Challenges in Analytical Quality Assurance

Manfred Reichenbächer • Jürgen W. Einax

Challenges in Analytical Quality Assurance

Springer

Dr. Manfred Reichenbächer
Friedrich-Schiller-Universität Jena
Institut für Anorganische und
Analytische Chemie
Lehrbereich Umweltanalytik
D-07743 Jena
Germany
manfred-reichenbaecher@t-online.de

Prof. Dr. Jürgen W. Einax
Friedrich-Schiller-Universität Jena
Institut für Anorganische und
Analytische Chemie
Lehrbereich Umweltanalytik
D-07743 Jena
Germany
juergen.einax@uni-jena.de

Additional material to this book can be down loaded from http://extras.springer.com

ISBN 978-3-642-16594-8 e-ISBN 978-3-642-16595-5
DOI 10.1007/978-3-642-16595-5
Springer Heidelberg Dordrecht London New York

Library of Congress Control Number: 2011922327

Cover design: eStudio Calamar S.L.

Printed on acid-free paper

Springer is part of Springer Science+Business Media (www.springer.com)

Preface

Analytical chemistry plays an important role in many branches of chemistry, biochemistry, pharmacy, life science and food production, as well as in monitoring of our environment, our health, etc. Many decisions are based on the results of quantitative chemical analysis, and it is important to be aware of the quality of the results whenever analytical chemistry methods are used. The development of analytical chemistry is thus increasingly characterized by the introduction of analytical quality assurance principles. The harmonization of European and international markets is triggering this process, and analytical laboratories in the chemical and pharmaceutical industries, as well as analytical routine laboratories in other disciplines such as environmental and food analysis, have generally accepted and introduced the appropriate standards, norms, and principles in the analytical process.

Nowadays, the analyst is not only expected to understand modern instrumental methods, they are also expected to understand and follow the regulatory requirements: for example, good laboratory practice (GLP) used in pharmaceutical analysis and elsewhere. This is a wide field, starting with the planning and selection of methods and sampling protocols. Next, the analyst has to validate the method and to test whether the approach is fit-for-purpose. This means they must use appropriate, calibrated equipment for the analytical measurements and must complete documentation at the end of the process, according to the stated requirements. Moreover, using principles of internal quality assurance, the analyst must be able to prove that the analytical methods are fit-for-purpose at any time. In addition, the work of the laboratory should be checked by interlaboratory comparison.

Despite its increasing importance, analytical quality assurance is hardly covered in university education. The beginner working in a chemical analytical laboratory will therefore face many issues for which they have not been trained. This book tries to help overcome this deficiency. Approaches are introduced and explained in detail on the basis of challenges as they appear to an analytical chemist in analytical practice. Most of the examples result from research in cooperation with industry and non-university laboratories. They have also been successfully applied in practical student courses in analytical quality assurance at our university.

Objective decisions require statistical tests. Therefore, all the challenges are solved by appropriate statistical tests which must be applied according to the regulatory requirements or which are recommended to establish the analyst's decision. Considerable weight is placed on solutions obtained according to these regulatory requirements.

Clearly, nowadays there are software packages for most of the problems, but we present each solution in detail in order to recalculate the results from first principles, because we believe that the analyst should know what the software program calculates. There are software packages, for example, for the calculation of the limit of detection. However, is it calculated on the basis of the German norm (DIN) or the IUPAC recommendations? Here, the analyst will obtain different results and therefore, in case of doubt, should be able to check the calculations. Besides the solutions given immediately following the challenges, MS Excel® spreadsheet functions can be found on the internet for solving the challenges, and these can also be applied to the reader's own problems.

As mentioned above, analytical quality assurance is a wide field which includes, besides the experimental requirements, the creation of documents according to regulatory requirements such as standard operation procedures (SOPs). Therefore, we had to make a selection of topics, and omitted this important documentation, which the analyst will learn, for example, in special workshops. We have only briefly introduced the extensive field of method development and tool qualification. However, the reader will find good books written by specialists in these fields.

Method validation is one of our main objectives. As all decisions must be taken with the help of statistical tests, the reader will find a comprehensive overview of method validation, taking into account all regulatory requirements. Thus the analyst will find, for example, all six methods for checking the trueness of analytical methods, and all the tests for linearity. The reader will find suitable methods for their own analytical approaches, as each test is supported by practical challenges.

We also point out that some frequently used procedures in statistics might not be the correct approach in analytical chemistry. For example, linearity is almost always checked by the correlation coefficient r. However, we argue that this is false, and discuss it in detail.

For further information on each chapter, there are many good books written by specialists. We apologize to colleagues whose work we could not cite because of space limitations. We wanted to introduce the reader to the wide field of analytical quality assurance in the style of a textbook rather than present a monograph with an exhaustive bibliography.

A comment on the symbols: we endeavored to apply unified symbols but we sometimes used the symbols suggested in documents such as DIN in order to retain compliance with the regulations. Last but not least, we have presented about 80 complex challenges. As there may be mistakes remaining, the authors will be grateful for any readers' comments.

Thus, we hope that beginners will find these inspiring challenges a positive and helpful introduction to the experimental work of analytical quality assurance.

Advanced analysts will also find suggestions and statistical tests necessary to ensure objectivity in their decisions.

Finally, we would like to express our thanks to the staff at Springer for all their help and courtesy, especially with regard to correction of the English.

Jena, January 2011 Manfred Reichenbächer and Jürgen W. Einax

Challenges

Excel-worksheets for solving the challenges are available at http://extras.springer.com/2011/978-3-642-16594-8. For further information, please consult the file README_978-3-642-16594-8.txt.

(*continued*)

Contents

List of Important Symbols

Some important symbols and abbreviations are listed below. Further symbols and abbreviations the use of which is restricted to special sections are defined in those sections.

A	Absorbance (without unit); peak area (in counts)
a_0	Regression coefficient (intercept)
a_1	Regression coefficient (slope)
a_2	Regression coefficient (quadratic term)
As	Asymmetry factor
c	Concentration
$\hat{C}$	Calculated value of the Cochran test on homogeneity of variances
CI	Confidence interval
CV	Coefficient of variation
df	Degrees of freedom
e	Error; difference between measured and estimated values; residual
F	Values of the F-distribution
H_0	Null hypothesis
H_1, H_A	Alternative hypothesis
I	Intensity
k	Number of groups
k'	Retention factor
l	Distance
L_0	Limit or threshold value
LAL	Lower action limit
LD	Limit of detection
LOF	Lack-of-fit
LQ	Limit of quantification
LWL	Lower warning limit
MS	Mean square
m	Number of features; mass
M	Molecular mass
med, $\tilde{x}$	Median

n	Number of objects, experiments, or observations
P	Two-sided probability
$\bar{P}$	One-sided probability
PE	Pure error
PI	Prediction interval
Q	Quantile
$\hat{Q}$	Calculated value of the Dixon outlier test
QC	Quality coefficient
$\hat{q}_r$	Calculated value of the David test on normal distribution
r	Correlation coefficient; repeatability limit; number of reflections
R	Range; reproducibility limit
R_s	Resolution
$\hat{r}_m$	Calculated value of the Grubbs outlier test
$Rr\,\%$	Recovery ratio
Reg	Regression
Rf	Response factor
RMS	Root mean squares
RSD	Relative standard deviation
s	Standard deviation
$s_{y.x}$	Residual standard deviation; calibration error
$s_{x.0}$	Process standard deviation; analytical error
s^2	Variance
Sens	Sensitivity
SS	Sum of squares
t	Value of the t-distribution
T	Transmission; temperature
t_r	Retention time
u	Uncertainty
U	Expanded uncertainty
u_{comb}	Combined uncertainty
UAL	Upper action limit
UV	Ultraviolet
UWL	Upper warning limit
V, v	Volume
VIS	Visible
w	Weighting factor; peak width
x	Variable
$\bar{\bar{x}}$	Grand mean value
y	Variable; response
z	Standardized variable
α	Probability of an error of the first kind; risk; absorptivity; selectivity factor
β	Probability of an error of the second kind
γ	Activity coefficient

λ	Wavelength
μ	True mean value
σ	True standard deviation
χ^2	Value of the chi-square distribution
$\hat{\chi}^2$	Calculated value of the Bartlett test for homogeneity of variances

Superscript Indices

^	Estimated value
–	Mean value; median
*	Outlier suspected value

Subscript Indices

a	Analysis
add	Stocked
bl	Blank
bw	Between
cal	Calibration
crit	Critical
i	Running index
in	Within
j	Running index
max	Maximum
min	Minimum
p	Pooled
r, rel	Relative
r	Repeatability
R	Reproducibility
sp	Spiked
st	Standard
tot	Total
val	Validation
w	Weighted

List of Abbreviations

AAS	Atomic absorption spectroscopy
ANOVA	Analysis of variance
API	Active pharmaceutical ingredient
AQA	Analytical quality assurance
CuSum	Cumulative sum of differences
DAD	Diode array detector
DIN	Deutsches Institut für Normung e. V. (German standards)
CRS	Chemical reference substance
CRM	Chemical reference material
ECD	Electron capture detector
ELISA	Enzyme-linked immunosorbent assay
FID	Flame ionization detector
GC	Gas chromatography
GLP	Good laboratory practice
HPLC	High performance liquid chromatography
HS	Headspace
IC	Ion chromatography
ICH	International conference on harmonization
ICP–OES	Optical emission spectroscopy with inductively coupled plasma
IS	Internal standard
ISO	International organization for standardization
MHE	Multiple headspace extraction
MS	Mass spectrometry
ODR	Orthogonal distance regression
OL	Outlier
OLS	Ordinary least squares
SPE	Solid phase extraction
SPME	Solid phase micro extraction
St	Stock solution
USP	United States Pharmacopeia

Chapter 1
Introduction

Quality of products and services has encompassed more and more areas of society, such as foods and drugs, environment, health, safety of the working population, and many others. According to DIN ISO 8402 [1], *quality* is "the totality of features and characteristics of product or service that bear on its ability to satisfy stated or implied needs", and the *assurance* of quality is defined as "all those planned and systematic actions necessary to provide adequate confidence that a product, process or service will satisfy quality requirements". Responsible for the compliance with these requirements is the *quality management system,* defined by ISO 9001:2008 [2] as "coordinated activities to direct and control an organization with regard to quality". The quality management system provides assurance through the following four tools [3, 4]:

- Quality planning:
  Planning activities are focused on setting quality objectives and specifying necessary operational processes and resources to fulfill quality objectives. They also include planning for quality assurance, quality control, and quality improvement activities.
- Quality assurance:
  Quality assurance includes all preventive activities which are focused on providing confidence that quality requirements will be fulfilled. It also includes proactive controls to prevent problems associated with customer dissatisfaction.
- Quality control:
  Quality control concerns activities focused on conforming to quality requirements so that customers receive only products that meet their requirements.
- Quality improvement:
  Quality improvement is the part of the management system which is focused on the continual improvement of activities increasing the ability to fulfill the requirements.

Analytical chemistry plays an important role in almost all parts of human life. Its results are essential for the output of industrial, pharmaceutical, and agricultural production, for research and development, and also for education. This includes the overwhelming majority of branches of chemistry, biochemistry, pharmacy, and life

M. Reichenbächer and J.W. Einax, *Challenges in Analytical Quality Assurance*,
DOI 10.1007/978-3-642-16595-5_1,

science, food production, material sciences, but also monitoring and control of our environment, human health, etc.

Analytical chemistry has a key position in the field of quality assurance. But it has a double function: on the one hand, the analysis must provide reliable data for customers; on the other hand, it must indicate that these data are valid. Its self-control is achieved by *validation*, which is defined in the international standard EN ISO/IEC 17025:2000 [5] as "validation is the confirmation by examination and the provision of objective evidence that particular requirements for a specific intended use are fulfilled". According to these requirements only properly validated methods may be applied and, additionally, using concepts of internal and external quality control, it must be documented that the test methods are capable of producing results that are fit-for-purpose every time.

This part of the extensive field of "quality" – *quality assurance* by means of instrumental analytical methods – is the subject of this book. In the field of harmonization of European and international markets, analytical laboratories of chemical and pharmaceutical industries as well as routine analytical laboratories in other disciplines such as environmental and food analysis have accepted and generally introduced appropriate standards, norms, and principles. These are mostly based on objective and statistically defined methods. They are the basis of numerous decisions not only in the development and production of pharmaceuticals and chemicals, but also in the field of environmental and consumer protection.

Because these decisions require the use of statistical methods, using statistics effectively is an important part of the analyst's job. This concerns all steps, beginning with the planning and realization of appropriate experiments for the validation of analytical methods, calibration of instruments, data acquisition under controlled conditions in order to make objective decisions, and ends with the analytical report as well as the archiving of the data. For most steps, statistical methods can or must be applied for objective decisions. Therefore, the modern analyst should not only have an excellent knowledge of instrumental methods and the chemistry of the analytes, they must also be able to understand and follow the regulatory requirements; for example, the principles of good laboratory practice (GLP) in the course of the analysis of pharmaceutical products, explosives, or pesticides, which also includes the application of statistical methods.

There are many good books which comprehensively present the theoretical basis of chemometrics and statistics; some of them are cited as references [6–11]. There are also a few books, mainly in German, applying the statistical treatment of analytical data to problems of analytical quality assurance [12–15].

The present book focuses on the procedures for solving practical and reasonable problems within Analytical Quality Assurance (AQA) which the analyst may meet in everyday work, using statistics as a help for decisions rather than concentrating on the theory. These problems are presented as "Challenges" and their solutions.

This book pursues the aim of learning by exercises. After a short theoretical and methodological introduction to the problems of analytical practice and its background, the reader should practice by means of the Challenges, which are mainly taken from daily problems met in an analytical laboratory. Although nowadays commercial software packages are mostly applied for the estimation of measured data, the answers will indicate the solution step-by-step on the basis of MS Excel spreadsheet functions which are used by international regulatory norms such as DIN/EN and others. Furthermore, the reader will also find simple Excel spreadsheets on the internet for the solution of most Challenges, with possible application to their own problems.

Each chapter builds on topics from previous chapters and the solution of Challenges given in subsequent chapters will mostly require the knowledge acquired in earlier chapters.

After a short listing of the possible kinds of errors in analytical measurements Chap. 2 is focused on the characteristics of *random errors* which can be evaluated by statistical methods. Important distributions of measured values, calculation of the *standard deviation* as the parameter of the distribution of measured values, and the *confidence interval* as the parameter of the analytical error and its relation to quality are briefly explained.

In Chap. 3 an outline is given of the *statistical tests* which are relevant in the field of AQA. After an introduction to the principles of hypothesis testing, the relevant tests for series of measurements are given which are important for the evaluation of the standard deviation. These are a simple test of *normal distribution* and a test of *trends* as well as *outlier* tests, which have to be used in AQA according to regulatory requirements. Furthermore, methods of comparison of standard deviations and mean values are presented, subdivided into methods for two standard deviations and mean values, respectively, and those for more than two parameters. The Cochran and Bartlett tests are two common methods for checking the significance of more than two standard deviations, and the analysis of variance (*ANOVA*) must be applied for testing more than two mean values. Challenges in the application of one-way and two-way ANOVA are presented. Note that the ANOVA design explained in Chap. 3 will be applied in further applications such as linearity test and internal or external laboratory tests.

Linear regression and its special case – *calibration* – is one of the main subjects of this book and it is extensively discussed in Chap. 4, which starts with a critical evaluation of the terms correlation, regression, and calibration (Sect. 4.1). A critical evaluation of the correlation is necessary because this parameter is often used erroneously in analytical practice. Challenges concerning the parameters of the linear calibration model are presented in Sect. 4.2, as well as its simplification if the intercept cannot be distinguished from zero (Sect. 4.3), and application of the quadratic regression model in Sect. 4.4.

In analytical practice, the choice of the proper *working range*, the difference between the highest and lowest values of the analyte in the sample, is an important parameter if the analytical method is to be fit for a specific purpose, because predicted values outside of the working range are not statistically guaranteed. Therefore, special attention has to be paid to the choice of the correct working range using some practical examples which also include preparation steps of the samples (Sect. 4.5).

Chapter 5 provides an overview of all the parameters which must be validated on the basis of the regulatory requirements within the scope of AQA. Many Challenges taken from problems of research in the areas of environmental and pharmaceutical analysis should enable the reader to employ this knowledge in their own problems. The parameter *precision* (Sect. 5.2) is separated into instrumental or system precision and the precision of the analytical procedure: its repeatability, intermediate precision, and reproducibility. A considerable part of this book is dedicated to tests for *linearity* of the regression line (Sect. 5.3). Besides visual methods, all statistical tests for checking linearity are presented and proved as Challenges: the *Mandel test*, the *lack-of-fit test by ANOVA* applied to replicated measurements, and the test of significance of the *quadratic regression coefficient* a_2. As explained above, the correlation coefficient most frequently used as the argument for linearity is often not appropriate for the linearity test and should be avoided in the analysis.

Section 5.4 provides challenges for testing *outliers* in the regression line by means of the F- and t-tests. Sections 5.5 and 5.6 describe the test of the *homogeneity of variances* within the regression line and the alternative *weighted linear least squares* estimation method which can be applied by heteroscedasticity of variances. Section 5.7 provides challenges for testing *outliers* in the regression line by means of the F- and t-test. Section 5.7 presents all methods and Challenges for checking *trueness* within the scope of AQA: mean value t-test, recovery rate, recovery rate of stocked samples, recovery function, standard addition method, and method comparison. Furthermore, standard addition is presented as an alternative analytical method applied to samples which are influenced by the matrix. The Challenges of this section encompass a broad field of problems so that the reader will find an appropriate method for solving their own specific analytical problem.

Many problems concern the analysis of samples with very low concentrations of analytes, e.g. for trace analytical methods in environmental analysis or for the determination of byproducts in substances, for which the parameters *limit of detection* and *limit of quantification* are defined. However there are differences in the definition of these parameters between the IUPAC definition and the German DIN regulatory documents; the latter are used in this book (Sect. 5.8).

Robustness or *ruggedness* of an analytical method is not explicitly given in the list of the required validation parameters, in documents such as International Conference on Harmonization (ICH), but it is recommended as part of method development to establish the critical measurement parameters which can be influenced by sample preparation and by the measurement conditions. The evaluation of the robustness of an analytical method is described in Sect. 5.9 and verified by a Challenge from pharmaceutical analysis.

After the reader has learnt step-by-step how to validate a new analytical method, Sect. 5.10 provides an application of the validation steps on the basis of the validation of an analytical method by the example of the determination of nitrite-N in iron-containing waste water in terms of monitoring the limit value. Thus, the reader can test their knowledge of method validation.

Chronologically, the choice and development of an appropriate analytical method for a specific analytical purpose is the first stage in the validation of a method. But method development is a wide field with specific investigations for each method, and therefore within the scope of this book only some aspects of one analytical method can be considered. Thus, *method development* is focused on chromatographic methods described in Chap. 6. After the definition of the relevant performance parameters of a chromatogram (Sect. 6.1), the important validation parameters *selectivity* and *specificity* are explained and verified on the basis of Challenges obtained by the analytical practice given in Sect. 6.2. Finally, the method development of headspace gas chromatography (HS-GC) is chosen as an example of one of the most important methods for the determination of organic compounds in liquid and in solid samples. Challenges provide solutions for various applications of HS-GC in AQA.

Besides the validity of the analytical methods, the reliability of all the instruments used for the experiments and measurements provide the fundamentals of analytical quality assurance. The performance verification of analytical instruments and tools is described for selected examples: UV–ViS spectrometers (Sect. 7.2), HPLC instruments (Sect. 7.3), and balances (Sect. 7.4).

After the reader has learnt about method development, tool qualification, and the steps of the validation of an analytical method, they must also know how to control the quality of the results in routine analysis. This is realized by methods of internal quality control, preferably verified by *control charts,* which are described in Chap. 8. Control charts are extremely valuable in providing a means of monitoring the total performance of the analyst, the instruments, and the test procedures, and can be utilized by any laboratory. There are a number of different types of control charts but the Shewhart (Sect. 8.2) and CuSum charts (Sect. 8.3) are those which are mainly used.

Chapter 9 concerns *interlaboratory studies,* which are organized into method-performance studies, material-certification studies, and proficiency studies. Note that participation in proficiency studies is necessary for laboratories in the process of their accreditation and participation in proficiency testing is obligatory for each accredited laboratory. A method performance study according to relevant ISO and DIN documents is described in Sect. 9.2 and is verified by the data obtained by an interlaboratory study for the determination of bromide in industrial waste water. Section 9.3 provides a Challenge for proficiency testing by the example of the determination of lead in flood sediment.

Finally, Chap. 10 attends to the problem of *measurement uncertainty,* which is a requirement of some regulatory documents; its estimation can be realized by various procedures. After explaining purpose, definition, and terminology in Sect. 10.1, the steps in evaluating measurement uncertainty according to ISO Guide 89:1995 (GUM) are described in Sect. 10.2. However the calculation of measurement uncertainty is best realized by the MS Excel spreadsheet method which is given in Sect. 10.3 together with some Challenges. A practical and understandable way of calculating measurement uncertainty is the Nordtest Report which is described and verified as a Challenge in Sect. 10.4. Note that this procedure is primarily written for environmental laboratories in the Nordic countries but is also applied in other countries, in particular in water analysis.

It is obvious that this book cannot claim a comprehensive coverage of the topic, but we think that the encapsulation of the essential requirements of AQA in about 80 Challenges offers a good starting point for the requirements of a modern analytical laboratory.

In spite of a careful audit of the numerous Challenges, the authors are grateful to hear of any mistakes which may remain. Note that all results presented were calculated by Excel functions; therefore, small differences to results calculated by a hand computer are possible.

References

1. DIN EN ISO 8402 (2005) Quality management and quality assurance – vocabulary. Beuth, Berlin
2. ISO 9001 (2008) Quality management systems – requirements. Beuth, Berlin
3. Kromidas S (ed) (1995) Qualität im analytischen Labor. Wiley-VCH, Weinheim
4. Kromidas S (2000) Handbuch Validierung in der Analytik 1. Aufl. Wiley-VCH, Weinheim
5. EN ISO/IEC 17025 (2000) General requirements for the competence of testing and calibration laboratories. European Committee for Standardization, Brussels
6. Brown SD, Tauler R, Walczak B (eds) (2009) Comprehensive chemometrics: chemical and biochemical data analysis (4-volume set). Elsevier, Amsterdam, Oxford
7. Brereton RGB (2007) Applied chemometrics for scientists. Wiley, Chichester
8. Einax JW, Zwanziger HW, Geiß S (1997) Chemometrics in environmental analysis. Wiley-VCH, Weinheim
9. Massart DL, Vandeginste BGM, Buydens LMC, De Jong S, Lewi PJ, Smeyers-Verbeke J (1997) Handbook of chemometrics and qualimetrics. Part A. Elsevier, Amsterdam
10. Otto M (2007) Chemometrics, 2nd edn. Wiley-VCH, Weinheim
11. Danzer K (2007) Analytical chemistry – theoretical and metrological fundamentals. Springer, Berlin
12. Doerffel K (1990) Statistik in der analytischen Chemie, 5. Aufl. Deutscher Verlag for Grundstoffindustrie, Leipzig
13. Ellison SLR, Berwick VJ, Duguid Farrant TJ (2009) Practical statistics for the analytical scientist, 2nd edn. RSC, Cambridge, UK
14. Funk W, Dammann V, Donnevert G (2005) Qualitätssicherung in der Analytischen Chemie, 2. Aufl. Wiley-VCH, Weinheim
15. Gottwald W (2000) Statistik für Anwender. Wiley-VCH, Weinheim

Chapter 2
Types of Errors in Instrumental Analysis

2.1 Overview

Even under constant experimental conditions (same operator, same tools, and same laboratory, short time intervals between the measurements), repeated measurements of series of identical samples always lead to results which differ among themselves and from the true value of the sample. Therefore, quantitative measurements cannot be reproduced with absolute reliability.

According to their character and magnitude, the following types of deviation can be distinguished [1–4]:

Random Errors. Random errors are the components of measurement errors that vary in an unpredictable manner in replicated measurements. They reflect the distribution of the results around the mean value of the sample which are randomly distributed to lower and higher values. Random errors characterize the *reproducibility* of measurements, and, therefore, their *precision*. They are caused by effects such as measuring techniques (e.g. noise), sample properties (e.g. inhomogeneities), and chemical effects (e.g. equilibrium). Even under carefully controlled conditions random errors cannot, in principle, be avoided, they can only be minimized and evaluated with statistical methods.

Systematic Errors. Systematic deviations (errors) displace the results of analytical measurements to one side, to higher or lower values which lead to false results. Such an effect is described by the performance characteristic *trueness,* which is defined as "the closeness of agreement between the expectation of a test result or measurement result and a true value" [1]. Measurement trueness is not a *quantity* and cannot be expressed numerically [2], but measures for closeness of agreement can be given. Thus the trueness can be quantified as *bias* which is defined as the difference between the average of several measurements on the same sample $\hat{\bar{x}}$ and its (conventionally) true value μ:

$$\mathrm{Bias}(x) = \hat{\bar{x}} - \mu \qquad (2.1\text{-}1)$$

or if expressed as a percentage

M. Reichenbächer and J.W. Einax, *Challenges in Analytical Quality Assurance*,
DOI 10.1007/978-3-642-16595-5_2, © Springer-Verlag Berlin Heidelberg 2011

$$\text{Bias}\ \% = \frac{(\hat{\bar{x}} - \mu)}{\mu} \cdot 100 \qquad (2.1\text{-}2)$$

or as the recovery ratio

$$Rr\% = \frac{\hat{\bar{x}}}{\mu} \cdot 100. \qquad (2.1\text{-}3)$$

In contrast to random errors, systematic errors can and *must* be avoided or eliminated if their origins become known, because they yield false results. Note that systematic errors cannot be statistically evaluated.

Systematic errors are always combined with random errors as shown in Fig. 2.1-1.

The measurement *accuracy* is defined as the "closeness of agreement between a measured quantity value and a true quantity value of a measurand" [2]. The measurement accuracy is not given a numerical value, but it is a qualitative performance characteristic which expressed the closeness of agreement between a measurement result and the value of the measurand, and thus it describes the precision as well as the trueness [5]. Therefore, the term "measurement accuracy" should not be used for measurement trueness.

The performance parameter of accuracy is the measurement uncertainty.

Measurement Uncertainty. The uncertainty of measurements is defined as "a parameter associated with the result of a measurement that characterizes the dispersion of the value that could reasonably be attributed to the measurand" [2]. The uncertainty concept divides the errors into two uncertainty components:

- Those that can be characterized by the experimental *standard deviations* (uncertainty components from Type A).
- Those that can be evaluated from *assumed probability distributions* based on experimental or other information (uncertainty components from Type B).

The combined uncertainty from both components is calculated by the law of propagation of errors (see Chap. 10).

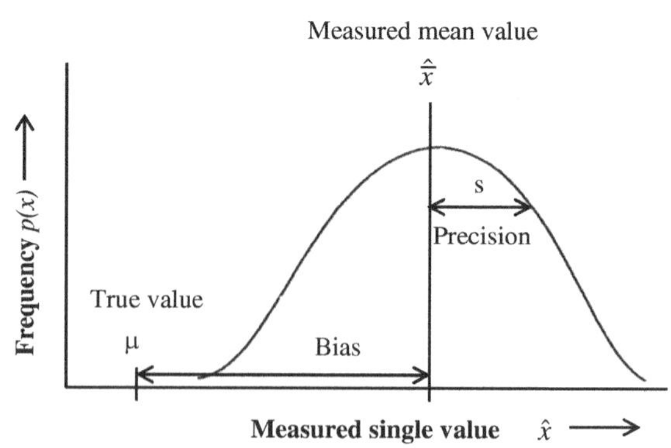

Fig. 2.1-1 Graphical representation of the terms precision and trueness

Outliers. Outliers are individual measurement values which considerably differ from the mean value. Outliers would falsify the estimation of parameters such as the mean value and the standard deviation, and therefore they must be detected by statistical methods and eliminated from the data set or, if this is not possible, one must work with methods resistant to outliers (robust methods [6]).

Trend. A data set shows a trend when the chronologically ordered values move steadily downwards or upwards. Such a data set is not under statistical control; therefore, after it has been recognized statistically, the trend must be eliminated. Note that a data set which shows a trend is to be rejected.

Gross Errors. Gross errors result from human mistakes, or have their origins in instrumental or computational errors. Frequently, they are easy to recognize and the origins must be eliminated.

Challenge 2.1-1

Table 2.1-1 shows series of data sets obtained by the five methods A–E. Which kinds of errors can be *visually* detected? The true content of the sample is $\mu = 100$.

Table 2.1-1 Hypothetical analytical results obtained with five methods

Method	x_1	x_2	x_3	x_4	x_5	x_6	$\bar{x}$
A	99	101	98	100	100	102	100
B	106	98	104	95	94	103	100
C	106	102	102	99	98	93	100
D	99	101	98	82	100	102	97
E	115	117	116	116	112	114	115

Solution to the Challenge 2.1-1

The data set in series *C* obviously shows a trend downwards, i.e. a trend is present. Though the calculated mean value is correct the data are not appropriate for the analytical result. The data set in *C* has to be rejected.

The value 82 in series *D* is obviously an outlier which leads to a false mean value. After elimination of this value a correct value (100) can be calculated.

Method E clearly yields a false mean value $\bar{x} = 115$. The result is obviously too high because a systematic error is present.

Methods A and B yield correct mean values but the individual results show a higher dispersion around that mean in series B than in A. This means that the precision in series A is better.

This exercise can be regarded as a *plausibility control,* which is an import step in analytical quality assurance. Plausibility control means checking data series, analytical results, and others, without statistical tests, to see if the data *can be corrected.* This procedure has to be carried out before the release of data for further processes or for the documentation of analytical results. Thus, for example, the check of series D reveals the existence of an outlier ($x_4 = 82$) for which an outlier test must be carried out (see Challenge 3.3-1), and the trend in series C is also obvious.

Note that errors cannot always be clearly recognized; statistical methods are mostly necessary, but this is the subject of the following chapters.

2.2 Random Errors

2.2.1 Distribution of Measured Values

When one wants to view the distribution of many available data, it is useful to group the n data into k classes with n_j variables in each class and visualize their frequency density or probability distribution $p(x)$ with a *histogram,* which is a graphical display of tabulated frequencies presented as bars [7–9]. Figure 2.2.1-1 shows an example for the frequency density $p(x)$ of measured values x. The bars must be adjacent and the intervals (or bands) are generally of the same size. The rule of thumb $k = \sqrt{n}$ provides an appropriate number of classes k for the construction of a histogram with n data.

If the number of repeated measurements is increased to infinity and one reduces the width of the classes towards zero, a symmetrical bell-shaped distribution of measurement values is usually obtained, which is called *Gaussian* or *normal distribution* (see curve ND in Fig. 2.2.1-2).

The frequency density $p(x)$ is described by the function

$$p(x) = \frac{1}{\sigma\sqrt{2\pi}} \exp\left\{-\frac{1}{2}\left(\frac{x-\mu}{\sigma}\right)^2\right\}. \tag{2.2.1-1}$$

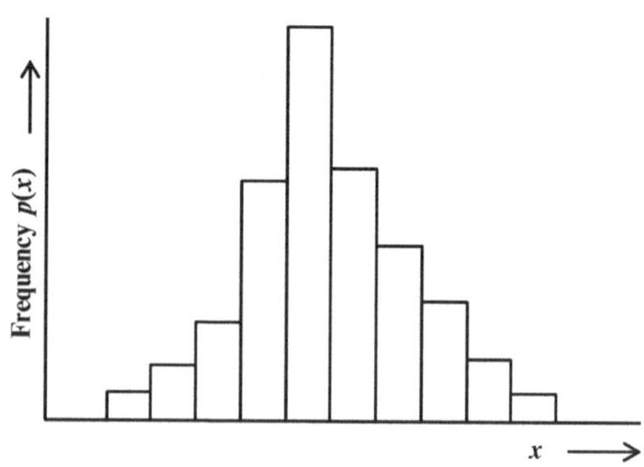

Fig. 2.2.1-1 An example for the frequency density $p(x)$ of measured values x

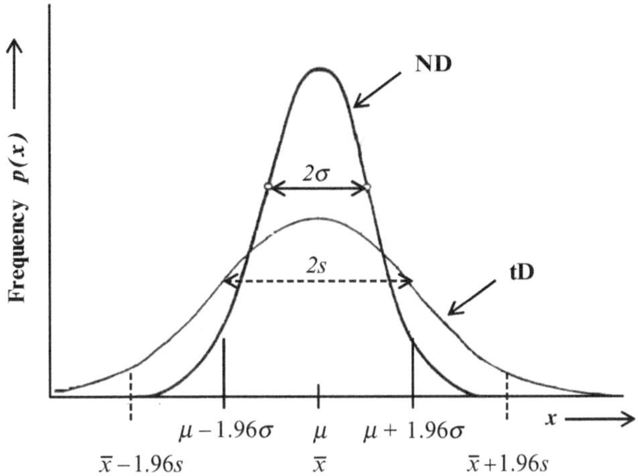

Fig. 2.2.1-2 Gaussian (normal) distribution (ND) and *t*-distribution (tD) of measured values x

The parameters of the normal distribution are:

- Mean value μ

$$\mu = \frac{\sum_{i=1}^{n} x_i}{n}. \tag{2.2.1-2}$$

- Variance σ^2

$$\sigma^2 = \sum_{i=1}^{n} \frac{(x_i - \mu)^2}{n}. \tag{2.2.1-3}$$

To avoid the scale effect, standardized values with

$$z = \frac{x - \mu}{\sigma} \tag{2.2.1-4}$$

are often used. Equation (2.2.1-1) is transformed into (2.2.1-5):

$$p(x) = \frac{1}{\sqrt{2\pi}} \exp\left(-\frac{z^2}{2}\right). \tag{2.2.1-5}$$

Equation (2.2.1-5) holds true for the *standardized normal distribution.*

In the literature one can find some tables for z-values [7]. Table A.1 gives the areas between the boundary $z = 0$ and a chosen value z (see Fig. 2.2.1-3).

Because of the symmetry of the normal distribution the table gives p-values only for positive values of z. With this table one can ask, for example, what percentage of determinations will fall between two chosen boundaries (see Challenge 2.2.1-2).

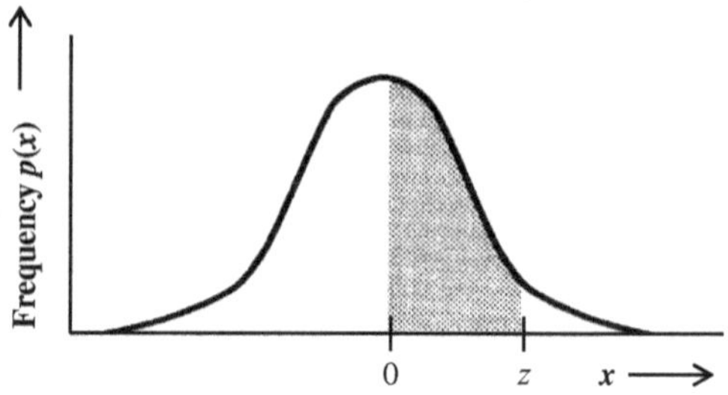

Fig. 2.2.1-3 The *shaded area* describes the probability $p(x)$ of finding a value between 0 and z

In analytical practice, random samples of the basic population are investigated. The parameters μ and σ are substituted by the estimated values $\bar{x}$ and s for n measurements, which are calculated by (2.2.1-6) and (2.2.1-7), respectively:

$$\bar{x} = \frac{\sum_{i=1}^{n} x_i}{n}, \tag{2.2.1-6}$$

$$s = \sqrt{\frac{\sum_{i=1}^{n} (x_i - \bar{x})^2}{n-1}}. \tag{2.2.1-7}$$

Both values can be obtained by MS Excel functions = AVERAGE(Data) and =STDEV(Data), respectively.

Note the calculation of the mean value $\bar{x}$ as well as the standard deviation s is based on the normal distribution of the data set.

However, there are data sets for which no assumptions about the distribution of the population can be made. These data sets are handled by so-called *robust methods* [9, 10]. The central tendency is expressed by the *median* $\tilde{x}$ instead of the mean value $\bar{x}$. The median is resistant to outlying observations which have a large effect on the mean and the standard deviation.

After ranking the n data, the median $\tilde{x}$ is the middle value of the given numbers in ascending order.

The median of a ordered data set $x_1, x_2, \ldots, x_n$ is

$$\tilde{x} = x_{\frac{n+1}{2}} \tag{2.2.1-8}$$

when the size of the distribution is *odd*, and

$$\tilde{x} = \frac{1}{2}\left(x_{\frac{n}{2}} + x_{\frac{n}{2}+1}\right) \tag{2.2.1-9}$$

when the size of the distribution is *even*.

In practice, the median is calculated by the Excel function = MEDIAN(Data) without ranking of the data set.

Challenge 2.2.1-1

The mean values of 40 batches of an intermediate product of a synthesis for an active pharmaceutical ingredient (API), calculated as the content relative to a standard, are given in Table 2.2.1-1.

Create the histogram for the data set with an appropriate number of classes!

Can the data set be considered normally distributed?

Table 2.2.1-1 Mean values $\bar{x}$ in % (w/w) of 40 batches of an intermediate product of a synthesis

n	$\bar{x}$	n	$\bar{x}$	n	$\bar{x}$	n	$\bar{x}$
1	103.9	11	96.2	21	99.7	31	96.9
2	102.7	12	99.9	22	100.6	32	108.0
3	101.0	13	92.3	23	107.5	33	105.8
4	94.8	14	101.2	24	90.5	34	94.6
5	105.2	15	100.8	25	108.8	35	102.8
6	100.4	16	99.0	26	101.9	36	104.2
7	97.0	17	100.8	27	102.5	37	99.9
8	101.6	18	104.0	28	97.4	38	106.4
9	109.0	19	99.2	29	107.0	39	103.5
10	90.8	20	109.7	30	104.5	40	96.7

Solution to Challenge 2.2.1-1

The mean values $\bar{x}$ arranged in increasing size are listed in Table 2.2.1-2.

With the rule of thumb for the choice of the number of classes $k = \sqrt{40} = 6.3$, seven classes are chosen. The number of mean values n_j which belong to the k classes are given in Table 2.2.1-3. The histogram is visualized in Fig. 2.2.1-4 from the data of Table 2.2.1-3. Figure 2.2.1-4 shows that the mean values $\bar{x}$ may be regarded as normally distributed, which is demonstrated by the bell-shaped curve in Fig. 2.2.1-5. (A statistical test for normal distribution is presented in Sect. 3.2.1.)

Table 2.2.1-2 Mean values $\bar{x}$ in % (w/w) arranged in increasing size

n	$\bar{x}$	n	$\bar{x}$	n	$\bar{x}$	n	$\bar{x}$
24	90.5	16	99.0	14	101.2	30	104.5
10	90.8	19	99.2	8	101.6	5	105.2
13	92.3	21	99.7	26	101.9	33	105.8
34	94.6	12	99.9	27	102.5	38	106.4
4	94.8	37	99.9	2	102.7	29	107.0
11	96.2	6	100.4	35	102.8	23	107.5
40	96.7	22	100.6	39	103.5	32	108.0
31	96.9	15	100.8	1	103.9	25	108.8
7	97.0	17	100.8	18	104.0	9	109.0
28	97.4	3	101.0	36	104.2	20	109.7

Table 2.2.1-3 Classes k with their width as well as the number of the mean values n_j for each class k

k	Width	n_j
1	90–93	3
2	93–96	2
3	96–99	6
4	99–102	12
5	102–105	8
6	105–108	6
7	108–111	3

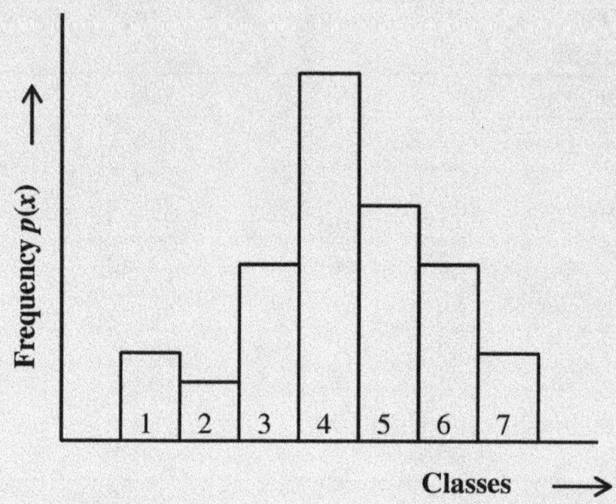

Fig. 2.2.1-4 Histogram generated with the data of Table 2.2.1-3

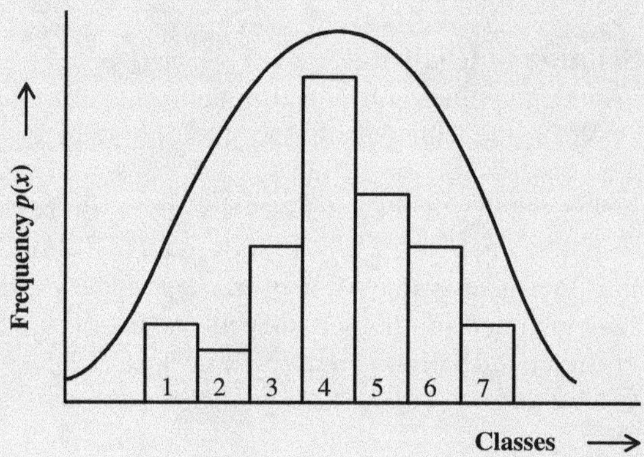

Fig. 2.2.1-5 Bell-shaped curve of the histogram in Fig. 2.2.1-4

Challenge 2.2.1-2

Calibration standards were prepared in the range 90–110% (w/w) for the determination of the content of the API with the same method as given in Challenge 2.2.1-1.

(a) What percentage of determinations will fall in this range?

(b) What percentage of determinations would fall in the range 99–101% (w/w)?

Solution to Challenge 2.2.1-2
According to the results in Challenge 2.2.1-1, the data set of the mean values $\bar{x}_i$ is normally distributed. Using the data set given in Table 2.2.1-2, the grand mean is $\bar{\bar{x}} = 101.2\%$ (w/w) and the standard deviation is $s = 4.857\%$ (w/w). These values are used for the calculation of the z-values according to (2.2.1-4).

(a) The z-value of the lower limit is $z_{ll} = -2.31$ and that of the upper limit is $z_{ul} = 1.81$. According to Table A.1 the probability p of finding an value between 0 and z is 0.4896 for the lower and 0.4649 for the upper limit, which gives the sum 0.9545. Thus, 95.5% of the results will fall within the range of the calibration standards, and only 4.5% will fall outside.
(b) For the range 99–101%(w/w), only 19.3% of the values are included and 80.7% fall outside, which is calculated by the intermediate quantities: $z_{ll} = -0.46$, $p(z_{ll}) = 0.1772$, $z_{ul} = -0.04$, $p(z_{ul}) = 0.0160$, $p(z_{ll} + z_{ul}) = 0.1932$.

Challenge 2.2.1-3
The screening of atrazine on a field by ELISA has yielded the mean values of 12 samples (n) obtained by triplicates given in Table 2.2.1-4.

Table 2.2.1-4 Data set obtained by screening of atrazine using ELISA

n	$\bar{x}_{atrazine}$ ppb (w/w)	n	$\bar{x}_{atrazine}$ ppb (w/w)	n	$\bar{x}_{atrazine}$ ppb (w/w)
1	2.5	5	4.6	9	13.8
2	0.9	6	0.5	10	1.2
3	1.1	7	8.6	11	0.8
4	7.9	8	3.1	12	6.4

(a) Calculate the mean value $\bar{x}$, the standard deviation s, and the median $\tilde{x}$; (b) Calculate the same parameters after addition of a further value 100 ppb (w/w) to the data set. Evaluate the results.

Solution to Challenge 2.2.1-3
(a) The mean value $\bar{x}$ and the standard deviation s calculated by (2.2.1-6) and (2.2.1-7), respectively, are $\bar{x} = 4.28$ ppm (w/w) and $s = 4.145$ ppm (w/w). In order to calculate the median, the data set has to be ordered (Table 2.2.1-5). Note that the median can also be calculated by the Excel function = MEDIAN(Data) without ordering of the data set.

Because the rank n is even the median is obtained by (2.2.1-9); the median is the mean of the values of rank 6 and 7: $\tilde{x} = 2.80$ ppm (w/w).

(continued)

Table 2.2.1-5 Analytical values of Table 2.2.1-4 in their ascending order n

n	x_{atrazine} ppb (w/w)	n	x_{atrazine} ppb (w/w)	n	x_{atrazine} ppb (w/w)
1	0.5	5	1.2	9	6.4
2	0.8	6	2.5	10	7.9
3	0.9	7	3.1	11	8.6
4	1.1	8	4.6	12	13.8

(b) After addition of the value 100 ppb (w/w) to the data set the mean value is $\bar{x} = 11.65$ ppb (w/w)and the standard deviation is $s = 26.842$ ppb (w/w). Because the rank is now odd ($n = 13$) the median is the observation with rank $(13 + 1)/2 = 7$, according to (2.2.1-8): $\tilde{x} = 3.1$ ppb (w/w).

Whereas the addition of a single but outlying observation causes a large effect on the mean value as well as on the standard deviation, the median is hardly changed. The mean value increases from 4.28 to 11.65 ppb (w/w) whereas the median increases only from 2.8 to 3.1 ppb (w/w), which shows that the median is a better representative of the central tendency after addition of only one value to the data set.

2.2.2 *Standard Deviation*

The standard deviation s is calculated from n replicate measurements of the same sample by (2.2.1-7). The number of *degrees of freedom* is $\text{df} = n - 1$ which corresponds to the number of control measurements.

The standard deviation obtained from replicates of different samples with varying content is calculated by (2.2.2-1)

$$s = \sqrt{\frac{\sum_{i=1}^{n_A} \sum_{j=1}^{m} (x_{ij} - \bar{x}_i)^2}{n - m}} \tag{2.2.2-1}$$

with $\text{df} = n - m$, in which m is the number of samples, n_A is the number of replicates for each sample, and n is the total number of determinations, $n = m \cdot n_A$.

Equation (2.2.2-2) should be used for the computation of s:

$$s = \sqrt{\frac{\sum_{j=1}^{m} \text{SS}_i}{n - m}}. \tag{2.2.2-2}$$

SS_i is the sum of squares of the sample i, which is a calculator function and also a MS Excel function = DEVSQ(Data).

In the special case of paired replicates, each determination is carried out in duplicate. The standard deviation is calculated according to (2.2.2-3):

$$s = \sqrt{\frac{\sum (x'_j - x''_j)^2}{2 \cdot m}}. \qquad (2.2.2\text{-}3)$$

The degrees of freedom df $= m$, x'_j and x''_j are the paired values of double measurements for each sample, and m is the number of samples.

The *variance* (*var*) is the square of the standard deviation:

$$\text{var} = s^2. \qquad (2.2.2\text{-}4)$$

The *relative standard deviation* s_r is given by

$$s_r = \frac{s}{\bar{x}} \qquad (2.2.2\text{-}5a)$$

and when it is expressed as a percentage by

$$s_r\% = 100 \cdot s_r \qquad (2.2.2\text{-}5b)$$

which is, for example, an appropriate parameter for the comparison of precision of various analytical methods.

The *standard deviation of the means* (SEM) $s(\bar{x})$ is called the standard error of the mean and is calculated using the equation

$$s(\bar{x}) = \frac{s}{\sqrt{n}}. \qquad (2.2.2\text{-}6)$$

The standard error of the mean is the standard deviation of the sample mean estimate of a population. It represents the variation associated with a mean value. The SEM is the expected value of the standard deviation of means of several samples.

Challenge 2.2.2-1

The process standard deviations for the determination of sulphur in steels according to the volumetric titration of SO_2 after burning of the samples were obtained by two different methods:

Method A: Repeated measurements of the *same* steel standard. The results are listed in Table 2.2.2-1.

(continued)

Table 2.2.2-1 Analytical values for a steel standard

Replicate	x in % (w/w)
1	0.0259
2	0.0238
3	0.0257
4	0.0242
5	0.0267
6	0.0239
7	0.0248
8	0.0259
9	0.0262
10	0.0241
11	0.0240

Table 2.2.2-2 Analytical values for ten steel standards

Standard	x' in % (w/w)	x'' in % (w/w)
1	0.0252	0.0236
2	0.0096	0.0110
3	0.0298	0.0282
4	0.0430	0.0448
5	0.0274	0.0281
6	0.0326	0.0294
7	0.0456	0.0480
8	0.0156	0.0135
9	0.0352	0.0330
10	0.0362	0.0374

Method B: Double measurements with ten *different* steel standards always of different content. The results are given in Table 2.2.2-2.

Calculate the standard deviation and give the degrees of freedom for both methods.

Note that the statistical test for normal distribution, the requirement for standard deviation, is given in Sect. 3.2.1.

Solution to Challenge 2.2.2-1

Method A: The standard deviations for the data set in Table 2.2.2-1 are calculated by (2.2.1-5), but this is a function on every hand calculator and an Excel function = STDEV(Data).

The standard deviation is $s = 0.0010\%$ (w/w) S, obtained by df = 10 degrees of freedom.

Method B: The standard deviation is $s = 0.00137\%$ (w/w) S calculated by (2.2.2-3) with the intermediate quantities $\sum (x' - x'')^2 = 0.0000375$ and df $= m = 10$.

Note that the degrees of freedom df are equal for both methods!

Challenge 2.2.2-2

An analytical laboratory has to determine manganese in steels with contents between 0.35 and 1.15% (w/w) Mn. For the determination of the standard deviation of the analytical method, five steel standards were analyzed by the volumetric method. The results are presented in Table 2.2.2-3.

Table 2.2.2-3 Analytical values for the five steel standards in % (w/w) Mn

Standard 1	0.31	0.30	0.29	0.32
Standard 2	0.59	0.57	0.58	0.57
Standard 3	0.71	0.69	0.71	0.71
Standard 4	0.92	0.92	0.95	0.95
Standard 5	1.18	1.17	1.21	1.19

Calculate the standard deviation.

Solution to Challenge 2.2.2-2

The standard deviation for the data set in Table 2.2.2-3 is calculated by (2.2.2-1). The sums of squares SS_i obtained by the MS Excel function = DEVSQ(Data) are:

Standard	1	2	3	4	5
SS_i	0.0005	0.000275	0.0003	0.0009	0.000875

The standard deviation is $s = 0.014\%$ (w/w) Mn which is obtained with $\sum SS_i = 0.00285$, $n = 20$, $m = 5$, and df = 15.

Note that the calculation of the standard deviation by (2.2.2-1) is only allowed if the variances of groups are homogeneous, which will be tested later (see Challenge 3.4-1).

2.2.3 Confidence Interval

Measured values which follow a normal distribution can occur in the whole range defined as $-\infty < x < \infty$. Therefore, it is useful to define dispersion ranges which include a certain number of measured values with a given high level of significance P, usually $P = 95\%$ or $P = 99\%$. The integration interval for $P = 95\%$ is $\mu \pm 1.96 \cdot \sigma$ and its limits are called *confidence limits* at the significance level $P = 95\%$. The range between the limits is called the *confidence interval*. Note that the integration between the limits $\pm 1.96 \cdot \sigma$ covers 95% of the values x_i. Thus, there is a probability of 95% that a measured value $\bar{x}$ will fall in the range $\mu \pm 1.96 \cdot \sigma$ under the assumption the values x_i of the mean belong to the same population.

For small sample sizes with n samples, the normal distribution $n(\sigma, \mu)$ is substituted by the *t-distribution* $n_t(s, \bar{x}, n)$. Figure 2.2.1-2 shows the relation between the normal distribution of a given population and the t-distribution for small samples. One can recognize that the t-distribution (curve tD) is broader at the base and the confidence interval is also broader. The confidence limits are given for the various n and degrees of freedom df, respectively, in the t-table (Table A.2) or by Excel function =TINV(α, df). Note that α is the risk, which is connected with the significance level by the relation $\alpha = 1 - P$.

Note that only two-tailed values are directly available from this function. In order to obtain a one-tailed critical value for the significance level α and df degrees of freedom the function =TINV(2α, df) is used. (One- and two-tailed values are explained in detail in Chap. 3.)

The confidence interval is calculated by (2.2.3-1):

$$\Delta\bar{x} = \frac{s_x \cdot t(P, \mathrm{df})}{\sqrt{n}}. \tag{2.2.3-1}$$

The t-values, i.e. the critical values of the t-distribution, are taken from Student's t-table for a certain significance level P (usually 95 or 99%) and the degrees of freedom df refers to the data set from which the standard deviation s_x is obtained.

The analytical result is expressed in the form:

$$\bar{x} \pm \Delta\bar{x}, \text{ given in the units of measurement.} \tag{2.2.3-2}$$

The mean value $\bar{x}$ is calculated by (2.2.1-2). But there is still a question: how big may the difference of two or more measurement values be for the formation of the mean value? Can all values x_i obtained be utilized or there are limits?

The *critical difference* Δ_{crit} between the highest and the lowest measurement values in a set of repeated determinations is given by *Pearson's criterion*:

$$\Delta_{\mathrm{crit}} = |x_{\max} - x_{\min}| < D(P, n_j) \cdot s_x. \tag{2.2.3-3}$$

The Pearson factors $D(P, n_j)$ for $P = 95\%$ and the number of repeated determinations n_j are given in Table 2.2.3-1.

Table 2.2.3-1 Pearson factors $D(P, n_j)$ for the critical difference between the highest and lowest measurement value of repeated determinations n_j with the significance level $P = 95\%$

n_j	2	3	4
$D(P, n_j)$	2.77	3.31	3.65

For example, the difference between the two measurement values may not exceed the limit $2.77 \cdot s_x$ for a double determination.

However, sometimes the simple relation $\Delta_{crit} < 2 \cdot s_x$ is used; therefore, the limit criterion used should be given in the documents.

Challenge 2.2.3-1

Let us come back to the determination of sulphur in steel (Challenge 2.2.2-1).

Calculate the confidence interval for the mean value of sulphur using method A and method B at the significance level $P = 95\%$ for

(a) Double determinations
(b) Fourfold determinations

Solution to Challenge 2.2.3-1

The confidence interval is calculated by (2.2.3-1).

The results are listed in Table 2.2.3-2. The values of s_x and df were determined in Challenge 2.2.2-1.

Table 2.2.3-2 Intermediate quantities and results of the calculation of the confidence interval $\Delta\bar{x}$ for the determination of sulphur in steel by different methods

Parameter	Method A	Method B
s_x	0.0011 in % (w/w)	0.0014 in % (w/w)
df	10	10
$t(P = 95\%, \text{df})$	2.228	2.228
a. $\Delta\bar{x}$ for $n = 2$	0.0017 in % (w/w)	0.0022 in % (w/w)
b. $\Delta\bar{x}$ for $n = 4$	0.0012 in % (w/w)	0.0015 in % (w/w)

Challenge 2.2.3-2

(a) According to the procedure given in Challenge 2.2.2-2, the double determination of manganese in a steel sample yields the values $x_1 = 0.65\%$ (w/w) Mn and $x_2 = 0.63\%$ (w/w) Mn.
    Present the analytical result in the form $x \pm \Delta\bar{x}\%$ (w/w) Mn.
    Give a verbal interpretation of the result.
(b) The analytical results obtained with triplicates of a sample of manganese steel are:

% (w/w) Mn	0.65	0.63	0.68

Test whether the calculation of the mean value is permitted.
What should one do if the limit is exceeded?

Solution to Challenge 2.2.3-2

(a) The confidence interval is $\Delta\bar{x} = 0.021\%$ (w/w) Mn calculated for $n_j = 2$(double determination) with the data obtained by Challenge 2.2.2-2: $s_x = 0.014\%$ (w/w) Mn and $t(P = 95\%, \text{df} = 15) = 2.131$.

The analytical result is 0.64 ± 0.02% (w/w) Mn.

The true value of the content of manganese in the steel sample lies in the range 0.62–0.66% (w/w) Mn. But this is true only for the significance level $P = 95\%$, with the risk $\alpha = 5\%$ that the true value will lie outside this range.

(b) For $n_j = 3$ the Pearson factor is 3.31. With $s_x = 0.014\%$ (w/w) Mn, the critical difference is 0.046% (w/w) Mn, but the difference in the experimental values is $x_{max} - x_{min} = 0.05\%$ (w/w) Mn. The calculation of the mean value is not permitted.

One should at best make a further analysis.

2.2.4 *Confidence Interval and Quality*

The quality control of products in environmental compartments and elsewhere requires decisions on the basis of analytical results, which means deciding whether a limit value is transgressed or not. Such a limit or threshold value stipulated in official documents can be an upper limit (e.g. in the case of environmental compartments) or a lower limit (e.g. for the potassium content of a fertilizer).

Let us take an example: the specified threshold for the content of the monomer styrene in industrially produced polystyrene for a certain application is ≤0.8% (w/w). Analytical quality assurance yields a content of 0.75% (w/w) for a batch pf polystyrene. Is the limit value exceeded or not, i.e. is this batch has to be discarded or is the quality standard fulfilled? How is it to be recognized easily: this decision is of great economic interest?

But, as Fig. 2.2.4-1 shows, the decision cannot be made without knowledge of the confidence interval of the analytical result.

The same mean value $\bar{x}$ was obtained with two methods which are different in regard to their precision. The quality criterion is fulfilled in the upper case I, because the limit value L_0 falls outside the confidence interval CI. One says that L_0 does not belong to the parent basic population of the sample. But in case II with the larger standard error, L_0 is included in the basic population of the sample which means, in a statistical sense, there is no difference between $\bar{x}$ and L_0. Therefore, the limit value is exceeded and, for example, the product cannot be delivered for sale.

For the control of limiting values, as well as some other problems, only the *one-sided* limit of the confidence interval is important. This is the upper limit in the

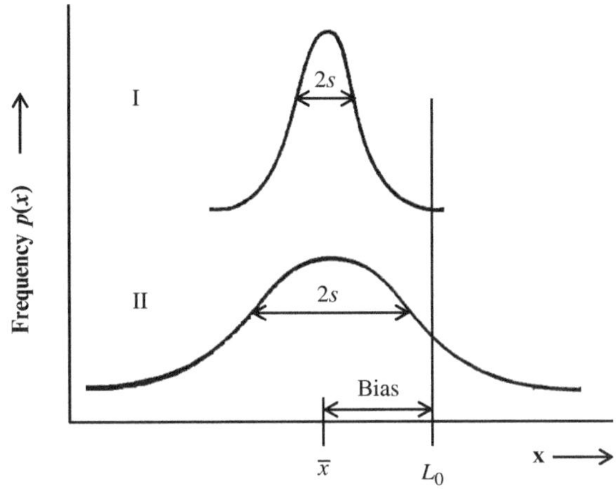

Fig. 2.2.4-1 Influence of the precision s on the transgression of the threshold value L_0 for the same analytical result $\bar{x}$

case of Fig. 2.2.4-1. The significance level of one-sided confidence intervals is also taken from the t-table or the MS Excel spreadsheet, but with another value for the statistical significance level. It is worth knowing that for the usual significance levels $t(\bar{P}_{\text{one-sided}} = 95\%, \text{df}) \approx t(P_{\text{two-sided}} = 90\%, \text{df})$.

An analytical mean value fulfils the quality standard for a required maximal threshold value L_0 if

$$\bar{x} + \frac{s \cdot t(\bar{P}_{\text{one-sided}}, \text{df})}{\sqrt{n_a}} \leqslant L_0. \qquad (2.2.4\text{-}1)$$

The degrees of freedom df refer to the number of replicates with which the standard deviation of the analytical method s has been determined, and n_a is the number of replicates in the routine analysis.

As Figure 2.2.4-1 reveals, an analytical method with a small confidence interval is desirable because the experimentally determined mean value can be closer to the limit value without it being exceeded. If one inspects (2.2.3-1), the confidence interval for a given significance level, usually $P = 95\%$, is determined by the standard deviation of the method s and the degrees of freedom df for its determination as well as the number n_a of the replicates in the routine analysis. The larger the number n the smaller will be the value $\Delta\bar{x}$. But the influence of n on the value of the confidence interval falls exponentially, as demonstrated in Fig. 2.2.4-2 [9]. Many replicates in routine quality control quickly increases costs, but the effect is only small. Double determinations are often sufficient.

However, the standard deviation of the analytical method s is direct proportional to $\Delta\bar{x}$. Thus, it has the biggest influence on the magnitude of $\Delta\bar{x}$. The determination of s is a unique procedure, and therefore a larger number of replicates should be made. On the other hand, the greater the number of replicates for the determination of s, the smaller the t-value.

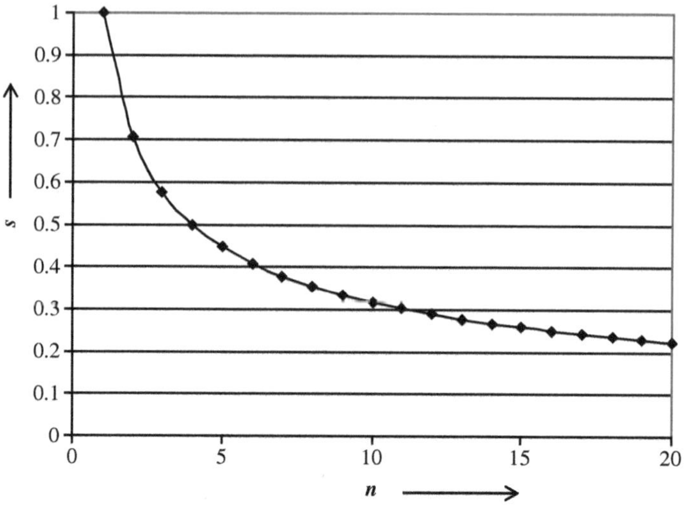

Fig. 2.2.4-2 Relation between s and the number of repeated measurements n [9]

Challenge 2.2.4-1

According to a company specification the content of benzene (bz) in technical n-hexane may not be greater than $L_O = 0.80\%$ (v/v). The analytical quality control will be carried out by GC with n-octane as an internal standard (IS).

The process standard deviation was determined with varying numbers of replicates of the same sample:

Method A: 12 individual samples

Method B: 6 individual samples.

The relative peak areas A_{bz}/A_{IS} obtained from the chromatograms are given in Table 2.2.4-1.

(continued)

Table 2.2.4-1 The relative peak areas A_{bz}/A_{IS} obtained from the chromatograms

Replicate	A_{bz}/A_{IS}	Replicate	A_{bz}/A_{IS}
Method A			
1	0.855	7	0.866
2	0.834	8	0.873
3	0.862	9	0.819
4	0.860	10	0.854
5	0.854	11	0.886
6	0.843	12	0.875
Method B			
1	0.788	4	0.796
2	0.772	5	0.747
3	0.769	6	0.758

A_{bz} is the peak area of benzene and A_{IS} is the peak area of the internal standard n-octane

Which mean value of benzene $\bar{x}_{bz}$ may not be exceeded if

(a) Double determinations or
(b) Fourfold determinations will be carried out in the quality control?

Evaluate the results.

Solution to Challenge 2.2.4-1
The critical mean value of benzene $\bar{x}_{crit,bz}$ which may not be exceeded is calculated according to (2.2.4-1):

$$\bar{x}_{crit.,bz} \leqslant L_0 - \frac{s \cdot t(\bar{P}_{one-sided}, df)}{\sqrt{n_a}}. \quad (2.2.4\text{-}2)$$

The intermediate quantities and the critical mean value of benzene $\bar{x}_{crit,bz}$ calculated according to (2.2.4-2) are given in Table 2.2.4-2 for the various conditions.

Table 2.2.4-2 Intermediate quantities and the limit value of benzene $\bar{x}_{crit,bz}$ calculated according to (2.2.4-2)

Parameter	Method A	Method B
Determination of the standard deviation s		
n	12	6
df	11	5
s in % (v/v)	0.0184	0.0182
$t(\bar{P}_{one-sided} = 95\%, df)$	1.796	2.015
Routine quality control		
n_a	2	2
$\Delta\bar{x}$ in % (v/v)	0.023	0.026
$\bar{x}_{crit,bz}$ in % (v/v)	0.777	0.774
n_a	4	4
$\Delta\bar{x}$ in % (v/v)	0.017	0.018
$\bar{x}_{crit,bz}$ in % (v/v)	0.783	0.782

As Table 2.2.4-2 shows, the critical mean value of benzene $\bar{x}_{crit,bz}$ differs only minimally with the various conditions. Double determinations in the routine quality control and determination of the standard deviation of the analytical method with twelve replicates yields the critical value $\bar{x}_{crit,bz} = 0.78\%$ (v/v). Increasing the numbers of replicates for the determination of s as well as in the routine quality control does not have a practical influence on the critical mean value.

Challenge 2.2.4-2
A company produces polystyrene for a certain application. The content of the residual monomer may not exceed 0.60% (w/w) styrene. The monomer will

(continued)

Table 2.2.4-3 Analytical results x in % (w/w) styrene obtained by two replicates with six polystyrene samples by MHE-HS-GC

Sample	1	2	3	4	5	6
x'	0.573	0.654	0.916	0.439	0.753	0.848
x''	0.525	0.691	0.972	0.489	0.812	0.892

be analyzed by MHE-HS-GC (see Chap. 7). The standard deviation of the analytical method was determined by two replicates with six samples. The results are listed in Table 2.2.4-3.

In the routine quality control of a sample the following analytical results were obtained:

% (w/w) styrene	0.562	0.591	0.559

Will the sample meet the quality requirement?

Solution to Challenge 2.2.4-2

The standard deviation of the analytical method s_x calculated according to (2.2.2-3) with $\sum (x' - x'')^2 = 0.014726\ (\%(\text{w/w}))^2$ and $m = 6$ is $s = 0.03503\%\ (w/w)$.

The confidence interval calculated by (2.2.3-1) with $\bar{x} = 0.5707\%$ (w/w), $t(\bar{P}_{\text{one-sided}} = 95\%, \text{df} = 6) = 1.943$, and $n_a = 3$ is $\Delta\bar{x} = 0.0393\%$ (w/w). Thus, the upper confidence limit is $\bar{x} + \Delta\bar{x}_{\text{one-sided}} = 0.61\%$ (w/w) styrene. This value exceeds the documented quality limit of $L_0 = 0.60\%$ (w/w), and therefore the sample does not fulfil the quality requirements. It cannot be delivered for sale.

2.2.5 *Propagation of Errors*

When the final result is obtained from more than one independent measurement, or when it is influenced by two or more independent sources of errors, these errors can be accumulated or compensated. This is called the propagation of errors.

In the case of independent variables $x_1, x_2, \ldots, x_n$, i.e. if there is no correlation between the x-values, i.e. the covariances

$$\text{cov}(x_1, x_2, \ldots, x_n) = \frac{1}{n-1} \cdot \left[\sum (x_{1i} - \bar{x}_1) \cdot (x_{21} - \bar{x}_2) \cdot \cdots \cdot (x_{ni} - \bar{x}_n)\right] = 0, \qquad (2.2.5\text{-}1)$$

the total error can be estimated according to the Gaussian *law of error propagation:*

$$\sigma_x^2 = \left(\frac{\partial f}{\partial x_1}\right)^2 \cdot \sigma_{x_1}^2 + \left(\frac{\partial f}{\partial x_2}\right)^2 \cdot \sigma_{x_2}^2 + \cdots + \left(\frac{\partial f}{\partial x_n}\right)^2 \cdot \sigma_{x_n}^2. \tag{2.2.5-2}$$

For addition or subtraction the variances are additive:

$$\sigma_x^2 = \sigma_{x_1}^2 + \sigma_{x_2}^2 + \cdots + \sigma_{x_n}^2. \tag{2.2.5-3}$$

For multiplication or division the squared relative standard deviations are additive:

$$\left(\frac{\sigma_x}{x}\right)^2 = \left(\frac{\sigma_{x1}}{x_1}\right)^2 + \left(\frac{\sigma_{x2}}{x_2}\right)^2 + \cdots + \left(\frac{\sigma_{x_n}}{x_n}\right)^2. \tag{2.2.5-4}$$

Note that, as mentioned above, these equations are correct only when the variables are independent, i.e. if they are not correlated!

Challenge 2.2.5-1
The content of a pharmaceutical product will be detected by HPLC.

The percentage content of the active pharmaceutical ingredient (API) $x_{\mathrm{API}}\%$ (w/w) is calculated by (2.2.5-5).

$$x_{\mathrm{API}}\%\ (\mathrm{w/w}) = \frac{\bar{A}_{\mathrm{s}}\ \text{in counts} \cdot 100}{(c_{\mathrm{s}}\ \text{in g L}^{-1})(\mathrm{Rf}\ \text{in counts L g}^{-1})} \tag{2.2.5-5}$$

$\bar{A}_{\mathrm{s}}$ is the mean peak area of the sample obtained by the chromatogram, c_{s} is the concentration of the sample, and Rf is the response factor which is determined with a solution of chemical reference substance (CRS) according to (2.2.5-6):

$$\mathrm{Rf} = \frac{\bar{A}_{\mathrm{CRS}}\ \text{in counts}}{(c_{\mathrm{CRS}}\ \text{in g L}^{-1}) \cdot (x_{\mathrm{CRS}}\ \text{in \% (w/w)}) \cdot 0.01}. \tag{2.2.5-6}$$

$\bar{A}_{\mathrm{CRS}}$ is the mean of the peak area of CRS, c_{CRS} is the concentration of CRS, and $x_{\mathrm{CRS}}\%$ (w/w) is the certified content of CRS.

According to the United States Pharmacopeia (USP) the relative standard deviation of the precision of injection of the sample should be $s_{\mathrm{r}}\% \leqslant 1.0$.

Testing the precision of injection, as usual in pharmaceutical analyses, a sample CRS was measured with six replicates. The peak areas obtained from the HPLC chromatograms are presented in Table 2.2.5-1.

The experimental data for the determination of the content of the API $x_{\mathrm{API}}\%$ (w/w) are given in Table 2.2.5-2.

Table 2.2.5-1 Peak areas A obtained from the HPLC chromatograms of a CRS solution

Replicate	A in counts
1	678,458
2	670,554
3	678,458
4	664,119
5	680,246
6	672,179

Table 2.2.5-2 Experimental data for determination of the API

Solutions	
Determination of Rf	$c_{\mathrm{CRS}} = 0.813 \text{ g L}^{-1}$
Determination of x_{API}	$c_{\mathrm{s}} = 0.803 \text{ g L}^{-1}$
Certified content of the CRS x_{CRS}	99.15% (w/w)
Peak areas A in counts obtained by the HPLC chromatograms	
Rf	Sample
114,856	112,969
115,681	111,781
114,836	111,876
113,592	113,006

(a) Test whether the claimed precision of injection is achieved;
(b) Calculate the API content of the sample with its confidence interval $\bar{x} \pm \Delta x\%$ (w/w) API.

Solution to Challenge 2.2.5-1

(a) The data set of Table 2.2.5-1 gives $s = 6,190.1$ counts, $\bar{A}_{\mathrm{CRS}} = 674,002.3$ counts, and $s_{\mathrm{r}}\% = 0.92$.

The relative standard deviation $s_{\mathrm{r}}\%$ is smaller than the limit value given in USP. This means the injection precision of the HPLC method is achieved.

(b) The intermediate quantities are:

- $\mathrm{Rf} = 142,343.1$ counts L g^{-1} calculated by (2.2.5-6) with $\bar{A}_{\mathrm{Rf}} = 114,741.3$ counts and further data given in Table 2.2.5-2
- $s^2_{A_{\mathrm{Rf}}} = 742,016.9$ counts^2,
- $\bar{A}_{\mathrm{s}} = 112,408.0$ counts,
- $s^2_{A_{\mathrm{s}}} = 449,492.7$ counts^2.

The content of the sample calculated by (2.2.5-5) is $\bar{x}_{\mathrm{API}}\%$ (w/w) $= 98.34$. The total variance s^2_{tot} for the determination of the API derived according to (2.2.5-2) is

(continued)

$$s_{\text{tot}}^2 = \left(\frac{100}{c_s \cdot \text{Rf}}\right)^2 s_{A_s}^2 + \left(\frac{100 \cdot \bar{A}_s}{c_s \cdot (\text{Rf})^2}\right)^2 s_{A_{\text{Rf}}}^2. \qquad (2.2.5\text{-}7)$$

s_{tot}^2 calculated by (2.2.5-7) is $s_{\text{tot}}^2 = 0.3576\ \text{g}^2\ \text{L}^{-2}$ and the standard deviation is $s_{\text{tot}} = 0.5980\ \text{g L}^{-1}$, respectively.

The confidence interval

$$\Delta\bar{x}_{\text{API}}\%\ (\text{w/w}) = \frac{s_{\text{tot}} \cdot t(P, \text{df})}{\sqrt{n}} \qquad (2.2.5\text{-}8)$$

is $\Delta\bar{x}\%\ (\text{w/w}) = 0.73$ calculated with $\text{df}_{\text{total}} = \text{df}_{\text{RF}} + \text{df}_s = 6$, $t(P = 95\%, \text{df} = 6) = 2.447$, and $n = 4$.

Result: The content of the sample is $\Delta\bar{x}\%\ (\text{w/w}) = 98.34 \pm 0.73$. The true value lies in the range 97.61–99.07% (w/w) at the significance level $P = 95\%$ and with the risk $\alpha = 5\%$ that the true value may be found outside this range.

Challenge 2.2.5-2

Let us now estimate the errors in photometric analysis which is an important method in AQA. As example, we will choose IR spectrophotometric analysis which must often be applied in AQA (see for example Challenge 3.3-3).

The spectrophotometric analysis is based on *Lambert–Beer's law*

$$A = \alpha \cdot c \cdot l \qquad (2.2.5\text{-}9)$$

where α is the absorptivity (a constant which is usually given in L mol^{-1} cm^{-1} or in m^2 mol^{-1}), c is the concentration in mol L^{-1}, and l is the optical path length, i.e. the diameter of the cuvette.

In IR spectrophotometry the optical path length lies in the μm range, and therefore it is determined by the interference method. The order of the interferences n which are obtained if the empty cuvette is traversed by IR light is calculated by

$$r = 2 \cdot l \cdot \nu_{\text{max}} \qquad (2.2.5\text{-}10)$$

where ν_{max} is the maximum of the interference, r is the number of the reflection, and l is the optical path length. According to (2.2.5-10) l is obtained from the slope of the function $n = f(2\nu)$.

Using standard calibration solutions with amount m in volume V, the absorptivity α is calculated by (2.2.5-11). Usually, in IR spectrophotometry the constant α is given in the units L g^{-1} cm^{-1} according to (2.2.5-11):

(continued)

$$\alpha = \frac{A \cdot V}{l \cdot m}, \tag{2.2.5-11}$$

which will also be used in the following.

Finally, let us turn to the absorbance A which is measured by the spectrophotometer.

The absorbance is defined by

$$A = \frac{\log I_0}{\log I} = \log I_0 - \log I, \tag{2.2.5-12}$$

where I_0 and I are the intensity of the reference beam and the intensity of the sample beam, respectively.

The error of the measurement of the absorbance is given by the law of error propagation according to (2.2.5-2):

$$\sigma_A^2 = \left(\frac{\partial A}{\partial I_0}\right)^2 \cdot \sigma_{I_0}^2 + \left(\frac{\partial A}{\partial I}\right)^2 \cdot \sigma_I^2. \tag{2.2.5-13}$$

From (2.2.5-12) follows

$$\sigma_A^2 = \left(\frac{\partial(\log I_0 - I)}{\partial I_0}\right)^2 \cdot \sigma_{I_0}^2 + \left(\frac{\partial(\log I_0 - I)}{\partial I}\right)^2 \cdot \sigma_I^2, \tag{2.2.5-14}$$

which gives (2.2.5-15), and with log $e = 0.43$ (2.2.5-16), respectively

$$\sigma_A^2 = \left(\frac{\log e}{I_0}\right)^2 \cdot \sigma_{I_0}^2 + \left(\frac{\log e}{I}\right)^2 \cdot \sigma_I^2 \tag{2.2.5-15}$$

$$\sigma_A^2 = \left(\frac{0.43}{I_0}\right)^2 \cdot \sigma_{I_0}^2 + \left(\frac{0.43}{I}\right)^2 \cdot \sigma_I^2. \tag{2.2.5-16}$$

The Challenges are:

(a) Derive the equation for variance of the absorptivity σ_α^2 from the law of propagation of errors.

(b) A problem in spectrophotometry is the magnitude of the chosen *absorbance A*. Decide if the relative error of the measurement of the absorbance is constant or variable. Derive the relation for this relative error and create a graph for the relative error of the measurement of the absorbance using values in the range 0.025–2.5 in appropriate steps. Estimate the result with regard to the choice of an optimal range for the measurement of A.

(continued)

(c) A further parameter in IR spectrophotometry is the magnitude of the slit width and the *slit width program*, respectively. Decide which slit width program should be chosen.
A tip: consider the fact that I_0 grows with the square of the slit width.
(d) Calculate the absorptivity α in L $cm^{-1}g^{-1}$ with its random error for the carbonyl band of lactic acid ester (LAE) at 1,735 cm^{-1} from the following data:

Sample solution	m = 50.28 mg LAE in 10 mL n-hexane		
Absorbance A	0.391525	0.391701	0.392668
	0.393124	0.392147	0.391010

The diameter of the cuvette is determined by the interference maxima given in Table 2.2.5-4.

The random errors of mass m, diameter of the cuvette l, and volume V are estimated from the data sets given in Tables 2.2.5-3 and 2.2.5-5.

Note that the influence of temperature, the tolerance of the volumetric flask, and other factors are neglected here. This is the subject of Chap. 10. Calculate the percentage of the individual variances in the variance of the absorptivity.
(e) Test whether a fivefold increase in sample volume will appreciably diminish the random error σ_A. The procedure for the determination of

(continued)

Table 2.2.5-3 Estimation of the random error of the balance

Number	Gross weight in g	Tare weight in g	Number	Gross weight in g	Tare weight in g
1	6.19740	6.09748	6	6.13155	6.03159
2	6.09595	5.99596	7	6.22193	6.12196
3	6.13175	6.03178	8	6.09995	6.00000
4	6.13467	6.03472	9	6.07420	5.97420
5	6.06939	5.96935	10	6.08567	5.98577

Table 2.2.5-4 Estimation of the random error of the diameter of the cuvette l from interference maxima I_{max} measured with the empty cuvette

Order number r	I_{max}	Order number r	I_{max}
1	793	11	1,128
2	829	12	1,160
3	861	13	1,191
4	895	14	1,226
5	927	15	1,259
6	960	16	1,292
7	993	17	1,327
8	1,027	18	1,358
9	1,060	19	1,394
10	1,092	20	1,425

Table 2.2.5-5 Estimation of the random error of the volume $V = 10$ mL. A 10 mL volumetric flask was filled up with water and the mass was determined.

Number	m in g	Number	m in g
1	9.964761	6	9.962722
2	9.974138	7	9.989396
3	9.983647	8	9.972522
4	9.985056	9	9.983176
5	9.997446	10	9.969295

Table 2.2.5-6 Estimation of the random error of the volume with $V = 50$ mL. A 50 mL volumetric flask was filled up with water and the mass was determined.

Number	m in g	Number	m in g
1	50.063150	6	50.061528
2	50.090792	7	50.116459
3	50.051704	8	50.116849
4	50.066276	9	50.031270
5	50.052144	10	50.097729

volume error is the same as for 10 mL volumes (Table 2.2.5-5). The experimental values are listed in Table 2.2.5-6.

Solution to Challenge 2.2.5-2

(a) Using the law of propagation of errors, equation (2.2.5-2), the error of the absorptivity α is given by (2.2.5-17):

$$\sigma_\alpha^2 = \underbrace{\left(\sigma_A \cdot \frac{V}{l \cdot m}\right)^2}_{\text{I}} + \underbrace{\left(\sigma_V \cdot \frac{A}{l \cdot m}\right)^2}_{\text{II}} + \underbrace{\left(\sigma_l \cdot \frac{-A \cdot V}{l^2 \cdot m}\right)^2}_{\text{III}} + \underbrace{\left(\sigma_m \cdot \frac{-A \cdot V}{l \cdot m^2}\right)^2}_{\text{IV}} \quad (2.2.5\text{-}17)$$

Term I represents the error in the measurement of the absorbance, term II that for the volume of the measuring solution, term III that of the optical path length, and term IV that for the weight of the mass of the sample for the preparation of the measuring solution.

(b) The equation for the relative error in the measurement of the absorbance σ_A/A is obtained from (2.2.5-15) with $\sigma_{I_0} = \sigma_I = 1$:

$$\frac{\sigma_A}{A} = \frac{\lg e}{A} \cdot \sqrt{\frac{1}{I_0^2} + \frac{1}{I^2}}. \quad (2.2.5\text{-}18)$$

Next, I is substituted as follows:

From (2.2.5-12) one obtains for I

$$\log I = \log I_0 - A \quad (2.2.5\text{-}19)$$

(continued)

and

$$I = 10^{(\log I_0)-A} = \frac{I_0}{10^A}. \quad (2.2.5\text{-}20)$$

Equation (2.2.5-20) in (2.2.5-18) gives:

$$\frac{\sigma_A}{A} = \frac{\log e}{A} \cdot \sqrt{\frac{1}{I_0^2} + \frac{10^{2A}}{I_0^2}} = \frac{\log e}{I_0} \cdot \frac{\sqrt{1+10^{2A}}}{A}. \quad (2.2.5\text{-}21)$$

Equation (2.2.5-21) shows that the relative error for the measurement of the absorbance is not constant, but is a function of the absorbance.

The relative errors for the measurement of the absorbance for a chosen data set calculated by (2.2.5-21) with $I_0 = 1$ and $\log e = 0.43$ are listed in Table 2.2.5-7 and the graph is presented in Fig. 2.2.5-1.

As the graph shows, the minimum of the relative errors of the measurement of the absorbance is in the range from about 0.3 to about 1.0. Therefore, this is an optimal range for spectrophotometry. For solutions with absorbance lower than 0.1 or greater than 2.0 the relative error rises rapidly.

(c) As (2.2.5-18) shows, the relative error of the measurement of the absorbance diminishes with increasing I_0. Because I_0 grows with the square of the slit width, one should take the largest slit width or the split program.

(d) The diameter of the cuvette l is determined by the regression analysis of the data set in Table 2.2.5-4 and is the slope of the function $n = f(2v)$ according to (2.2.5-10), $l = 0.01505$ cm. The standard error of the diameter corresponds to the standard deviation of the slope calculated by (5.2-13), which is $\sigma_l = 1.833 \times 10^{-5}$ cm obtained with $SS_{xx} = 2{,}934{,}826.2$ and $s_{y.x} = 0.0314027$.

(continued)

Table 2.2.5-7 Relative errors of the measurement of the absorbance $s_{r,A} = \sigma_A/A$ calculated by (2.2.5-21)

A	$\frac{\sigma_A}{A}$	A	$\frac{\sigma_A}{A}$	A	$\frac{\sigma_A}{A}$
0.020	31.13	0.400	2.91	0.850	3.62
0.025	25.06	0.450	2.86	0.900	3.83
0.050	12.93	0.500	2.85	0.950	4.06
0.100	6.91	0.550	2.88	1.000	4.32
0.150	4.96	0.600	2.94	1.250	6.13
0.200	4.03	0.650	3.03	1.500	9.07
0.250	3.51	0.700	3.14	1.750	13.82
0.300	3.20	0.750	3.27	2.000	21.50
0.350	3.01	0.800	3.43	2.250	33.99

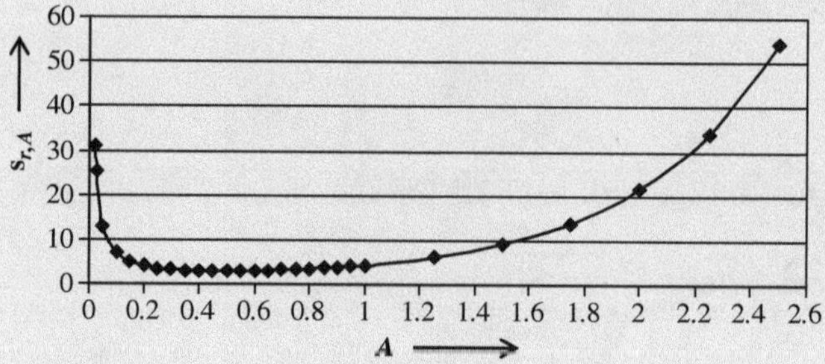

Fig. 2.2.5-1 Relative errors of the measurement of the absorbance $s_{r,\,A}$ as a function of the absorbance A

The mean value of the absorbance $\bar{A}$ is 0.392029. The absorptivity α is calculated according to (2.2.5-11):

$$\alpha = \frac{0.392029 \cdot 0.01\ \text{L}}{0.01505\ \text{cm} \cdot 0.05028\ \text{g}} = 5.18\ \text{L cm}^{-1}\text{g}^{-1}. \qquad (2.2.5\text{-}22)$$

The standard deviation of the measurement of the absorbance is $\sigma_A = 7.773 \times 10^{-4}$.

The standard deviation of the net values (difference between gross and tare of the data in Table 2.2.5-3) is $\sigma_m = 3 \cdot 98 \cdot 10^{-5}$ g.

The standard deviation of the filling of the volumetric flask is calculated using the data set in Table 2.2.5-5: $\sigma_V = 0.01128$ mL for a 10 mL volumetric flask.

The random error of the absorptivity α is calculated by (2.2.5-17) with the parameters $V = 0.01$ L, $m = 0.05028$ g, $l = 0.01505$ cm, $A = 0.392029$, $\sigma_V = 1.128 \cdot 10^{-5}$ L, $\sigma_m = 3.979 \cdot 10^{-5}$ g, $\sigma_l = 1.833 \cdot 10^{-5}$ cm, and $\sigma_A = 0.0007773$.

The result is $\sigma_\alpha^2 = 0.0001962$ and $\sigma_\alpha = 0.0140$.

The absorptivity α calculated according to (2.2.5-11) is $\alpha = 5.18$ L g^{-1}cm^{-1}. Thus, the relative standard deviation calculated by (2.2.2-5a) is $s_r\% = 0.27$.

The percentages of the individual variances in the total variance σ_α^2 are given in Table 2.2.5-8.

As Table 2.2.5-8 shows, the measurement of the absorbance has the greatest influence on the random error of the absorptivity α, but one has to consider that all the uncertainties of Type B were rejected here.

(e) The standard deviation of the filling of the volumetric flask is calculated with the data set in Table 2.2.5-6: $\sigma_V = 2.908 \cdot 10^{-5}$ L for a 50 mL volumetric flask. The variance of the absorptivity is $\sigma_\alpha^2 = 0.000155$ calculated with the fivefold-increased mass $m = 2.514$ g according to

(continued)

Table 2.2.5-8 Percentages of the individual variances in the total variance of the absorptivity σ_α^2

σ_m^2 mass	σ_V^2 volume	σ_l^2 cuvette	σ_A^2 absorbance
8.6%	17.4%	20.3%	53.8%

(2.2.5-17). The standard deviation of the absorptivity is $\sigma_\alpha = 0.01245$ and $s_r\% = 0.24$.

The increase in sample volume does not improve the precision in practice, but would only incur higher costs for the sample and the solvent.

References

1. ISO 3534-2 (2006) Statistics – vocabulary and symbols – part 2: applied statistics. International Organization for Standardization, Geneva
2. Joint Committee for Guides in Metrology (JCGM) (2008) International vocabulary of metrology – basic and general concepts and associated terms (VIM). International Organization for Standardization, Geneva
3. ISO Guide 98 (1995) Guide to the expression of uncertainty in measurement. International Organization for Standardization, Geneva
4. Danzer K (2007) Analytical chemistry – theoretical and metrological fundamentals. Springer, Berlin
5. Menditto A, Patria M, Magnusson B (2007) Understanding the meaning of accuracy, trueness and precision. Accred Qual Assur 12:45–47
6. Danzer K (1989) Robuste Statistik in der analytischen Chemie. Fresenius Z Anal Chem 335:869–875
7. Massart DL, Vandeginste BGM, Buydens LMC, De Jong S, Lewi PJ, Smeyers-Verbeke J (1997) Handbook of chemometrics and qualimetrics – part A. Elsevier, Amsterdam
8. Ellison SLR, Berwick VJ, Duguid Farrant TJ (2009) Practical statistics for the analytical scientist, 2nd edn. RSC, Cambridge
9. Doerffel K (1990) Statistik in der analytischen Chemie, 5 Aufl. Deutscher Verlag für Grundstoffindustrie, Leipzig
10. Danzer K (1989) Robuste Statistik in der analytischen Chemie. Fresenius Z Anal Chem 335:869–875

Chapter 3
Statistical Tests

3.1 General Remarks

Hypothesis testing is a very important part of statistics, and can be used to investigate whether the mean values of two or more series of measurement are equal, whether the variances of two or more data sets are identical, etc.

The hypothesis tests used in AQA are carried out using the following steps:

1. *Stating the null and alternative hypotheses*
   Statistical tests are based on the null and alternative hypotheses. The *null hypothesis* H_0 is that there is no difference between the values being compared. For example, when the mean values of two data sets are compared, the null hypothesis is that the population means are equal or, in other words, the mean values μ_1 and μ_2 belong to the same population.
   The shorthand notation is
   H_0: $\mu_1 = \mu_2$.
   In the case that the null hypothesis is not true, one needs to formulate an *alternative hypothesis* H_1 (or H_A):
   H_1: $\mu_1 \neq \mu_2$.
   The alternative hypothesis is that the population means are not equal, i.e. the mean values μ_1 and μ_2 differ significantly and belong to different populations. The alternative hypothesis is confirmed if the null hypothesis has to be rejected. The alternative hypothesis H_1: $\mu_1 \neq \mu_2$ does not make any assumptions about the sign of the difference, but sometimes this can be important.
   The following cases can be distinguished:

   H_0: $\mu_1 = \mu_2$ (the population means are equal),
   H_1: $\mu_1 > \mu_2$ (μ_1 is significantly higher than μ_2), or another alternative hypothesis:
   H_1: $\mu_1 < \mu_2$ (μ_1 is significantly lower than μ_2).

2. *Checking the distribution of the data*
   Significance tests obtained in AQA mostly assume that the data are approximately normally distributed. Appropriate tests for normal distribution will be given in Sect. 3.2.1. Significance tests can give misleading results if the assumptions are not appropriate for the data sets.

M. Reichenbächer and J.W. Einax, *Challenges in Analytical Quality Assurance*,
DOI 10.1007/978-3-642-16595-5_3,

3. *Selection and calculation of the appropriate test*
   Each test has to be carried out by using a particular equation for the calculation of the test value, which is marked by a particular sign, for example $\hat{t}$ or $\hat{F}$. This is the subject of the following chapters.
4. *Comparing the calculated test value with the critical value*
   In order to decide whether the null hypothesis can be accepted or must be rejected, the calculated test value, for example $\hat{t}$ or $\hat{F}$, has to be compared with the critical value. If the calculated test value is greater than the critical value, in this case $t(P, \mathrm{df})$ and $F(P, \mathrm{df}_1, \mathrm{df}_2)$, respectively, the null hypothesis has to be rejected and the alternative hypothesis is valid.

   The appropriate critical value is determined by

- a level of significance P
- the number of "tails" (one- or two-sided tests)
- the degrees of freedom df.

Each test result is only valid for a certain freely chosen level of significance P. For the majority of tests a significance level $P = 95\%$ is used. Note this corresponds to the risk $\alpha = 5\%$.

The confidence interval (CI) is related to the risk α as follows:

$$\mathrm{CI}\% = 100(1 - \alpha). \tag{3.1-1}$$

Thus, a risk of $\alpha = 0.05$ is equivalent to a confidence interval of CI $= 95\%$.

In cases where great consequence is attached to the test result, a higher confidence interval $P = 0.99$ and $P = 99\%$, respectively, with the lower risk $\alpha = 0.01$ must be chosen. If H_0 for the significance level $P = 99\%$ is rejected, then the difference is *highly* significant.

The alternative hypothesis given above H_1: $\mu_1 \neq \mu_2$ means only that there is a difference between the means in *either direction*, i.e. μ_1 may be greater or less than μ_2. This is known as a *two-sided, two-tail*, or *two-tailed hypothesis test.*

But there are situations where we are concerned only in knowing whether the mean of one data set is "greater than" or "smaller than" that of the other. These alternative hypotheses tests are H_1: $\mu_1 < \mu_2$ and H_1: $\mu_1 > \mu_2$, respectively. In both cases we are only interested in whether there is a difference between the means in *one direction*. This hypothesis test is called a *one-sided, one-tail*, or *one-tailed hypothesis test.*

The distinction between one- and two-sided tests is important because of the various significance limits as shown in Fig. 3.1-1 for the normal distribution of a mean $\bar{x}$ around μ at the risk $\alpha = 0.05$. The distance a is the interval within which H_0 would be accepted for the two-sided test and b is the interval in which H_0 would be accepted for a one-sided test. 95% of the data lie within the limits -1.96σ and $+1.96\sigma$ for the two-sided test (distance a) and within the limits $-\infty$ and $+1.65\sigma$ for the one-sided test (distance b).

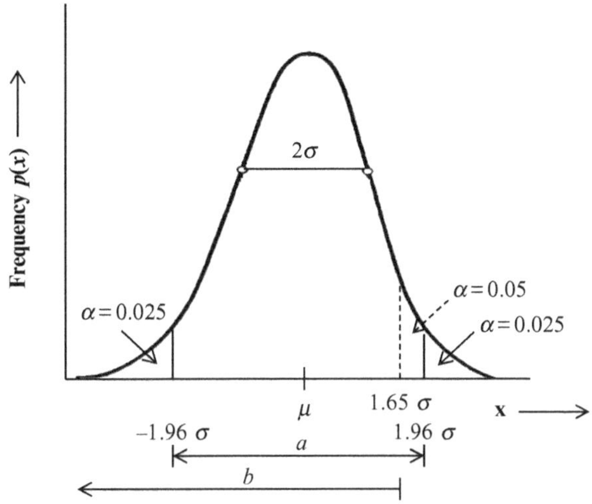

Fig. 3.1-1 One-sided decision limit (at $\mu + 1.65\sigma$) compared to two-sided limit (between $\mu - 1.96\sigma$ and $\mu + 1.96\sigma$). (**a**) Interval within which H_0 would be accepted for the two-sided test. (**b**) Interval within which H_0 would be accepted for a one-sided test

Table 3.1-1 Kinds of errors and the reality

Decision of the test	Actual condition	
	H_0 is true	H_1 is true
H_0 is not rejected	True decision with the significance level $P = 1 - \alpha$	Error of second kind (Type II) β-error
H_0 is rejected	Error of first kind (Type I) α-error	True decision

In statistics, two types are used to describe potential errors made in a statistical decision process:

- Type I error (*α-error*): H_0 is rejected although it is true (false positive decision or false alarm).
- Type II error (*β-error*): H_0 is erroneously not rejected although the alternative hypothesis is true (false negative decision).

The relation between the null hypothesis and the actual condition – the reality – is summarized in Table 3.1-1.

3.2 Tests for Series of Measurements

The *precision* of the analytical method given by the standard deviation is an important validation parameter. But the following three requirements have to be fulfilled for the calculation of the standard deviation:

- The data set must be *normally distributed.*
- The data set must be free of *outliers*.
- The data may not show a *trend* with the respect to their time of measurement.

The fulfilment of these requirements can be verified by statistical tests.

3.2.1 Rapid Test for Normal Distribution (David Test)

Several tests can be used to verify whether data are based on a normal distribution (e.g. χ^2 test, Kolmogorov–Smirnov test, Shapiro–Wilk test) [1], but in AQA the rapid test by David is usually preferred. The test value $\hat{q}_r$ is given by (3.2.1-1):

$$\hat{q}_r = \frac{x_{max} - x_{min}}{s}. \tag{3.2.1-1}$$

The parameters x_{max} and x_{min} are the greatest and the lowest values in the series of measurement, respectively, and s is the standard deviation.

The data are normally distributed according to David if the calculated value $\hat{q}_r$ is inside the boundaries of the David table for a given significance level P (see Table A-8).

Challenge 3.2.1-1
The data set given in Table 2.2.1-1 was tested for normal distribution by the histograms in Challenge 2.1-1.

Which result does the statistical test yield?

Solution to Challenge 3.2.1-1
With $x_{max} = 109.7$, $x_{min} = 90.5$, and $s = 4.857$ the test value calculated by (3.2.1-1) is $\hat{q}_r = 3.95$. This value is inside the boundaries of the David table for $n = 40$ at $P = 95\%$ $(3.67 < 3.95 < 5.16)$ and $P = 99\%$ $(3.47 < 3.95 < 5.56)$, respectively.

3.2.2 Test for a Trend

A trend is a progressively decreasing or increasing drift of measured values in chronological order. A trend is an indicator that a process is not under *statistical control*. If this is the case, statistical parameters cannot be calculated. Therefore, it is evident that trends must be avoided and data sets with a trend have to be rejected.

Sometimes, the presentation of the data in chronological order, as for example in control charts (see Chap. 8), can visually indicate the trend in a data set. But without control charts a statistical test has to be used for the detection of a trend unless a drift can be visually recognized for definite. The simple test by Neumann may be used in AQA [2].

In the Neumann test for a trend, the test value is the ratio of the variances of the $(n-1)$ pairs of consecutive data of a measurement series in chronological order $x_1, x_2, \ldots, x_n$ (Δ^2) to the variance of the data values s^2 themselves:

$$\Delta^2 = \frac{\sum\limits_{i=1}^{n} (x_i - x_{i+1})^2}{n-1}, \qquad (3.2.2\text{-}1)$$

$$\frac{\Delta^2}{s^2} = \frac{\sum\limits_{i=1}^{n} (x_i - x_{i+1})^2}{\sum\limits_{i=1}^{n} (x_i - \bar{x})^2}. \qquad (3.2.2\text{-}2)$$

$\sum\limits_{i=1}^{n} (x_i - \bar{x})^2$ is the sum of squares SS_i which may be calculated by the Excel function =DEVSQ(Data).

Consecutive values are considered independent at a significance level P if the test value calculated by (3.2.2-2) is *larger* than a critical limit tabulated by Neumann (Table A-10) for a given sample size n.

In analytical practice a trend will often be detected with this simple relation: a trend must be considered if $\Delta^2 < 2s^2$, and a consecutive series of measurements can be assumed to vary randomly and not show a trend if $\Delta^2 \geq 2s^2$.

Challenge 3.2.2-1

The analytical results for the determination of benzene in three samples of waste water with HS-GC are given in Table 3.2.2-1.

Test whether the mean value may be calculated for both samples and evaluate the results.

Table 3.2.2-1 Analytical results (given in µg L^{-1}) for the determination of benzene of waste water in chronological order

Sample	x_1	x_2	x_3	x_4	x_5	x_6
1	3.13	3.19	3.18	3.24	3.25	3.28
2	3.13	3.19	3.18	3.24	3.25	3.26
3	3.14	3.12	3.15	3.13	3.12	3.17

Solution to Challenge 3.2.2-1

The inspection of the data sets in Table 3.2.2-1 shows that the measured values for sample 3 are randomly distributed, but the chronological order of the values for samples 1 and 2 reveals a trend because the values rise steadily. Therefore, the data must be checked by a trend test. The results calculated by (3.2.2-2) are listed in Table 3.2.2-2.

The exact (Table 3.2.2-2) as well as the rough test (Table 3.2.2-3) deliver an unequivocal result only for sample 3: in sample 3 no trend is detected then the critical values are smaller than the calculated one at each significance level, and the rough test $\Delta^2 \geq 2s^2$ also fulfils the conditions for a trend-free data set.

The situation for samples 1 and 2 is different. A trend is found by the exact test in both samples at the significance level $P = 95\%$, but for $P = 99\%$ only the data set for sample 1 shows a trend, whereas the calculated test value is somewhat greater than the critical value of sample 2.

Strictly speaking, no trend is proved, but according to the result of the rough test as well as the result at the significance level $P = 95\%$, the data set of sample 2 should also be rejected. The origin of the drift in the data set of sample 2 should be sought, and the determination should be repeated after the cause is removed.

Of course, the measured values of sample 3 must be also checked for normal distribution.

The test value $\hat{q}_r = 2.576$ calculated by (3.2.1-1) lies between the lower (2.28) and upper limits (3.012) of the David table for $P = 95\%$, which means that the data may be regarded as normally distributed. The standard deviation, the mean value, and further parameters may be calculated with the data of sample 3.

Table 3.2.2-2 Results for the Neumann test

Sample	$\sum(x_i - x_{i+1})^2$	$\sum(x_i - \bar{x})^2$	Test value (3.2.2-2)
1	0.0083	0.015083	0.5503
2	0.0075	0.012683	0.5913
3	0.0043	0.001883	2.2832
Critical value for $P = 95\%, n = 6$			0.8902
Critical value for $P = 99\%, n = 6$			0.5615

Table 3.2.2-3 Results for the rough test

Sample	$2{\cdot}s^2$	Δ^2	Result
1	0.00603	0.00166	$\Delta^2 < 2{\cdot}s^2$
2	0.00507	0.00150	$\Delta^2 < 2{\cdot}s^2$
3	0.00075	0.00086	$\Delta^2 > 2{\cdot}s^2$

3.2.3 Test for Outliers

There are some statistical tests for outliers in series of measurements, but in AQA the tests by Dixon and Grubbs are usually applied [3]. Note that an outlier value x^* must always be only the highest or the lowest value in a series of measurements.

According to DIN EN 53 804-1 [3], the *Dixon* test must be used for measurement series up to $n \leq 29$. The measured values are sorted in ascending or descending order, depending on whether the suspected outlier value is the lowest or the highest value. The test value $\hat{Q}$ calculated by (3.2.3-1)

$$\hat{Q} = \frac{|x_1^* - x_b|}{|x_1^* - x_k|} \tag{3.2.3-1}$$

depends on the magnitude of the data set n. The indices b and k have the following values:

$b = 2$	for $3 \leq n \leq 10$	$b = 3$	for $11 \leq n \leq 25$
$k = n$	for $3 \leq n \leq 7$	$k = n - 1$	for $8 \leq n \leq 13$
$k = n - 2$	for $14 \leq n \leq 29$		

The test value $\hat{Q}$ is compared with the critical limit $Q(P, n)$ given in Table A-7. An outlier value is statistically detected at the significance level P if $\hat{Q}$ is greater than the critical value $Q(P, n)$.

In practice, the statistical outlier test according to *Grubbs* is used for nearly all measurement series but DIN EN 53 804-1 [3] recommends at least 30 replicates for a reliable performance of this statistical test. Note that some regulatory documents demand the Grubbs test independent of the data size. With the subdivision of the test according to n, different results obtained by both tests are avoided.

The test value of the Grubbs test $\hat{r}_m$ is calculated by (3.2.3-2).

$$\hat{r}_m = \frac{|x^* - \bar{x}|}{s}. \tag{3.2.3-2}$$

An outlier value is statistically detected with the significance level $\bar{P}_{\text{one-sided}}$ if the test value $\hat{r}_m$ is greater than the critical value $r_m(\bar{P}_{\text{one-sided}}, n)$ given in Table A-6.

Note that each outlier value detected with a statistical test in a series of measurements has to be rejected from this series.

Because the tests by Dixon and Grubbs often do not yield the same result, the outlier test by *Hampel* can be used in addition. This test is based on robust statistics using the median, which is more robust than the mean value $\bar{x}$. An outlier must always be the highest or the lowest value. When one of these values is removed as a suspected outlier, both the mean value and the standard deviation become smaller, which results in changes to the values found by the Grubbs test. But after removing an outlier checked by the Hampel test, the re-calculated test values will usually remain the same. In contrast to other tests where only one outlier can be discarded or the outliers are discarded sequentially, the Hampel test makes no assumptions about the potential outlier(s).

The test values are calculated by the following steps [4]:

1. Calculation of the median $\tilde{x}$.
2. Calculation of the absolute residuals of the median:

$$r_i = |x_i - \tilde{x}|. \tag{3.2.3-3}$$

3. Calculation of the median of the absolute deviations (MAD) according to (2.2.1-8) and (2.2.1-9), respectively.
4. Calculation of the test values $\hat{H}_i$ for all observations i:

$$\hat{H}_i = \frac{r_i}{5.06\,\mathrm{MAD}}. \tag{3.2.3-4}$$

If the test value $\hat{H}_i$ is greater than 1, this observation is regarded as an outlier at the significance level $P = 95\%$.

Box and whisker plots:
The box and whisker plot (also called "box plot") is a type of graph which is used to show the shape of the distribution, its central value, and its spread, which allows a visual representation of the data. It is helpful for indicating whether a distribution is skewed and whether there are any unusual observations (outliers) in the data set.

Box plots are constructed as following:

1. Calculate the median according to (2.2.1-8) and (2.2.1-9) or by the Excel function =MEDIAN(Data).
2. Calculate the *first* (lower) and the *third* (upper) *quartiles* *Q1* and *Q3*, respectively. Quartiles, by definition, separate a quarter of data points from the rest. The first quartile *Q1* is the value *under* which 25% of the data lie and the third quartile *Q3* is the value *over* which 25% of the data are found. Note the second quartile *Q2* is the median itself. The calculation of quartiles can be verified by the Excel function = QUARTILE(Data, 1), and = QUARTILE(Data, 3), respectively.
3. Calculate the *interquartile range* (IQR) which is the difference between *Q3* and *Q1*:

$$\mathrm{IQR} = Q3 - Q1 \tag{3.2.3-5}$$

4. Calculate the *lower* and *upper whisker lines*, L_W and U_W, respectively:

$$L_W = Q1 - 1.5 \cdot \mathrm{IQR}, \tag{3.2.3-6}$$

$$U_W = Q3 + 1.5 \cdot \mathrm{IQR}. \tag{3.2.3-7}$$

5. Construct the graph of the *box plot* with ends corresponding to *Q1* and *Q3* in which the median is represented by a horizontal bar.

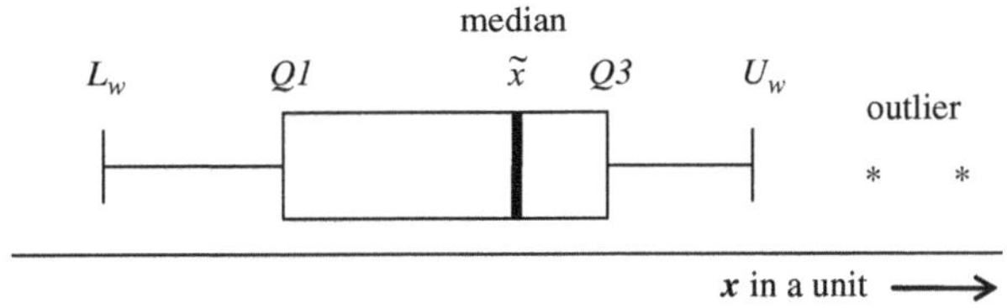

Fig. 3.2.3-1 Example of a box and whisker plot

6. Draw a vertical line from each end to the lower and the upper whisker line (shown by a small horizontal line) which is the most remote data points that are not outliers.
7. *Outliers* are indicated by points outside the whiskers.

Figure 3.2.3-1 shows an example of a box and whisker plot.

Box and whisker plots allow a visual interpretation of data sets. The median shows the central line for each group, the length of the box indicates the dispersion of the data, its range is characterized by the whiskers, and outliers are shown as points which lie outside of the whiskers.

The bar of the median situated outside the box indicates a skew distribution of the data.

Furthermore, box plots are also useful for the comparison of different groups of data.

Challenge 3.2.3-1

(a) In Table 2.2.2-3 the analytical results are given for the determination of Mn in steels. The largest value for sample 5 $x = 1.21\%$ (w/w) may be a suspect value. Test whether it is an outlier using the appropriate method.

(b) According to the visual inspection of the analytical data for method *D* in Table 2.1-1, value 82 was regarded as an outlier. This assumption is to be confirmed by a statistical test.

Solution to Challenge 3.2.3-1

(a) For a short data set with $n = 4$ the Dixon test must be used according to [3].

Because the suspect value is the highest in the series of measurement, the values have to be sorted in descending order. Note that the suspect value is always x_1 (Table 3.2.3-1).

The calculation for the Dixon test is

$$\hat{Q} = \frac{|x_1^* - x_2|}{|x_1^* - x_n|} = \frac{(1.21 - 1.19)}{(1.21 - 1.17)} = 0.500.$$

(continued)

Table 3.2.3-1 Ordered values of the data of sample 5 in Table 2.2.2-3

x_1	x_2	x_3	$x_4 = x_n$	
1.21	1.19	1.18	1.17	% (w/w)

Table 3.2.3-2 Ordered values of method *D* in Table 2.1-1

x_1	x_2	x_3	x_4	x_5	x_6	x_7
82	97	98	99	100	101	102

The critical value for $n = 4$ at the significance level $P = 95\%$ obtained from Table A-7 is $Q = 0.765$, which is greater than the test value $\hat{Q}$, and therefore the measured value 1.21 is not an outlier.

Although the Dixon test must be applied according to DIN EN 53 804-1 [3], we are interested in the result of the Grubbs test.

With $s = 0.01708$ and $\bar{x} = 1.187$, the Grubbs test value calculated by (3.2.3-2) is $\hat{r}_m = 1.317$. The test value is smaller than the critical value $r_m(\bar{P} = 95\%, n = 4) = 1.463$, and therefore the value 1.21 is not an outlier, which is the same result as obtained by the Dixon test.

(b) The suspect value is the lowest one, and therefore sorting the measured values in ascending order is necessary, which is given in Table 3.2.3-2.

The calculation for the Dixon test for $n = 6$ with $b = 2$ and $k = n$ is:

$$\hat{Q} = \frac{|x_1^* - x_2|}{|x_1^* - x_n|} = \frac{|82 - 97|}{|82 - 102|} = 0.750. \qquad (3.2.3\text{-}6)$$

The critical value $Q\,(P = 95\%,\ n = 7) = 0.507$ is smaller than the test value $\hat{Q}$, and thus the suspected outlier can be confirmed at the significance level $P = 95\%$.

The same result is obtained by the Grubbs test with $s = 6.831$ and $\bar{x} = 97$, the test value calculated by (3.2.3-2) is $\hat{r}_m = 2.196$. The test value exceeds the limit of the critical value $r_m(\bar{P} = 95\%,\ n = 7) = 1.938$.

Challenge 3.2.3-2

The analytical results of the determination of benzene in waste water with HS-GC are given in Table 3.2.3-3.

In order to calculate the mean value, the data sets have to be checked for outliers. Test both samples for outliers. Evaluate the result.

Table 3.2.3-3 Analytical results (given in μg L^{-1}) of the determination of benzene in waste water by HS-GC

Replicate	Sample	
	1	2
1	1.234	1.234
2	1.251	1.251
3	1.226	1.226
4	1.238	1.238
5	1.531	1.531
6	1.278	1.278
7	1.363	
8	1.214	

Solution to Challenge 3.2.3-2

According to DIN EN 53 804-1 [3], the Dixon test must be applied for the data sizes $n = 8$ and $n = 6$, respectively.

The test value is calculated for sample 1 with $n = 8$ by (3.2.3-7)

$$\hat{Q} = \left| \frac{x_1^* - x_2}{x_1^* - x_{n-1}} \right| \tag{3.2.3-7}$$

and for sample 2 with $n = 6$ by (3.2.3-8)

$$\hat{Q} = \frac{|x_1^* - x_2|}{|x_1^* - x_n|}. \tag{3.2.3-8}$$

The values x_1^*, x_2, and x_{n-1} are obtained by Excel functions which are given together with intermediate quantities and the test values $\hat{Q}$ in Table 3.2.3-4.

Note that the data set of sample 2 is the same as replicates 1–6 of sample 1. While the highest value $x_{\max} = 1.531$ is not an outlier in the data set of sample 1, it is confirmed as an outlier in the data set of sample 2 because the test value $\hat{Q} = 0.830$ exceeds the critical value at the significance level $P = 95\%$.

Using the Grubbs test, $x_{\max} = 1.531$ is also confirmed as an outlier. The test value $\hat{r}_m = 2.226$ calculated with $s = 0.1074$ and $\bar{x} = 1.2919$ is higher than the critical value $r_m(P = 95\%, n = 8) = 2.032$.

Because it is possible that the results obtained by both statistical tests may be different, it is necessary in AQA to document the procedure used, or, better, if there is no established statistical test for an outlier in the regulatory documents, one should apply the outlier test according to DIN 51 848-3 and 53 804-1.

Table 3.2.3-4 Excel functions, intermediate quantities, and test values $\hat{Q}$ of the Dixon outlier test

Function	x_1^*	x_2	x_{n-1}	$\hat{Q}$
Sample 1				
$x_1^* = x_{max}$				
= MAX(Data)	1.531			
= LARGE(Matrix, 2)		1.363		
= SMALL(Matrix, 2)			1.226	0.551
$x_1^* = x_{min}$				
= MIN(Data)	1.214			
= SMALL(Matrix, 2)		1.226		
= LARGE(Matrix, 2)			1.363	0.081
$Q(P = 95\%, n = 8)$	0.554			
Sample 2				
$x_1^* = x_{max}$				
= MAX(Data)	1.531			
= LARGE(Matrix, 2)		1.278		
= MIN(Data)			1.226	0.830
$x_1^* = x_{min}$				
= MIN(Data)	1.226			
= SMALL(Matrix, 2)		1.234		
= MAX(Data)			1.531	0.026
$Q(P = 95\%, n = 6)$	0.560			

Challenge 3.2.3-3

Check if the maximum value of the analytical results of atrazine $x = 13.8\,\text{ppb}\,(\text{w/w})$ given in Table 2.2.1-4 can be regarded as an outlier according to the Grubbs, Dixon, and Hampel tests.

Solution to Challenge 3.2.3-3

Dixon test

The test value is calculated by (3.2.3-1) with $b = 3$ and $k = n - 1$. The test value is $\hat{Q} = 0.454$ calculated with $x_1^* = 13.8$, $x_3 = 7.9$, and $x_{n-1} = 0.8$. The critical value $Q(P = 95\%, n = 12) = 0.546$ is greater than the test value $\hat{Q}$ which means that the maximum value of 13.8 ppb (w/w) cannot be regarded as an outlier.

Grubbs test

The test value $\hat{r}_m = 2.296$ obtained with $\bar{x} = 4.283$ and $s = 4.145$ according to (3.2.3-2) exceeds the critical value $r_m(P = 95\%, n = 12) = 2.285$. Thus, the suspect value $x_{max} = 13.8\,\text{ppb}\,(\text{w/w})$ is confirmed as an outlier at the

(continued)

Table 3.2.3-5 Absolute residuals of the median $\tilde{x} = 2.80$ obtained with the data set given in Table 2.2.1-4

Sample n_i	$r_i = \|x_i - 2.80\|$	$\hat{H}_i$
1	0.3	0.030
2	1.9	0.193
3	1.7	0.172
4	5.1	0.517
5	1.8	0.182
6	2.3	0.233
7	5.8	0.588
8	0.3	0.030
9	11.0	1.115
10	1.6	0.162
11	2.0	0.203
12	3.6	0.365

significance level $P = 95\%$.This contradicts the result obtained by the Dixon test.

Which result will the *Hampel* test give?

The median is obtained with the Excel function = MEDIAN(Data): $\tilde{x} = 2.80$. The values of the absolute residuals of the median $r_i = |x_i - 2.80|$ are given in Table 3.2.3-5.

The MAD of the r_i values is also obtained with the Excel function = MEDIAN(Data): MAD = 1.95.

The test values $\hat{H}_i$ calculated according to (3.2.3-4) are also listed in Table 3.2.3-5.

As Table 3.2.3-5 shows, the test value $\hat{H}_9$ is greater than 1, therefore the maximum analytical value of 13.8 ppb (w/w) for $n = 9$ is confirmed as an outlier at the significance level $P = 95\%$. This is in accordance with the Grubbs test. In order to calculate the average atrazine content, this value should be discarded from the data set.

Challenge 3.2.3-4

The determination of aroma compounds in white wine should be carried out by headspace–solid phase micro extraction–gas chromatography (HS-SPME-GC) [5] in routine analysis. The problem is the choice of an appropriate fiber for the extraction step. In order to check some fibers, a test solution of 33 μg L^{-1} linalool (as an example of terpenoids), 50 μg L^{-1} ethyl butyrate (as an example of the substance class of esters), and 30 μg L^{-1} hexanoic acid (as an example of aliphatic acids) in 10% (v/v) ethanolic solution was analyzed using various fibers. The results obtained by the strongly polar polyacrylate fiber are given in Table 3.2.3-6.

Construct the box and whisker plots and evaluate the results.

Table 3.2.3-6 Analytical results (given in μg L^{-1}) of the determination of test analytes linalool, ethyl butyrate, and hexanoic acid obtained by HS-SPME-GC

Replicate	Linalool	Ethyl butyrate	Hexanoic acid
1	32.5	45.7	35.6
2	34.8	56.3	21.6
3	35.6	33.5	10.8
4	33.9	51.8	22.8
5	33.7	39.8	27.5
6	39.8	52.7	28.9
7	33.3	41.2	23.6

Solution to Challenge 3.2.3-4

The median $\tilde{x}$ and the quartiles $Q1$ and $Q3$ are obtained by the corresponding Excel functions = MEDIAN(Data), = QUARTILE(Matrix, 1), and = QUARTILE(Matrix, 3), respectively. The interquartile range (IQR) is calculated by (3.2.3-5) and the whiskers by (3.2.3-6) and (3.2.3-7). The results are summarized in Table 3.2.3-7 and the box and whisker plots are shown in Fig. 3.2.3-2.

Figure 3.2.3-2 yields the following results:

- The data sets for linalool and hexanoic acid show skewness because the medians are not situated in the centre of the boxes. Furthermore, in both data sets an outlier is present.
- The data for linalool show only a small distribution and the true value μ lies inside the box. This means that the fiber and the headspace SPME technique are appropriate for the extraction of terpenoids.
- The data for the ester are widely distributed, but the true value μ is situated inside the box. The strongly polar polyacrylate fiber is obviously not appropriate for the extraction of esters with reasonable precision. A less polar fiber should be tested.
- The headspace extraction of polar acids yields false results. Obviously, the headspace technique is not appropriate for the extraction of strongly polar analytes because of their low volatility. The direct injection SPME technique should be tried for the extraction of organic acids.

Table 3.2.3-7 Intermediate quantities for the calculation of the box and whisker plots

	Linalool	Ethyl butyrate	Hexanoic acid
$\tilde{x}$	33.90	45.70	23.60
Q1	33.50	40.50	22.20
Q3	35.20	52.25	28.20
IQR	1.70	11.75	6.00
L_W	30.95	22.88	13.2
U_W	37.75	69.88	37.2
$\bar{x}$	34.80	45.86	24.40

Fig. 3.2.3-2 Box and whisker plots of the data presented in Table 3.2.3-7, the true values μ, the medians $\tilde{x}$, and the outliers*

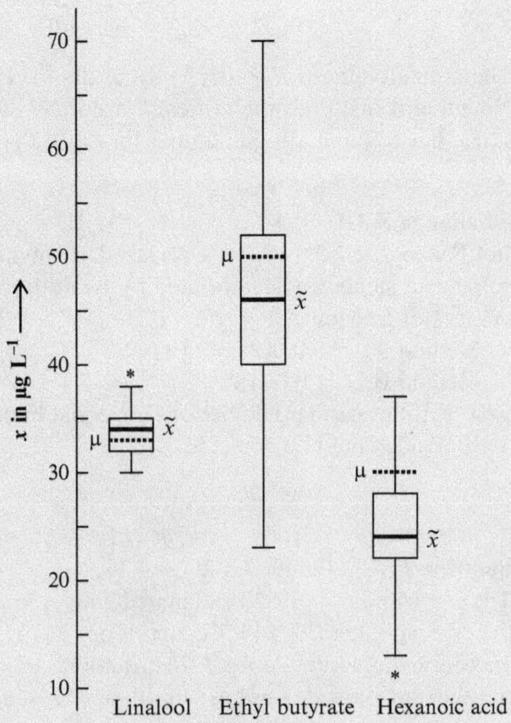

3.3 Comparison of Two Standard Deviations

Two standard deviations s_1 with degrees of freedom df_1 and s_2 with degrees of freedom df_2 are compared by means of the F-test. The test value $\hat{F}$ is calculated by (3.3-1), usually with $s_1 > s_2$:

$$\hat{F} = \frac{s_1^2}{s_2^2}. \tag{3.3-1}$$

The test value $\hat{F}$ is compared with the corresponding quantiles of the F-distribution for a certain significance level $F(P, df_1, df_2)$ which are given in Tables A-3 and A-4 for $P = 95\%$ and $P = 99\%$. The critical value is found at the intersection of the column df_1 corresponding to s_1^2 and the row df_2 corresponding to s_2^2. Note that confusing these values produces mistakes!

Sometimes, comparison is necessary between the laboratory standard deviation s_{Lab} obtained with the degrees of freedom df_{Lab} and a standard deviation σ from a document, such as a handbook or a DIN. In this case, given no degrees of freedom, infinity is chosen for df if no other information is given about the degrees of freedom of the documented standard deviation σ. The test value is calculated by

$$\hat{F} = \frac{s_1^2}{s_2^2} = \frac{s_{\text{Lab}}^2}{\sigma^2}. \tag{3.3-2}$$

The critical value is $F(P, \text{df}_1 = \text{df}_{\text{Lab}}, \text{df}_2 = \infty)$.

Note that if one finds the critical F-value by the Excel function = FINV(α, df_1, df_2), one has to input a high number for df_2, 1,000 or so.

Challenge 3.3-1

In Challenge 2.2.2-1 the process standard deviations for the determination of sulphur in steels was determined by two different procedures with the same degrees of freedom, df = 10.

Method A: $s = 0.00108\%$ (w/w) S

Method B: $s = 0.00137\%$ (w/w) S

Test if the standard deviations are equal in the statistical sense, or whether a difference could be detected.

Solution to Challenge 3.3-1

The test value $\hat{F} = 1.620$ calculated with $s_1 = s_B$ and $s_2 = s_A$ is smaller than the critical value $F(P = 95\%, \text{df}_1 = \text{df}_2 = 10) = 2.987$. Because $\hat{F}$ does not exceed the quantiles of the F-distribution, no difference is detected between the two standard deviations; in other words, s_A and s_B belong to the same population, or the null hypothesis H_0: $s_A = s_B$ is true.

Challenge 3.3-2

Let us once again consider Challenge 2.2.2-2, the determination of manganese in steel.

According to the handbook (hb) for steel analysis the process standard deviation for the determination of Mn is $s_{r,\text{hb}} = 0.000708\%$ (w/w) Mn.

Test whether there is a difference between $s_{r,\text{hb}}$ and the standard deviation determined in the laboratory with the data set given in Table 2.2.2-3.

Solution to Challenge 3.3-2

Starting with the results of Challenge 2.2.2-2, $s_{\text{Lab}} = 0.0137\%$ (w/w) Mn determined by df = 15, the test value is calculated to be $\hat{F} = 379.04$, which is very much higher than the table value $F(P = 95\%, \text{df}_1 = 15, \text{df}_2 = \infty) = 1.666$. Because the test value is greater than the critical F-value, the null hypothesis

$H_0: s_{\text{Lab}} = s_{r,\text{hb}}$ must be rejected, and the alternative hypothesis

$H_1: s_{\text{Lab}} \neq s_{r,\text{hb}}$ is true.

(continued)

The laboratory standard deviation s_{Lab} differs from that of the handbook, which means that precision is not yet reached for the routine quality control of Mn in steel in the laboratory.

Strictly spoken, the value 0.69 (Standard 3) has to be eliminated because it is checked being an outlier. But the result is the same because the new calculated test value $\hat{F} = 363.367$ exceeds the table value $F(P = 95\%, df_1 = 14, df_2 = \infty) = 1.693$.

Challenge 3.3-3

Quality control of fuel oils is carried out by DIN 51603-1 [6]. The specification of the maximal threshold value for FAME (fatty acid methyl ester) is 0.5% (v/v). The determination of FAME is accomplished by IR spectrophotometry. The repeatability limit r is specified by the (3.3-3):

$$r = 0.0126x + 0.0079, \tag{3.3-3}$$

with x as the mean value of the measurement values in % (v/v).

The quality control of such fuel oils must be introduced in an analytical laboratory. The process standard deviation for the determination of FAME is calculated using six standard oil samples with two replicates each. The results are listed in Table 3.3-1.

(a) Test if the experimentally determined process standard deviation s_r fulfils the DIN requirement using the data set given in Table 3.3-1. Give the precision of the laboratory as the relative standard deviation $s_r\%$.
(b) If the required precision is achieved, routine quality control can be started. Let us evaluate the results of four oil samples also given in Table 3.3-1. The Pearson criterion should be applied to the test to see if calculation of the mean value is allowed.

As described in Sects. 3.6 and 5.2, the repeatability limit r obtained by interlaboratory trials is calculated by (3.3-4):

$$r = t(P, df_{in}) \cdot s_{repeat} \cdot \sqrt{2}. \tag{3.3-4}$$

Table 3.3-1 Analytical results given in % (v/v) for the determination of FAME in fuel oils

Determination of the process standard deviation s_r						
Sample	1	2	3	4	5	6
x'	0.386	0.397	0.379	0.411	0.436	0.478
x''	0.378	0.385	0.371	0.419	0.429	0.471
Routine quality control						
Oil 1	0.492	0.491	0.486			
Oil 2	0.491	0.558	0.487			
Oil 3	0.467	0.447				
Oil 4	0.495	0.496				

Solution to Challenge 3.3-3

This complex Challenge is best solved by the following steps.

(a) 1. Calculation of the repeatability limit r

The repeatability limit r calculated by (3.3-3) with the mean values $\bar{x} = 0.4117\%$ (v/v) obtained with six oil samples is $r = 0.0131\%$ (v/v).

2. Calculation of the precision of the analytical process

$$s_{\text{repeat}} = \frac{r}{\sqrt{2}\,t(P = 95\%, \text{df} = \infty)}. \qquad (3.3\text{-}5)$$

The standard deviation is $s_{\text{repeat}} = 0.00472\%$ (v/v) calculated with $r = 0.0131\%$ (v/v) and $t(P = 95\%, \text{df} = \infty) = 1.96$.

3. Estimation of the precision of the laboratory

The precision of the laboratory s_{Lab} is determined by the results of the six oil samples given in Table 3.3-1. The intermediate quantities are presented in Table 3.3-2.

The standard deviation calculated by (2.2.2-3) is $s_{\text{Lab}} = 0.00601\%$ (v/v).

4. Comparison of the precision of the laboratory with the precision required by DIN with an F-test

The test value is $\hat{F} = 1.623$ calculated with $s_1 = s_{\text{Lab}} = 0.00601\%$ (v/v) and $s_2 = 0.00472\%$ (v/v). The critical value is $F(P = 95\%, \text{df}_1 = \text{df}_{\text{Lab}} = 6, \text{df}_2 = \text{df}_{\text{DIN}} = \infty) = 2.099$, which is greater than the test value $\hat{F}$. Thus, the null hypothesis H_0: $s_{\text{DIN}} = s_{\text{Lab}}$ is true. The precision required by DIN is accomplished.

Because the required precision is achieved, the routine quality control can start.

5. Calculation the precision of the analytical process expressed as $s_r\%$

The laboratory precision is $s_r\% = 1.46\%$ calculated according (2.2.2-5a) with $s = s_{\text{Lab}} = 0.00601\%$ (v/v) and $\bar{x} = 0.411\%$ (v/v).

(b) 1. Checking the data for outliers

The value 0.558% (v/v) in sample 2 is a suspect value which is tested as an outlier by the Dixon test. The test value is $\hat{Q} = 0.944$, calculated by (3.2.3-1) with $x_1^* = x_{\max} = 0.558$, $x_2 = 0.491$ and $x_n = 0.487$. The test value $\hat{Q}$ is greater than the critical value

(continued)

Table 3.3-2 Intermediate quantities and results for the calculation of the precision of the laboratory

Sample	1	2	3	4	5	6
$10^5(x'-x'')^2$	6.4	14.4	6.4	6.4	4.9	4.9
$\Sigma(x'-x'')^2$	0.000434	m	6			

$Q(P = 95\%, n = 3) = 0.941$, which means that the value 0.558% (v/v) is confirmed as an outlier at the significance level $P = 95\%$.

Thus, it has to be removed from the data set.

2. Estimation of the calculation of the mean values

   The criterion of whether the calculation of the mean is allowed or not is determined by the repeatability r which is given in the DIN. The criterion is $\Delta_{\text{exp}} = x_{\max} - x_{\min} < r$. Thus, the calculation of the means is only allowed if the difference between the highest and lowest value does not exceed the repeatability r. As Table 3.3-3 shows, the difference Δ_{exp} of sample 3 is greater then r, and thus the analysis of sample 3 has to be repeated. For the other three samples Δ_{exp} is smaller than r, and the means can be calculated.

   Note that if one does not reject the outlier in the data set of sample 2, $\Delta_{\text{exp}} = 0.071\%$ (v/v) exceeds the limit and the analysis of sample 2 would also have to be repeated.

3. Analytical quality control of the means of samples 1, 2, and 4

   The critical mean value $\bar{x}_{\text{crit}}$ is given by the difference between a fixed limit L_0, in this case the DIN value $L_0 = 0.5\%$ (v/v), and the critical confidence interval CI_{crit} which is calculated from the experimental data from the laboratory (see Sect. 2.2.4):

$$\bar{x}_{\text{crit}} = 0.5\% \ (\text{v/v}) - \text{CI}_{\text{crit,one-sided}} \tag{3.3-6}$$

with

$$\text{CI}_{\text{crit,one-sided}} = \frac{s_{\text{Lab}} \cdot t(\bar{P}_{\text{one-sided}}, \text{df}_{\text{Lab}})}{\sqrt{n_j}}. \tag{3.3-7}$$

The values in Table 3.3-4 were calculated with the outlier-free data set, $s_{\text{Lab}} = 0.00601\%$ (v/v), and $t(\bar{P} = 95\%, \text{df}_{\text{Lab}} = 6) = 1.943$.

As Table 3.3-4 shows, the mean values $\bar{x}$ of the samples 1 and 2 do not exceed the critical threshold value $\bar{x}_{\text{crit}}$ but the mean value of sample 4 is greater than $\bar{x}_{\text{crit}}$. Thus, the threshold value L_0 of the allowed concentration of FAME is exceeded for oil sample 4, and therefore this oil cannot be delivered.

(continued)

Table 3.3-3 The intermediate quantities and results for the calculation of the means using the outlier-free data set

Sample	n_j	$x_{\max}$	$x_{\min}$	$\Delta_{\text{exp}} = x_{\max} - x_{\min}$
1	3	0.492	0.486	0.006
2	2	0.491	0.487	0.004
3	2	0.467	0.447	0.020
4	2	0.495	0.495	0.001

Table 3.3-4 The intermediate quantities and results for quality control

Sample	n_j	$\bar{x}$ in % (v/v)	$CI_{crit,one\text{-}sided}$ in % (v/v)	$\bar{x}_{crit}$ in % (v/v)	Result
1	3	0.490	0.00675	0.493	$\bar{x} < \bar{x}_{crit}$
2	2	0.489	0.00826	0.492	$\bar{x} < \bar{x}_{crit}$
4	2	0.496	0.00826	0.492	$\bar{x} > \bar{x}_{crit}$

Only samples 1 and 2 fulfill the quality norm, but only after the rejection of the outlier in the analytical values of sample 2.

3.4 Comparison of More than Two Standard Deviations

To decide whether more than two variances differ randomly or significantly, two tests are usually employed in AQA:

(a) Cochran test [7,8]
The test value $\hat{C}$ is calculated by (3.4-1):

$$\hat{C} = \frac{s^2_{max}}{s_1^2 + s_2^2 + \cdots + s_k^2}, \tag{3.4-1}$$

where $s_1^2, s_2^2, \ldots, s_k^2$ are the variances of the measurement values with *equal* size ($n_1 = n_2 = \cdots = n_k$ and $df_1 = df_2 = \cdots = df_n$, respectively).

The test value $\hat{C}$ is compared with the value of the Cochran table (Table A-9) for k samples and df degrees of freedom at the significance level $P = 95\%$.

The test hypotheses for the Cochran test are

$$H_0: s^2_{max} = \sum_1^k s_k^2,$$

$$H_A: s^2_{max} \neq \sum_1^k s_k^2.$$

The null hypothesis H_0 is rejected if the test value $\hat{C}$ is greater than the critical value $C(P = 95\%, k, df)$.

(b) *Bartlett test* [1,9]
The homogeneity of variances from measurement values of *different* sizes is tested by the Bartlett test. The test value is calculated by (3.4-2) for k groups and the total number of measurement values n

$$\hat{\chi}^2 = \frac{2.3026}{c} \cdot \left(df \cdot \lg s^2 - \sum_{i=1}^k df_i \cdot \lg s_i^2 \right) \tag{3.4-2}$$

with the total number of degrees of freedom $\mathrm{df} = n - k$, and s_i^2 the variances of the ith group with degrees of freedom df_i.

The variance s^2 is calculated by (2.2.2-2). The correction factor c

$$c = \frac{\sum_{i=1}^{k} \frac{1}{\mathrm{df}_i} - \frac{1}{\mathrm{df}}}{3(k-1)} + 1 \qquad (3.4\text{-}3)$$

has to be considered only if the test value $\hat{\chi}^2$ is slightly greater than the table value $\chi^2(\bar{P}, \mathrm{df} = k - 1)$.The homogeneity of the variances is given at the significance level P if the test value $\hat{\chi}^2$ does not exceed the limits of the χ^2 distribution which are listed in Table A-5 for $P = 95\%$.

Challenge 3.4-1

Let us return to Challenge 2.2.2-2 in which the process standard deviation must be calculated for the determination of Mn in steel from the measurement values of five samples. However, the calculation of the process standard deviation requires the homogeneity of the variances of the data set listed in Table 2.2.2-3.

Test whether homogeneity is present.

Solution to Challenge 3.4-1

Because the qualitative inspection of the data set in Table 2.2.2-3 reveals that no outliers are obviously present, outlier tests will not be made, and therefore the number of measurement values is equal for all five samples. The Cochran test can be used with $k = 5$ (five samples) and $\mathrm{df} = 3$ (four replicates). The results are summarized in Table 3.4-1.

The test value $\hat{C} = 0.3158$ does not exceed the critical value $C(P = 95\%, k = 5, \mathrm{df} = 3) = 0.5981$.Therefore, homogeneity of the variances is present and the process standard deviation can be calculated.

Table 3.4-1 Results of the Cochran test for the measurement values given in Table 2.2.2-3

Sample	1	2	3	4	5
$10,000 \cdot s_k^2$	1.667	0.917	1.000	3.000	2.917
$10,000 \cdot s_{max}^2$	3.000	$\sum_{1}^{k} 10,000 \cdot s_k^2$	9.500		
$\hat{C}$	0.3158	$C(P = 95\%, k = 5, \mathrm{df} = 3)$			0.5981

Challenge 3.4-2

Because of the small injection volume of 1 μL or less in gas chromatography, the injection of the sample is a frequent source of errors; therefore, checking the precision of the syringe is an important operation in AQA.

For the testing of five syringes in the autosampler of a GC equipment, a stable test sample was splitlessly injected with nine replicates under the same conditions for all syringes. The peak areas A given in counts obtained from the GC chromatograms are listed in Table 3.4-2.

Check if all syringes are equal in their injection precision.

Table 3.4-2 Peak areas in counts of the GC chromatograms testing the injection precision of five syringes

Syringe	1	2	3	4	5
Replicate					
1	12,350	12,305	12,375	12,351	12,364
2	12,376	12,346	12,370	12,350	12,360
3	12,348	12,328	12,378	12,352	12,360
4	12,352	12,392	12,383	12,352	12,365
5	12,340	12,310	12,371	12,354	12,366
6	12,382	12,319	12,368	12,349	12,363
7	12,372	12,333	12,377	12,349	12,359
8	12,339	12,326	12,375	12,350	12,361
9	12,340	12,335	12,367	12,354	12,360

Solution to Challenge 3.4-2

The syringes work with the same precision if one cannot detect a difference between the variances of the peak areas obtained under the same conditions; this is the case if the variances are homogeneous. Before testing the homogeneity of the variances, a test for outliers in the data set is necessary, for which, in accordance with the DIN, we will choose the Dixon test.

According to (3.2.3-1) for $n = 9$, the test value $\hat{Q}$ is calculated by

$$\hat{Q}\left|\frac{x_1^* - x_2}{x_1^* - x_{n-1}}\right|. \tag{3.4-4}$$

The test values for each maximum and minimum value are summarized in Table 3.4-3. The critical value is $Q(P = 95\%, n = 9) = 0.512$.

The test value for the largest value of syringe 2 (12,392) exceeds the critical value. Therefore, the value 12,392 confirmed as an outlier and it must be rejected from the data set, with the consequence that the numbers of replicates are no longer equal. The Cochran test cannot be used, but the Bartlett test must be applied.

(continued)

Table 3.4-3 Intermediate quantities and results for the Dixon outlier test

Syringe	1	2	3	4	5
$x_1^* = x_{max}$					
x_{max}	12,382	12,392	12,383	12,354	12,366
x_2	12,376	12,346	12,378	12,354	12,365
x_{n-1}	12,340	12,310	12,368	12,349	12,360
$\hat{Q}_{max}$	0.143	0.561	0.333	0.000	0.167
$x_1^* = x_{min}$					
x_{min}	12,339	12,305	12,367	12,349	12,359
x_2	12,340	12,310	12,368	12,349	12,360
x_{n-1}	12,376	12,346	12,378	12,354	12,365
$\hat{Q}_{min}$	0.027	0.122	0.091	0.000	0.167

Table 3.4-4 Intermediate quantities for the test of homogeneity of the five syringes according to (3.4-2)

Syringe	1	2	3	4	5
n_i	9	8	9	9	9
n	44	k	5	df = n–k	39
SS_i	2,246.2	1,275.5	217.6	29.6	52.0
$\sum \text{SS}_i$	3,820.8				
s^2	97.970	$\log s^2$	1.9911		
$\text{df} \cdot \log s^2$	77.65				
s_i^2	280.78	182.21	27.19	3.69	6.50
$\log s_i^2$	2.448	2.261	1.434	0.568	0.813
df_i	8	7	8	8	8
$\text{df}_i \cdot \log s_i^2$	19.587	15.824	11.476	4.540	6.503
$\sum \text{df}_i \cdot \log s_i^2$	57.931				

The intermediate quantities and results for the homogeneity of the five syringes according to the Bartlett test are summarized in Table 3.4-4.

The test value $\hat{\chi}^2 = 45.41$ exceeds the critical value $\chi^2(\bar{P} = 99\%, \text{df} = k - 1 = 4) = 13.277$ substantially, and therefore the injection precision of the five syringes is not equal.

3.5 Comparison of Two Mean Values

The comparison of two mean values $\bar{x}_1$ and $\bar{x}_2$ of two different independent samples with n_1 and n_2 determinations is made by the t-test. The test value $\hat{t}$

$$\hat{t} = \frac{|\bar{x}_1 - \bar{x}_2|}{s_p} \sqrt{\frac{n_1 n_2}{n_1 + n_2}} \qquad (3.5\text{-}1)$$

with the pooled (average) standard deviation

$$s_p = \sqrt{\frac{(n_1 - 1)s_1^2 + (n_2 - 1)s_2^2}{n_1 + n_2 - 2}} = \sqrt{\frac{\mathrm{df}_1 \cdot s_1^2 + \mathrm{df}_2 \cdot s_2^2}{\mathrm{df}_1 + \mathrm{df}_2}} \tag{3.5-2}$$

must be compared with the critical value $t(P, \mathrm{df}_1 + \mathrm{df}_2)$ from the t-table. If $\hat{t}$ is smaller than the critical value, the null hypothesis $\mathrm{H}_0: \bar{x}_1 = \bar{x}_2$ is true; in other words, there is no difference between the two mean values at the significance P.

The calculation of the average standard deviation according to (3.5-2) is only allowed and the t-test can only be applied if the variances s_1^2 and s_2^2 do not differ significantly or, in other words, if they belong to the same population. This must be checked by the F-test according to (3.3-1).

If the variances s_1^2 and s_2^2 differ significantly the t-test according to *Welch* [9] can be applied. The test value $\hat{t}_\mathrm{W}$ is calculated by (3.5-3)

$$\hat{t}_\mathrm{W} = \frac{|\bar{x}_1 - \bar{x}_2|}{\sqrt{\frac{s_1^2}{n_1} + \frac{s_2^2}{n_2}}} \tag{3.5-3}$$

and, as above described, is compared with the critical value $t(P, \mathrm{df}_\mathrm{W})$. The degrees of freedom of the Welch test df_W are calculated according to (3.5-4):

$$\mathrm{df}_\mathrm{W} = \frac{\left(\frac{s_1^2}{n_1} + \frac{s_2^2}{n_2}\right)^2}{\frac{\left(\frac{s_1^2}{n_1}\right)^2}{n_1 - 1} + \frac{\left(\frac{s_2^2}{n_2}\right)^2}{n_2 - 1}}. \tag{3.5-4}$$

Note that the degrees of freedom calculated by (3.5-4) are non-integral numbers which are not given in the t-table. Either one has to interpolate or, better, the Excel function $=$ TINV(α, df_W) is used.

When it is necessary to decide if the mean value of a sample $\bar{x}$ differs randomly or significantly from a "true" value μ, which might be, for example, the theoretical value calculated from the stoichiometry of the chemical formula or a certified value from an interlaboratory trial, the t-test according to (3.5-5) has to be made:

$$\hat{t} = \frac{|\bar{x} - \mu|}{s} \cdot \sqrt{n}. \tag{3.5-5}$$

The null hypothesis H_0: $\bar{x} = \mu$ is true if $\hat{t}$ does not exceed the critical value $t(P, \mathrm{df})$.

If the mean values $\bar{x}_1$ and $\bar{x}_2$ differ from each other and one would like to know which value is false, both means must be tested *separately* against μ.

Challenge 3.5-1

The HPLC method for the assay of an API in a drug must be transferred from reference laboratory 1 to laboratory 2. The analysis should be conducted in accordance with good laboratory practice (GLP) rules [10].

According to the lab-to-lab transfer plan, one batch is selected and the determination of the assay is performed six times in each laboratory by the same procedure. The assay of the chosen batch is 98.0% (w/w). The acceptance criterion of the plan is fulfilled if the results from the two laboratories do not differ significantly. The analytical results from the laboratories are given in Table 3.5-1.

(a) Can you detect a significant difference between the results obtained by the laboratories?
(b) Are the results of both laboratories true?

Note the significance level for all tests is $P = 95\%$.

Table 3.5-1 Analytical results in % (w/w) obtained from two laboratories

Laboratory 1	98.0	98.4	98.7	98.4	97.5	98.6
Laboratory 2	98.7	97.5	97.0	97.7	97.6	97.4

Solution to Challenge 3.5-1

(a) The transfer of an analytical method from one laboratory to another is permitted if the independent sample means $\bar{x}_1$ and $\bar{x}_2$ are not significantly different, which is checked by the *t*-test. The *t*-test is based on the following assumptions:

1. The samples with means $\bar{x}_1$ and $\bar{x}_2$ are drawn from normal populations, which can be tested by the David test.
2. There must be no outliers in the data sets, for which the Dixon test will be applied.
3. There are no significant differences between the variances s_1^2 and s_2^2, which is tested by the *F*-test.

The test values for normal distribution (David test) and for outliers (Dixon test) are given in Table 3.5-2.

Results of the test for normal distribution:

For laboratory 1 and laboratory 2 the test values $\hat{q}_r$ lie within the boundaries of the David table: 2.28 and 3.012, respectively, for a sample size of $n = 6$ at the significance level $P = 95\%$. The analytical results of both laboratories can be regarded as normally distributed.

Results of the outlier test:

(continued)

Table 3.5-2 Results of the tests for normal distribution by the David test and for outliers by the Dixon test

David test according to (3.2.1-1)				
Laboratory	x_{max}	x_{min}	s	$\hat{q}_r$
1	98.7	97.5	0.446	2.69
2	98.7	97.0	0.568	2.99
Dixon test according to (3.2.3-1)				
Laboratory	$x_{2,max}$	$x_{2,min}$	$\hat{Q}(x_{max})$	$\hat{Q}(x_{min})$
1	98.6	98.0	0.083	0.417
2	97.7	97.4	0.588	0.235

Table 3.5-3 Intermediate quantities for the F-test

Laboratory	n	$\bar{x}$	s	df
1	6	98.27	0.4457	5
2	5	97.44	0.2702	4

The critical value is $Q(P = 95\%, n = 6) = 0.560$. The test value of the highest value x_{max} of laboratory 2 is larger than the critical value. As a consequence, x_{max} in laboratory 2 is regarded as an outlier at the significance level $P = 95\%$. Therefore, it must be removed from the data set.

Comparison of the two means $\bar{x}_1$ and $\bar{x}_2$:

The test values of the homogeneity of the variances by means of the F-test with the outlier-free data set are listed in Table 3.5-3.

Do the laboratories work statistically with the same precision?

The test value $\hat{F} = 2.721$ calculated by (3.3-1) is smaller than the critical value $F(P = 95\%, \text{df}_1 = 5, \text{df}_2 = 4) = 6.256$. Note that for the F-test s_1^2 must always be higher than s_2^2.

The null hypothesis H_0: $s_1^2 = s_2^2$ is accepted. Both laboratories work with equal precision, and therefore the t-test may be carried out.

Comparison of the means:

The test value calculated by (3.5-1) and (3.5-2) with $s_p = 0.3779$ is $\hat{t} = 3.613$. This value exceeds the limit of the critical value $t(P = 95\%, \text{df}_{total} = 9) = 2.262$, and as a consequence there is a difference between the results of the two laboratories.

Note that if one does not remove the outlier, the t-test value calculated is $\hat{t} = 2.091$. In this case, $\hat{t}$ is smaller than the critical value which means that the mean values in the two laboratories are equal, but this result is not correct.

(b) The t-test can also be used as a check for trueness. For each laboratory the test value $\hat{t}$ is calculated using (3.5-5) with the "true" value $\mu = 98.0\%$ (w/w), the known assay of the chosen batch. The analytical result is regarded as true if the $\hat{t}$-value calculated is smaller than the critical

(continued)

t-value at the significance level P and the number of degrees of freedom df of the data set in each laboratory. With the data set listed in Table 3.5-3 the following t-values are obtained:
Laboratory 1 (the reference laboratory):
$\hat{t}_1 = 1.465 \; t(P = 95\%, df_1 = 5) = 2.571$

Laboratory 2:
$\hat{t}_2 = 4.635 \quad t(P = 95\%, df_2 = 4) = 2.776$
The analytical result of the reference laboratory 1 is true, but the analytical result for the laboratory 2 is false. The lab-to-lab transfer is not yet successful.

Challenge 3.5-2
In an analytical laboratory the determination of Ni in the presence of a great amount of Fe in waste water must be introduced. For the choice of analytical method, a internal laboratory comparison of six different methods was carried out. A standard solution with 50.0 μg L^{-1} Ni and 500 μg L^{-1} Fe was analyzed by $n_j = 10$ replicates. The results are presented in Table 3.5-4.
Which methods provide a true value and which mean value is false? Check it at the significance level $P = 95\%$.

Table 3.5-4 Analytical results in μg L^{-1} Ni

Method	1	2	3	4	5	6
n_j						
1	49.0	50.2	50.2	49.3	49.3	49.9
2	50.1	50.5	49.3	49.3	49.9	50.2
3	49.4	49.2	49.8	49.7	49.6	50.5
4	49.1	49.9	50.3	49.5	49.0	50.0
5	50.2	50.1	50.1	50.2	49.3	49.6
6	49.8	50.3	49.4	49.8	49.1	50.1
7	49.9	50.6	51.2	49.9	49.6	49.7
8	49.5	49.8	49.8	49.5	50.2	49.5
9	50.3	50.5	49.9	49.9	49.2	50.4
10	49.7	50.4	50.1	49.9	49.4	50.1

Key for the analytical methods:
(1) volumetric analysis, (2) polarography, (3) photometry, (4) flame AAS (flame: N_2O/C_2H_2, $\lambda = 232$ nm), (5) flame AAS (flame: air/C_2H_2, $\lambda = 342$ nm), (6) ICP-OES

Solution to Challenge 3.5-2
The test values for checking the normal distribution according to the David test are listed in Table 3.5-5.

(continued)

Table 3.5-5 Results of checking the normal distribution according to the David test

Method	1	2	3	4	5	6
x_{max}	50.3	50.6	51.2	50.2	50.2	50.5
x_{min}	49.0	49.2	49.3	49.3	49.0	49.5
s	0.447	0.425	0.530	0.294	0.372	0.330
$\hat{q}_r$	2.907	3.295	3.584	3.057	3.228	3.030

Table 3.5-6 Intermediate quantities and test values for Dixon's outlier test

Method	1	2	3	4	5	6
$x_1^* = x_{max}$						
x_{max}	50.3	50.6	51.2	50.2	50.2	50.5
x_2	50.2	50.5	50.3	49.9	49.9	50.4
x_{n-1}	49.1	49.8	49.4	49.3	49.1	49.6
$\hat{Q}_{max}$	0.083	0.125	0.500	0.333	0.273	0.111
$x_1^* = x_{min}$						
x_{min}	49.0	49.2	49.3	49.3	49.0	49.5
x_2	49.1	49.8	49.4	49.3	49.1	49.6
x_{n-1}	50.2	50.5	50.3	49.9	49.9	50.4
$\hat{Q}_{min}$	0.083	0.462	0.100	0.000	0.111	0.111

All test values lie within the limits of the David table at the significance level $P = 95\%$: 2.67 and 3.685. Thus, the data sets are normally distributed.

The test values for Dixon's outlier test are calculated according (3.2.3-1) by

$$\hat{Q} = \left| \frac{x_1^* - x_2}{x_1^* - x_{n-1}} \right|. \tag{3.5-6}$$

The intermediate quantities obtained by the corresponding Excel functions as given in Table 3.2.3-4 are summarized in Table 3.5-6. The critical value $Q(P = 95\%, n = 10) = 0.477$ is exceeded by the maximal value 51.2 for method 3. Thus, this value must be rejected as an outlier. Further outliers cannot be checked.

The visual inspection of the data sets does not give any hint for using a statistical trend test.

The test for trueness is carried out by means of the t-test using (3.5-5) with $\mu = 50.0\ \mu g\ L^{-1}$ Ni. The test values $\hat{t}$ are listed in Table 3.5-7. If the test value $\hat{t}$ is smaller than the critical value $t(P = 95\%, df_j)$ then the analytical result is true, otherwise it is false. The results are also given in Table 3.5-7.

As the results in Table 3.5-7 show, the same mean value $\bar{x}$ can be "true" or "false" if the standard deviations are different. This is the case for

(continued)

$\bar{x} = 47.70\ \mu g\,L^{-1}$ Ni in methods 1 and 4. Note that the statement "true" or "false" cannot be made without knowledge of the precision (see Sect. 2.2.4).

Table 3.5-7 Results of testing the trueness of the determination of Ni in the presence of a high Fe content with six different methods, obtained with outlier-free data sets

Method	1	2	3	4	5	6
$\bar{x}$	49.70	50.15	49.88	49.70	49.46	50.00
n_j	10	10	9	10	10	10
df_j	9	9	8	9	9	9
s_j	0.447	0.425	0.346	0.294	0.372	0.330
$\hat{t}_j$; (3.5-5)	2.121	1.116	1.061	3.223	4.593	0.000
$t(P = 95\%, df_j)$	2.262	2.262	2.306	2.262	2.262	2.262
Result	true	true	true	false	false	true

Challenge 3.5-3

The photometric determination of methylene blue active detergents in waste water has to be introduced in a laboratory. To test the necessary sample preparation procedure, two samples were analyzed:

1. A waste water sample after a simple filtration (sample 1)
2. A waste water sample which was purified by solid phase extraction (SPE) (sample 2)

The analytical results of six replicates each are given in Table 3.5-8.

(a) Test whether there is a significant difference between the mean values of the two samples.
(b) The relative standard deviation $s_r\%$ for this analytical method should not exceed 5% routinely. Check whether this will be reached with the two procedures for the average sample amounts of 500 µg L^{-1}.

Note that the significance level for all tests is $P = 95\%$.
Evaluate the results.

Table 3.5-8 Analytical results (in µg L^{-1}) of the determination of methylene blue active detergents for two samples with different preparations with $n_i = 6$ replicates

n_i	1	2	3	4	5	6
Sample 1	438	512	478	490	515	438
Sample 2	456	478	469	493	476	456

Sample pre-treatment:
Sample 1 – simple filtration
Sample 2 – purification by SPE

Solution to Challenge 3.5-3

(a) The t-test is required for testing of differences between two mean values, but the calculation of the test values $\hat{t}$ according to (3.5-1) and (3.5-2) is only allowed if there is no significant difference between the respective standard deviations, which is checked by an F-test according to (3.3-1). If the two standard deviations do not belong to the same population, the Welch test has to be carried out. But firstly, the data must be checked to see whether the standard deviation can be calculated. The test values on normal distribution are $\hat{q}_{r,1} = 2.25$ and $\hat{q}_{r,2} = 2.60$, respectively. Thus, the data set of sample 1 exceeds the critical values at the significance level $P = 95\%$, which are 2.28 and 3.012, respectively. The deviation is only small; therefore the test result is ignored.

As the results of the Dixon test given in Table 3.5-9 shows, there is no outlier in either data set because the test values do not exceed the critical value $Q(P = 95\%, n = 6) = 0.560$.

As Table 3.5-9 shows, the test value $\hat{F}$ is greater than the critical value $F(P = 95\%, \mathrm{df}_1 = \mathrm{df}_2 = 5) = 5.050$ which means that the variances are different. Therefore, the Welch test is necessary, with the following results: $\hat{t}_W = 0.473$ calculated according to (3.5-3), $\mathrm{df}_W = 6.674$ calculated by (3.5-4), and the critical value $t(P = 95\%, \mathrm{df}_W) = 2.447$ obtained by Excel functions.

The test value $\hat{t}_W$ does not exceed the limit of the critical value, which means that there is no significant difference between the mean values obtained by two different sample pre-treatments.

Note that this result only gives the information that there is no difference between the means, but no information about the trueness of the analytical results. It is possible that both means are wrong. Information on the trueness would only be possible with a test against a known content μ according to (3.5-5), but μ is not given.

(b) The precision, calculated by (2.2.2-5a), is $s_r\% = 7.2$ for sample 1 and $s_r\% = 3.0$ for sample 2. The precision with the simple filtration is obviously worse than the required 5%; therefore, the SPE cleaning step has to be applied.

Table 3.5-9 Results for the Dixon and F-tests

Sample	Dixon test (3.2.3-1)					
	x_{max}	x_2	$\hat{Q}_{max}$	x_{min}	x_2	$\hat{Q}_{min}$
1	515	512	0.039	438	438	0
2	493	478	0.405	456	456	0
F-test, (3.3-1)						
Sample		s_i		df_i		$\hat{F}$
1		34.256		5		5.802
2		14.222		5		

3.6 Comparison of More than Two Mean Values: Analysis of Variance

Let us first consider some examples: in a laboratory the same sample is analyzed by k analysts under the same conditions with n_j replicates to check the mode of operation of the analysts. Another example is the testing of the performance of an accredited laboratory, which is also carried out in an interlaboratory study in which k laboratories participate. A partitioned sample is given to k laboratories to carry out the analysis with n_j replicates under the same conditions. A final example concerns the comparison of k analytical results obtained by k methods in the same laboratory.

The results obtained can be presented in the fashion given in Table 3.6-1 for all the examples mentioned above, but the replicates in the columns can also be different. In all these examples the question is, are the mean values $\bar{x}_k$ statistically equal or not? This question is answered by the *AN*alysis *O*f *VA*riance (ANOVA).

The fundamental technique of ANOVA is a partitioning of the total sum of squares (SS_{total}) into components related to the effects used in the model, for example,

$$SS_{total} = SS_{treatment} + SS_{error}. \tag{3.6-1}$$

The number of degrees of freedom can be partitioned in a similar way:

$$df_{total} = df_{treatment} + df_{error}. \tag{3.6-2}$$

There are some assumptions for ANOVA:

1. The populations from which the samples were obtained must be normally distributed.
2. The subjects are sampled randomly.
3. The population variances must be homogeneous.
4. The groups (cells) must be independent.
5. The null hypothesis H_0
   H_0: $\mu_1 = \mu_2 = \cdots = \mu_k$
   is rejected if at least one μ_i is not equal to another.

Table 3.6-1 General scheme for the one-way ANOVA layout with k series of measurements and n replicates

Group: Analyst; sample; laboratory; method; and others	1	2	...	k
	x_{11}	x_{12}	...	x_{1k}
	x_{21}	x_{22}	...	x_{2k}
	$\vdots$	$\vdots$	...	$\vdots$
	x_{i1}	x_{i2}	...	x_{ik}
	$\vdots$	$\vdots$	...	$\vdots$
	x_{n1}	x_{n2}	...	x_{nk}
Mean	$\bar{x}_1$	$\bar{x}_2$	...	$\bar{x}_k$
Variances	s_1^2	s_2^2	...	s_k^2

There are three conceptual classes of such models:

1. *Fixed-effects models* assume that the data came from normal populations which may differ only in their means (Model I).
2. *Random effects models* assume the data describe a hierarchy of different populations whose differences are constrained by their hierarchy (Model II).
3. *Mixed-effect models* describe situations in which both fixed and random effects are present (Model III).

If only a *single* effect is studied – the mode of operation of laboratory analysts or the influence of a method on the analytical result – this is called a *one-way* ANOVA.

The computational scheme of the required variances is summarized in an ANOVA table with a similar format to that for one-way experiments, given in Table 3.6-2.

SS_{bw}, SS_{in}, and SS_{tot} are the between-columns sum of squares, within-columns sum of squares, and total sum of squares, respectively; s^2_{bw} and s^2_{in} are the variances (also called mean squares) between and within the columns calculated with the degrees of freedom df_{bw} and df_{in}, respectively. $\bar{x}_i$ is the mean value of column i and $\bar{\bar{x}}$ is the grand mean value which is obtained from all values x_{ij} and the total number of values n:

$$\bar{\bar{x}} = \frac{\sum_{i=1}^{k}\sum_{j=1}^{n_j} x_{ij}}{n}. \tag{3.6-12}$$

The comparison of more than two mean values is traced back to the comparison of variances, which is performed by an F-test. The test value $\hat{F}$ is given by

Table 3.6-2 Computational scheme of *one-way* ANOVA

Source of error: *Between* columns		Equation
Sum of squares	$SS_{bw} = \sum_{i=1}^{k} n_i \cdot (\bar{x}_i - \bar{\bar{x}})^2$	(3.6-3)
Degrees of freedom	$df_{bw} = k - 1$	(3.6-4)
Variance	$s^2_{bw} = \frac{SS_{bw}}{df_{bw}}$	(3.6-5)
Source of error: *Within* columns		**Equation**
Sum of squares	$SS_{in} = \sum_{i=1}^{k}\sum_{j=1}^{n_j} (x_{ij} - \bar{x}_j)^2$	(3.6-6)
Degrees of freedom	$df_{in} = n - k$	(3.6-7)
Variance	$s^2_{in} = \frac{SS_{in}}{df_{in}}$	(3.6-8)
Source of error: *Total*		**Equation**
Sum of squares	$SS_{tot} = \sum_{i=1}^{k}\sum_{j=1}^{n_j} (x_{ij} - \bar{\bar{x}})^2$	(3.6-9)
	$SS_{tot} = SS_{bw} + SS_{in}$	(3.6-10)
Degrees of freedom	$df_{tot} = n - 1$	(3.6-11)

$$\hat{F} = \frac{s_{\text{bw}}^2}{s_{\text{in}}^2}. \tag{3.6-13}$$

The $\hat{F}$-value calculated by (3.6-13) must be compared with the critical F-value for $\text{df}_1 = \text{df}_{\text{bw}} = k - 1$ and $\text{df}_2 = \text{df}_{\text{in}} = n - k$ degrees of freedom at a significance level P.

If $\hat{F} > F(P, \text{df}_1, \text{df}_2)$, then the null hypothesis, that all mean values are equal in the statistical sense

H_0: $\mu_1 = \mu_2 = \cdots = \mu_k$,

must be rejected.

Because within-column variances are pooled, one must test whether these variances are equal. The homogeneity of more than two variances may be tested by the Cochran or Bartlett test (see Sect. 3.4). Furthermore, as a general requirement for the calculation of variances, the data sets must be normally distributed, which can be tested by the rapid David test (Sect. 3.2.1).

When the null hypothesis has been rejected, in the fixed effect model it is considered that at least one column has a mean value which is different from the others, but which column is this? It may be that the visual representation of the data can single out those columns for which it is most likely that differences exist. However, if this cannot be decided clearly, statistical tests are necessary.

There are some tests for such a problem in the literature, but the *Least Significant Difference* (LSD) method is fascinating in its simplicity. Any pair of means for which $|\bar{x}_j - \bar{x}_k| > \text{LSD}$ is considered different. LSD is calculated according to (3.6-14) and (3.6-15):

$$\text{LSD} = t(P, \text{df}_{\text{in}}) \cdot \sqrt{s_{\text{in}}^2 \cdot [(1/n_1) + (1/n_2)]}. \tag{3.6-14}$$

For equal sample size, i.e. $n_1 = n_2$, (3.6-14) can be simplified:

$$\text{LSD} = t(P, \text{df}_{\text{in}}) \cdot \sqrt{\frac{2 \cdot s_{\text{in}}^2}{n_j}}. \tag{3.6-15}$$

The t-value is obtained from the t-table for the significance level P, which is usually 95%, and the degrees of freedom df_{in}.

Two-way ANOVA:

Let us look at an example, as it very often occurs in analytical practice. To explore an appropriate method for the determination of the content of some metals in soil samples by atomic absorption spectroscopy (AAS), some experiments with two factors are necessary: the first factor is given by the various conditions for the solubility of the samples, and the second factor is the subsequent cleaning step with SPE or the techniques of AAS. The question is which factor will influence the result? This can be answered by the *two-way* ANOVA method because there are *two* independent variables.

Table 3.6-3 Two-way ANOVA design

Notations						
	Any	Last				
Score	i	n				
Factor A	j	p				
Factor B	k	q				
		Factor B				A Marginals
Factor A		b_1	b_2	b_k	b_q	
	a_1	x_{i11}	x_{i12}	x_{i1k}	x_{i1q}	$\bar{x}_{1\cdot}$
		x_{n11}	x_{n12}	x_{n1k}	x_{n1q}	
		$\bar{x}_{11}$	$\bar{x}_{12}$	$\bar{x}_{1k}$	$\bar{x}_{1q}$	
	a_2	x_{i21}	x_{i22}	x_{i2k}	x_{i2q}	$\bar{x}_{2\cdot}$
		x_{n21}	x_{n22}	x_{n2k}	x_{n2q}	
		$\bar{x}_{21}$	$\bar{x}_{22}$	$\bar{x}_{2k}$	$\bar{x}_{2q}$	
	a_j	x_{ij1}	x_{ij2}	x_{ijk}	x_{ijq}	$\bar{x}_{j\cdot}$
		x_{nj1}	x_{nj2}	x_{njk}	x_{njq}	
		$\bar{x}_{j1}$	$\bar{x}_{j2}$	$\bar{x}_{jk}$	$\bar{x}_{jq}$	
	a_p	x_{ip1}	x_{ip2}	x_{ipk}	x_{ipq}	$\bar{x}_{p\cdot}$
		x_{np1}	x_{np2}	x_{npk}	x_{npq}	
		$\bar{x}_{p1}$	$\bar{x}_{p2}$	$\bar{x}_{pk}$	$\bar{x}_{pq}$	
A Marginals		$\bar{x}_{.1}$	$\bar{x}_{.2}$	$\bar{x}_{.k}$	$\bar{x}_{.q}$	Grand mean $\bar{\bar{x}}$

A further assumption mentioned above has to be added: the groups must have the same sample size.

Table 3.6-3 shows the general two-way ANOVA design.

The two independent variables in a two-way ANOVA are called "factors", because the two variables (two factors) affect the dependent variables. Each factor will have two or more levels within it.

Treatment groups (cells) are formed from all possible combinations of the two factors; if the first factor has l levels and the second factor has k levels, then there will be $l \cdot k$ different treatment groups.

In many experimental systems, the effect of one factor depends on the level of the other. This is called *interaction*. The interaction effect is the effect that one factor has on the other. The computational scheme of the two-way ANOVA with consideration of the interaction effect is given in Table 3.6-4.

The computational scheme of ANOVA is also used for many other problems.

Of course there are also ANOVA applications with more than two factors, so called *multi-way* ANOVA, but these are used extremely rarely in AQA.

Challenge 3.6-1

To check the compatibility of four laboratory analysts in their analysis of nitrite in waste water by DIN EN ISO 10304-1 [11], a waste water sample had to be analyzed by the four analysts under the same conditions with the same ion chromatograph. The results obtained are given in Table 3.6-5.

(continued)

Table 3.6-4 Computational scheme of two-way ANOVA

Factor A		Equation
Degrees of freedom	$df_A = p - 1$	(3.6-16)
Sum of squares	$SS_A = n \cdot q \cdot \sum_{j=1}^{p} \left(\bar{x}_{j.} - \bar{\bar{x}}\right)^2$	(3.6-17)
Mean square	$s_A^2 = \frac{SS_A}{df_A}$	(3.6-18)
Test value $\hat{F}$	$\hat{F}_A = \frac{s_A^2}{s_R^2}$	(3.6-19)
Critical value	$F(P, df_A, df_R)$	
Factor B		**Equation**
Degrees of freedom	$df_B = q - 1$	(3.6-20)
Sum of squares	$SS_B = n \cdot p \cdot \sum_{k=1}^{q} (\bar{x}_{.k} - \bar{\bar{x}})^2$	(3.6-21)
Mean square	$s_B^2 = \frac{SS_B}{df_B}$	(3.6-22)
Test value $\hat{F}$	$\hat{F}_B = \frac{s_B^2}{s_R^2}$	(3.6-23)
Critical value	$F(P, df_B, df_R)$	
Interaction		**Equation**
Degrees of freedom	$df_{AB} = df_A \cdot df_B$	(3.6-24)
Sum of squares	$SS_{AB} = n \cdot \sum_{j=1}^{p} \sum_{k=1}^{q} \left(\bar{x}_{jk} - \bar{x}_{j.} - \bar{x}_{.k} + \bar{\bar{x}}\right)^2$	(3.6-25)
Mean square	$s_{AB}^2 = \frac{SS_{AB}}{df_A \cdot df_B}$	(3.6-26)
Test value $\hat{F}$	$\hat{F}_{AB} = \frac{s_{AB}^2}{s_R^2}$	(3.6-27)
Critical value	$F(P, df_{AB}, df_R)$	
Residual		**Equation**
Degrees of freedom	$df_R = df_T - df_A - df_B - df_{AB}$	(3.6-28)
Sum of squares	$SS_R = SS_T - SS_A - SS_B - SS_{AB}$	(3.6-29)
Mean square	$s_R^2 = \frac{SS_R}{df_R}$	(3.6-30)
Total		**Equation**
Degrees of freedom	$df_T = nkl - 1$	(3.6-31)
Sum of squares	$SS_T = \sum_{i=1}^{n} \sum_{j=1}^{p} \sum_{k=1}^{q} (x_{ijk} - \bar{\bar{x}})^2$	(3.6-32)

Table 3.6-5 Analytical results (in µg L^{-1} NO_2^-) obtained by ion chromatography with four analysts and three replicates n_j

Analyst	1	2	3	4
Replicates n_j				
1	10.2	11.2	10.3	10.5
2	10.4	10.9	10.4	10.7
3	10.0	10.9	10.7	10.4

(a) Check whether the mean values obtained by three replicates are statistically equal or whether a difference can be detected at the significance

(continued)

level $P = 95\%$. First, give a visual presentation of the analytical results of the four analysts.

(b) If you have detected a statistical difference between the means, check which mean(s) is/are different.

Solution to Challenge 3.6-1

(a) Any statistical test should always be preceded by visual inspection of the data. The visual presentation of the analytical results of the four analysts given in Table 3.6-5 is given in Fig. 3.6-1.

As the figure shows, neither the question of the difference between the means nor the question as to which mean(s) is/are different from the others can be decided visually, and therefore statistical tests are necessary to answer these questions.

First, we have to check the homogeneity of the variances in the four columns of Table 3.6-5. Because all columns are the same size, the Cochran test can be applied. The test value $\hat{C} = 0.3171$ is calculated by (3.4-1) with $\sum s_j^2 = 0.1367$ and $s_{\max}^2 = 0.04333$.

The critical value $C(P = 95\%, k = 4, \mathrm{df} = 2) = 0.7679$ is greater than $\hat{C}$, which means the variances of the groups are homogeneous, and ANOVA is allowed. The ANOVA data are summarized in Table 3.6-6. Because the test value $\hat{F}$ exceeds the tabulated value $F(P = 95\%, \mathrm{df_{bw}}, \mathrm{df_{in}})$, a difference between the means of the four analysts is detected: at least one analytical result is different from the others. But ANOVA has not told us which mean(s) is/are different, and furthermore the inspection of Fig. 3.6-1 does not give an answer either. Therefore, we will try to answer this question using the LSD test.

(b) The test value is LSD $= 0.3480$ calculated by (3.6-15) with $t(P = 95\%, \mathrm{df_{in}} = 8) = 2.306$, $s_{\mathrm{in}}^2 = 0.0342$, and $n_j = 3$. The absolute differences

(continued)

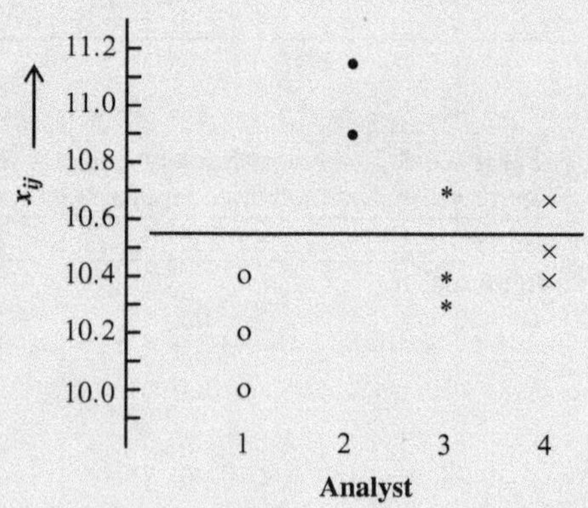

Fig. 3.6-1 Visual presentation of the analytical results of the four analysts given in Table 3.6-5

Table 3.6-6 Intermediate quantities for ANOVA of the data given in Table 3.6-5

n	12	k	4	
$\bar{\bar{x}}$ (3.6-12)	10.55			
Analyst	1	2	3	4
$\bar{x}_j$	10.20	11.00	10.47	10.53
n_j	3	3	3	3
$n_j \cdot (\bar{x}_j - \bar{\bar{x}})^2$	0.36750	0.60750	0.02083	0.00083
SS_{bw} (3.6-3)	0.99667	df_{bw} (3.6-4)	3	
s^2_{bw} (3.6-5)	0.3322			
$\sum_j (x_{ij} - \bar{x}_j)^2$	0.0800	0.0600	0.0867	0.0467
SS_{in} (3.6-6)	0.2733	df_{in} (3.6-7)	8	
s^2_{in} (3.6-8)	0.0342			
$\hat{F}$ (3.6-13)	9.724	$F(P = 95\%, df_{bw}, df_{in})$		4.066

Table 3.6-7 The comparison of the paired absolute difference of mean values together with the LSD value

Comparison of analysts	|Difference|	Result
1 with 2	0.8000	Greater
1 with 3	0.2667	Smaller
1 with 4	0.3333	Smaller
2 with 3	0.5333	Greater
2 with 4	0.4667	Greater
3 with 4	0.0667	Smaller

of all mean values together, given in Table 3.6-7, reveals that only the mean value of analyst 2 is different from the other mean values. Note that according to Fig. 3.6-1 the mean of analyst 1 also seems to differ significantly from the other means, but the statistical test gives another result.

Challenge 3.6-2

In an laboratory, a method for the routine analysis of Ni in industrial waste water is to be introduced. For this purpose, the AAS method according to DIN 38406-E 11 [12] was chosen. It is known that Fe in various concentrations can be present in the waste water. With this information three problems arise:

(a) Does the Fe concentration have an influence on the Ni determination?
(b) Which experimental conditions are appropriate for the AAS?
(c) Is there an interaction between (a) and (b)?

Here it should be easily recognized that these questions can be answered with a *two-way ANOVA*.

The *first* factor A is the influence of Fe on the determination of Ni, for which three types of test solutions were prepared:

(continued)

1. A test solution was used which does not contain Fe (denoted "without").
2. A test solution was used with a small concentration of 10 mg L^{-1} Fe (denoted "small").
3. A test solution was used with the highest possible concentration of 50 mg L^{-1} Fe (denoted "high").

The concentration of Ni is equal in all test solutions: 30 mg L^{-1} Ni.

The *second* factor B includes the conditions of the measurement by the AAS method:

1	Flame: N_2O/C_2H_2	$\lambda = 232$ nm	(Condition I)
2	Flame: Air/C_2H_2	$\lambda = 342$ nm	(Condition II)

The results of the AAS determinations of Ni with five replicates are summarized in Table 3.6-8.

Answer the questions given above on the basis of the experimental results.

Table 3.6-8 Results of the AAS determination of Ni (in mg L^{-1}) under two different conditions

Factor B: AAS	Factor A: Fe content		
	Without	Small	High
Condition I	20.1	20.7	22.0
	19.0	20.3	21.2
	20.5	20.9	22.0
	19.7	20.5	20.6
	20.3	19.6	22.3
Condition II	18.0	19.8	22.1
	19.3	20.1	21.2
	18.7	19.2	22.2
	21.0	19.6	22.0
	19.6	20.3	22.4

Solution to Challenge 3.6-2

Firstly, we have to check whether the variables in each cell are normally distributed and whether the variances of the populations are statistically equal (normality and homoscedasticity of all cells). Note that there is no reason for a test of outliers in the data set.

The test for *normal distribution* is carried out by the David test (see Sect. 3.2.1) and the *homogeneity of the variances* is tested by the Bartlett test (see Sect. 3.4).

The intermediate quantities and the results for the David and Bartlett tests are given in Tables 3.6-9 and 3.6-10.

(continued)

Table 3.6-9 Intermediate quantities and results for the check of normal distribution (David test)

	A_1	A_2	A_3
B_1			
x_{max}	20.5	20.9	22.3
x_{min}	19.0	19.6	20.6
s	0.5933	0.5000	0.7014
$\hat{q}_r$	2.53	2.60	2.42
B_2			
x_{max}	21.0	20.3	22.4
x_{min}	18.0	19.2	21.2
s	1.1212	0.4301	0.4604
$\hat{q}_r$	2.68	2.56	2.61

Table 3.6-10 Intermediate quantities and results for the check of homogeneity of variances (Bartlett test)

A_j	1	2	3	1	2	3
	B_1			B_2		
SS_j	1.408	1.000	1.968	5.028	0.740	0.848
$\sum SS_j$	10.992					
s^2 (2.2.2-2)		0.4580	df log s^2		−8.1392	
s_j^2	0.352	0.250	0.492	1.257	0.185	0.212
df_j	4	4	4	4	4	4
$df_j \log s_j^2$	−1.814	−2.408	−1.232	0.397	−2.931	−2.6947
$\Sigma df_j \log s_j^2$		−10.6828				
$\hat{\chi}^2$	*5.857*	$\chi^2(P = 99\%, df = 5)$				15.086
		$\chi^2(P = 95\%, df = 5)$				11.070

The lower and upper limit values of the David test are 2.15 and 2.753 at the significance level $P = 95\%$. Thus, a normal distribution is present in all cells.

As Table 3.6-10 shows, the variances are homogeneous at the significance level $P = 99\%$ as well as $P = 95\%$ because the test value $\hat{\chi}^2$ does not exceed the corresponding critical value. Thus, two-way ANOVA may be carried out.

The intermediate quantities for the calculation of the sum of squares, the variances, the test and the critical values for two-way ANOVA are given in detail in Table 3.6-11.

Since there are three questions about the analytical task, we have three decisions to make.

(a) Since the test value $\hat{F}_A = 28.649$ exceeds the critical value $F(P = 95\%, df_A, df_R) = 3.403$, the null hypothesis $H_0 = \mu_1 = \mu_2 = \cdots = \mu_6$ must be rejected. There is a very significant effect of the Fe content on the determination of Ni.

(continued)

Table 3.6-11 Intermediate quantities and results for the two-way ANOVA; the equations for the calculations are given in parentheses

p	3	q	2
n	5	$\bar{\bar{x}}$	20.507
Sum of squares			
A_j	1	2	3
$\bar{x}(A_j)$	19.62	20.10	21.80
$nq(\bar{x}_j \cdot - \bar{\bar{x}})^2$	7.862	1.654	16.727
SS_A (3.6-17)	26.243	df_A (3.6-16)	2
B_k	1	2	
$\bar{x}(B_k)$	20.647	20.367	
$nq(\bar{x}_k \cdot - \bar{\bar{x}})^2$	0.2940	0.2940	
SS_B (3.6-21)	0.588	df_B (3.6-20)	1
SS_T (3.6-32)	39.359	df_T (3.6-31)	29
$\bar{x}_{jk}$	19.92	20.40	21.62
	19.32	19.80	21.98
$(\bar{x}_{jk} - \bar{x}_{j\cdot} - \bar{x}_{\cdot k} + \bar{\bar{x}})^2$	0.0256	0.0256	0.1024
	0.0256	0.0256	0.1024
SS_{AB} (3.6.25)	1.536	df_{AB} (3.6-24)	2
SS_R (3.6-29)	10.992	df_R (3.6-28)	24
Variances			
s_A^2 (3.6-18)	13.121	s_{AB}^2 (3.6-26)	0.768
s_B^2 (3.6-22)	0.588	s_R^2 (3.6-30)	0.458
Critical values at $P = 95\%$		*Test values*	
$F(P = 95\%, df_A, df_R)$	3.403	$\hat{F}_A$ (3.6-19)	28.649
$F(P = 95\%, df_B, df_R)$	4.260	$\hat{F}_B$ (3.6-23)	1.284
$F(P = 95\%, df_{AB}, df_R)$	3.403	$\hat{F}_{AB}$(3.6-27)	1.677

(b) The influence of the conditions of AAS measurement on the result is not significant because the test value $\hat{F}_B = 1.284$ is smaller than the table value $F(P = 95\%, df_B, df_R) = 4.260$.

(c) The test value for the interaction $\hat{F}_{AB} = 1.677$ does not exceed the limits of the table value $F(P = 95\%, df_{AB}, df_R) = 3.403$, and thus factor A (the concentration of Fe) does not interact with factor B (the conditions of AAS measurement).

The two-way ANOVA yields the important result that the analytical method proposed for the determination of Ni in ferrous waste water cannot be applied without elimination of the influence of Fe.

References

1. Massart DL, Vandeginste BGM, Buydens LMC, De Jong S, Lewi PJ, Smeyers-Verbeke J (1997) Handbook of chemometrics and qualimetrics, part A. Elsevier, Amsterdam
2. Funk W, Dammann V, Donnevert G (2005) Qualitätssicherung in der analytischen Chemie, 2nd edn. Wiley-VCH, Weinheim

3. DIN EN 53 804-1 (2002) Statistical evaluation – part 1: continuous characteristics. Beuth, Berlin
4. Linsinger TPJ, Kandler W, Koska R, Grasserbauer M (1998) The influence of different evaluation techniques on the results of interlaboratory comparison. Accred Qual Assur 3:322–327
5. DeLa Calle Garcia D, Magnaghi S, Reichenbächer M, Danzer K (1996) Systematic optimization of the analysis of wine bouquet components by solid-phase-microextraction. J High Resol Chromatogr 19:257–262
6. DIN 51603-1 (2008) Liquid fuels – fuel oils, part 1: fuel oil specifications. Beuth, Berlin
7. Ellison SLR, Berwick VJ, Duguid Farrant TJ (2009) Practical statistics for the analytical scientist, 2nd edn. RSC, Cambridge
8. DIN ISO 57525-2 (2002) Accuracy (trueness and precision) of measurement methods and results- part 2: basic method for the determination of repeatability and reproducibility of a standard measurement method. Beuth, Berlin
9. Doerffel K (1990) Statistik in der analytischen Chemie, 5. Aufl. Deutscher Verlag für Grundstoffindustrie, Leipzig, Germany
10. EudraLex (2010) The rules for governing medicinal products in the European Union, vol 4, Brussels
11. DIN EN ISO 10304-1 (2009) Water quality – Determination of dissolved anions by liquid chromatography of ions, part 1: determination of bromide, chloride, fluoride, nitrate, nitrite, phosphate and sulphate. Beuth, Berlin
12. DIN 38406-E 11 (1991) German standard methods for the examination of water, waste water and sludge; cations (group E). Determination of nickel by atomic absorption spectrometry (AAS). Beuth, Berlin

Chapter 4
General Aspects of Linear Regression

4.1 Correlation, Regression, and Calibration

Correlation and regression analysis investigate the relationships between associated variables, but with different objectives depending on the nature of the variables.

Correlation analysis studies whether there is a linear relationship between two random variables x_i and y_i and how strong is it. The strength of the relationship between a pair of variables is quantified by the *correlation coefficient* r_{xy} (also called the Pearson correlation coefficient), which is calculated by (4.1-1):

$$r_{xy} = \frac{\mathrm{SS}_{xy}}{\sqrt{\mathrm{SS}_{xx} \cdot \mathrm{SS}_{yy}}} = \frac{s_{xy}^2}{s_x \cdot s_y}, \qquad (4.1\text{-}1)$$

with the covariance

$$s_{xy}^2 = \frac{\mathrm{SS}_{xy}}{\mathrm{df}}, \qquad (4.1\text{-}2)$$

the sums of squares

$$\mathrm{SS}_{xx} = \sum x_i^2 - \frac{\left(\sum x_i\right)^2}{n} = \sum (x_i - \bar{x})^2 \qquad (4.1\text{-}3)$$

$$\mathrm{SS}_{yy} = \sum y_i^2 - \frac{\left(\sum y_i\right)^2}{n} = \sum (y_i - \bar{y})^2 \qquad (4.1\text{-}4)$$

$$\mathrm{SS}_{xy} = \sum (x_i \cdot y_i) - \frac{\sum x_i \cdot \sum y_i}{n} = \sum (x_i - \bar{x}) \cdot (y_i - \bar{y}), \qquad (4.1\text{-}5)$$

and the degrees of freedom

$$\mathrm{df} = n - 1. \qquad (4.1\text{-}6)$$

M. Reichenbächer and J.W. Einax, *Challenges in Analytical Quality Assurance*,
DOI 10.1007/978-3-642-16595-5_4, © Springer-Verlag Berlin Heidelberg 2011

The square of the correlation coefficient between x_i and y_i, r^2 is called the *coefficient of determination*. It expresses the proportion of the sum of squares of regression SS_{reg} in the total sum of squares SS_{tot}:

$$r^2 = \frac{\sum (\hat{y}_i - \bar{y})^2}{\sum (y_i - \bar{y})^2} = \frac{SS_{reg}}{SS_{tot}}. \tag{4.1-7}$$

Note that in practice, the sum of squares SS_{xx}, SS_{yy}, the correlation coefficient r, and the coefficient of determination r^2 are obtained by the corresponding MS Excel functions =DEVSQ(Data x_i), =DEVSQ(Data y_i), =CORREL(Matrix 1, Matrix 2), and =RSQ(Matrix 1, Matrix 2), respectively.

One variable is not expressed as a function of the other since both are equivalent. There is neither a dependent nor an independent variable. Correlation questions can be, for example, the stability of a steel wire and its content of carbon or the concentration of certain metals in soil measured at different location near a plant.

The correlation coefficient r_{xy} is a dimensionless number in the range $-1 < r_{xy} < +1$. The values +1 or –1 indicate a perfect linear relationship between the variables x_i and y_i, and $r_{xy} = 0$ indicates that the variables are uncorrelated. Figure 4.1-1 illustrates some cases.

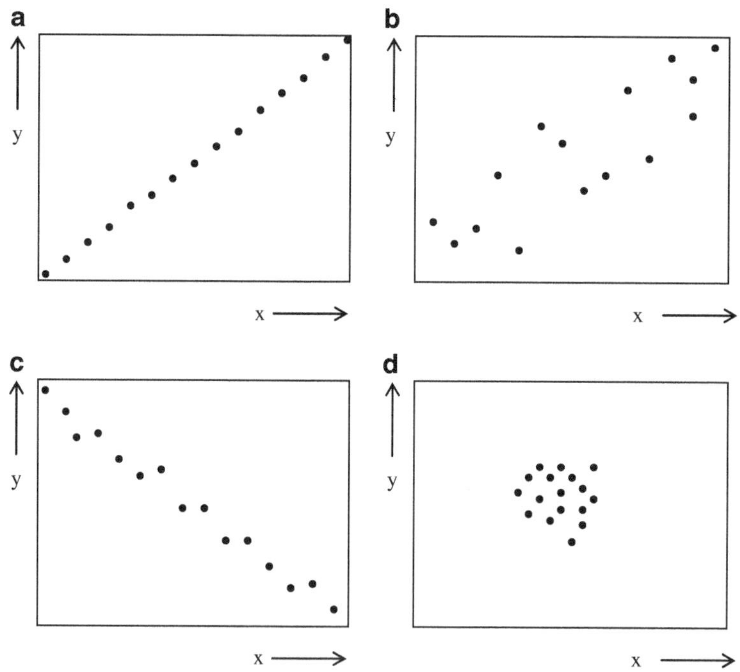

Fig. 4.1-1 Scatter plots of variables x and y with various degrees of correlation: (**a**) a nearly perfect positive correlation with $r_{xy} \approx +1$, (**b**) a moderate positive correlation with $r_{xy} < 1$, (**c**) a strong negative correlation with $r_{xy} \approx -1$, and (**d**) no correlation, $r_{xy} = 0$

Note that in the analytical calibration in which $x_i = c_i$ is fixed, there is no correlation problem because the function $y = f(x)$ is well-known, mostly from natural laws; for example, the Lambert–Beer law.

Regression analysis includes any techniques for modeling and analyzing several variables when the focus is on the relationship between one and more *dependent* variables and one or more *independent* variables. This relationship is expressed by a mathematical function. If this function is known it is possible to predict one and more variable. More specifically, regression analysis helps us to understand how the typical value of the dependent variable changes when any one of the independent variables is varied while the other independent variables are held fixed.

The determination of the mathematical relationship is carried out by *calibration*, a fundamental objective of instrumental analysis where an instrumental response (peak area, absorbance, and others) as the dependent variable is related to the known (given a priori) concentrations of the calibration standards as the independent variables.

There are some important conditions for calibration:

- *Certified reference material* (CRM) or *certified reference substances* (CRS) must be used for the preparation of standard solutions as independent variables.
- All calibration standards must be prepared *independently*, which means that each standard has to be prepared separately.
- The *measurement strategy* has to be fixed: the number of calibration standards, their distance (equally distributed or in larger numbers at the beginning or the end of the calibration range), the number of real replicate measurements must be fixed, which means that each standard must be prepared with the same treatments. If, for example, for HPLC analysis the preparation of the standard solutions means merely dissolving defined amounts of the CRS in the eluent, then two injections from the same vial are two independent measurements, because the error is mainly determined in the peak areas of HPLC analysis and not in the preparation of the standard solutions. If, however, HPLC analysis occurs with solutions which are produced, for example, with a pre-treatment (extraction of the sample or similarly), then two injections from the same vial would only produce two measured values from which a mean for this standard can be calculated, because the error of the pre-treatment is not involved.
- The type of *calibration model* has to be fixed: linear or non-linear, univariate or multivariate. However, the univariate linear model is the one usually applied in AQA.
- The y-values must be *normally distributed.*
- The variances of the y-values have to be equal throughout the range of x, i.e. there must be *homogeneity of variances* at each calibration point. If this is not the case, weighted calibration must be applied.

Challenge 4.1-1

In order to develop the method for the determination of some metals in solid samples in routine analysis, the extraction step must be optimized. For this purpose, the extraction of Cu, Ni, and Zn is tested under various conditions (temperature, microwave, etc.). The analytical results obtained by AAS analysis are summarized in Table 4.1-1 for ten different extraction conditions.

Check whether there is a correlation between the extraction of Cu and Ni as well as between Cu and Zn. Interpret your result.

Table 4.1-1 Analytical results of Cu, Ni, and Zn determined by AAS after the extraction of the same solid sample under various conditions

Extraction condition	Cu in $mg\ L^{-1}$	Ni in $mg\ L^{-1}$	Zn in $mg\ L^{-1}$
1	14.5	34.6	55.6
2	16.8	33.2	54.8
3	12.6	30.4	56.0
4	21.4	59.3	56.3
5	20.3	58.0	54.0
6	9.50	27.4	56.4
7	25.8	63.0	58.2
8	23.7	61.8	57.3
9	28.6	65.3	55.9
10	25.9	61.0	53.6

Solution for Challenge 4.1-1

Table 4.1-2 presents the intermediate quantities and result for the calculation of the correlation coefficient between the extracted amount of Cu and Ni, r_{xy}(Cu/Ni), according to (4.1-1)–(4.1-5). The other correlation coefficients obtained by Excel functions are r_{xy}(Cu/Zn) = 0.0502 and r_{xy}(Ni/Zn) = 0.0952.

The correlation coefficient r_{xy}(Cu/Ni) = 0.9465 is close to 1. This indicates a strong positive correlation which means that both elements are extracted with the same efficiency. For further investigations concerning the extraction step, only one element needs to be determined to know the extraction efficiency of the other.

The correlation coefficient between the extracted amounts of Cu and Zn, r_{xy}(Cu/Zn) = 0.0502, indicates that there is only a very weak correlation, which means that the extracted amount of Zn is independent of the extraction conditions. Under each condition the amount of Zn is completely extracted, perhaps because Zn exists as a slightly soluble compound.

Table 4.1-2 Intermediate quantities and the result for the calculation of the correlation coefficient r_{xy} between the amounts of extracted Cu (x-values) and Ni (y-values) according to (4.1-1)–(4.1-5)

$\bar{x}$(Cu) in mg L^{-1}	19.91	$\bar{y}$(Ni) in mg L^{-1}	49.40
Extraction condition	$(x_i - \bar{x})^2$	$(y_i - \bar{y})^2$	$(x_i - \bar{x}) \cdot (y_i - \bar{y})$
1	29.27	219.04	80.07
2	9.67	262.44	50.38
3	53.44	361.00	138.89
4	2.22	98.01	14.75
5	0.15	73.96	3.35
6	108.37	484.00	229.02
7	34.69	184.96	80.10
8	14.36	153.76	47.00
9	75.52	252.81	138.17
10	35.88	134.56	69.48
Sum	363.57 $= SS_{xx}$	2,224.54 $= SS_{yy}$	851.22 $= SS_{xy}$
r_{xy}	*0.9465*		

4.2 Linear Calibration Model

With independent concentrations of the standard x_i which fulfills the conditions given above, the dependent information values (response) y_i are obtained by measurement. The linear calibration function can be expressed by

$$y_i = a_0 + a_1 \cdot x_i + e_{y_i} \tag{4.2-1}$$

in which the model parameters a_0 and a_1 are the *intercept* and the *slope* of the true but unknown regression line, respectively.

The linear function can be fitted to the measured values by means of *Gaussian least squares estimation* or *ordinary least squares estimation* (OLS):

$$\hat{y}_i = a_0 + a_1 \cdot x_i. \tag{4.2-2}$$

The residuals e_{y_i} are the deviations of the measurement values y_i from their values predicted by the regression line:

$$e_{y_i} = y_i - \hat{y}_i = y_i - a_0 - a_1 \cdot x_i. \tag{4.2-3}$$

The graph for the linear calibration function and its parameters is shown in Fig. 4.2-1.

Fig. 4.2-1 Linear calibration line for the function $y = f(x_i)$ with the parameters intercept a_0, slope a_1, and the residuals e_i

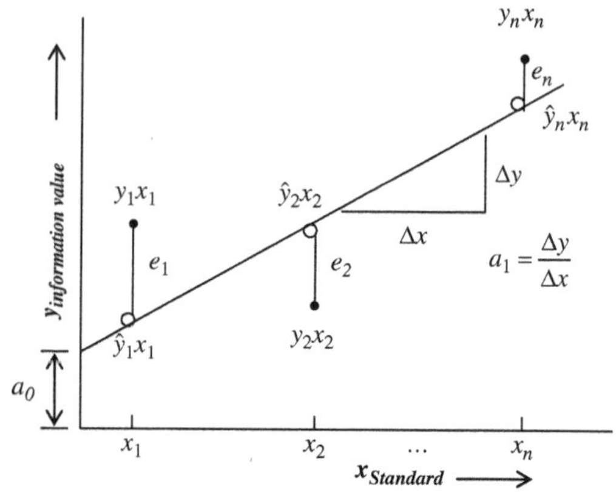

Parameters of the linear calibration model [1]:

1. *Intercept* a_0, and *slope* a_1
   The slope a_1 is the *sensitivity* of the analytical method, and the intercept a_0 is predominantly the *blank*.
   Although the calibration intercept and slope can be usually calculated by a hand calculator or by Excel functions = INTERCEPT(y_i, x_i) and = SLOPE(y_i, x_i), respectively, the equations for their calculation will be given:
   Intercept a_0:

$$a_0 = \bar{y} - a_1 \cdot \bar{x}. \tag{4.2-4}$$

Slope a_1:

$$a_1 = \frac{\sum (x_i - \bar{x}) \cdot (y_i - \bar{y})}{\sum (x_i - \bar{x})^2} = \frac{SS_{xy}}{SS_{xx}}. \tag{4.2-5}$$

In linear regression analysis, besides the intercept a_0 and the slope a_1, the following parameters are important; most of them are shown in Fig. 4.2-2.

2. *Residual standard deviation*, $s_{y.x}$
   The residual standard deviation $s_{y.x}$ indicates the *calibration error*. It is calculated by (4.2-6) or (4.2-7):

$$s_{y.x} = \sqrt{\left(\frac{1}{n_c - 2}\right) \cdot \left(SS_{yy} - \frac{SS_{xy}^2}{SS_{xx}}\right)}, \tag{4.2-6}$$

$$s_{y.x} = \sqrt{\frac{\sum_{i=1}^{n_c} (y_i - \hat{y}_i)^2}{df}}. \tag{4.2-7}$$

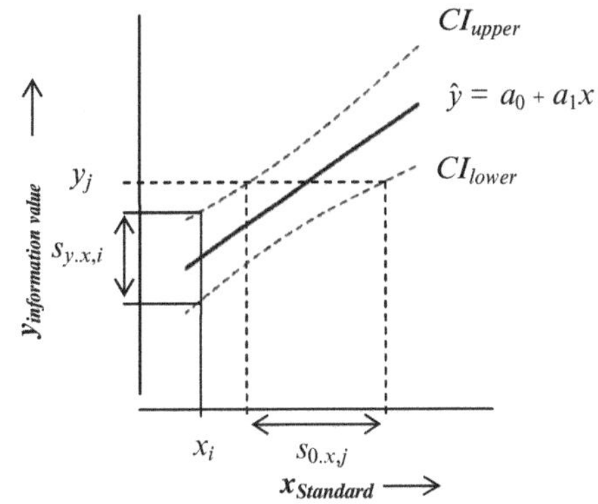

Fig. 4.2-2 Linear calibration function $\hat{y} = a_0 + a_1 \cdot x$ with its upper and lower confidence intervals $\mathrm{CI}_{\mathrm{upper}}$ and $\mathrm{CI}_{\mathrm{lower}}$, respectively; $s_{x.0,j}$ is the analytical error for the information value y_j, and $s_{y.x,i}$ is the calibration error for the standard x_i

The degrees of freedom df is given by df $= n_c - 2$ for the *linear* calibration function in which n_c is the number of independently measured values.

The sums of squares SS_{xx}, SS_{yy}, and SS_{xy} are calculated by (4.1-3), (4.1-4), and (4.1-5), respectively, but SS_{xx} and SS_{yy} can be obtained with the Excel functions DEVSQ(x_i) and DEVSQ(y_i).

The sum of the residuals can also be calculated by (4.2-8):

$$\sum_{i=1}^{n_c} (y_i - \hat{y}_i)^2 = \sum y_i^2 - a_0 \cdot \sum y_i - a_1 \cdot \sum x_i \cdot y_i. \qquad (4.2\text{-}8)$$

Note that in practice the residual standard deviation is obtained by Excel function =STEYX(y_i, x_i).

3. *Analytical error*, $s_{x.0}$

The analytical standard deviation $s_{x.0}$ indicates the random error of the analytical process:

$$s_{x.0} = \frac{s_{y.x}}{a_1}. \qquad (4.2\text{-}9)$$

The relative standard deviation $s_r\%$ of the analytical process is

$$s_r\% = \frac{s_{x.0}}{\bar{x}} \cdot 100. \qquad (4.2\text{-}10)$$

4. *Standard deviation of the intercept*, s_{a_0}

$$s_{a_0} = s_{y.x} \cdot \sqrt{\frac{1}{n_c} + \frac{\bar{x}^2}{\mathrm{SS}_{xx}}}. \qquad (4.2\text{-}11)$$

5. *Confidence interval of the intercept*, $\mathrm{CI}(a_0)$

$$\mathrm{CI}(a_0) = a_0 \pm t(P, \mathrm{df}) \cdot s_{a_0}. \qquad (4.2\text{-}12)$$

6. *Standard deviation of the slope*, s_{a_1}

$$s_{a_1} = \frac{s_{y.x}}{\sqrt{\mathrm{SS}_{xx}}}. \tag{4.2-13}$$

7. *Confidence interval of the slope*, $\mathrm{CI}(a_1)$

$$\mathrm{CI}(a_1) = a_1 \pm t(P, \mathrm{df}) \cdot s_{a_1}. \tag{4.2-14}$$

8. *Prediction of x from y*, $\hat{x}$
   The calibration line is used to predict the concentration of an analyte in a sample $\hat{x}$ using (4.2-15):

$$\hat{x} = \frac{\hat{\bar{y}} - a_0}{a_1}, \tag{4.2-15}$$

   in which $\hat{\bar{y}}$ is the mean of the information values with n_a determinations performed on the sample.

   The error of the predicted value $s(\hat{x})$ is calculated by (4.2-16):

$$s(\hat{x}) = \frac{s_{y.x}}{a_1} \cdot \sqrt{\frac{1}{n_c} + \frac{1}{n_a} + \frac{(\bar{y} - \hat{\bar{y}})^2}{a_1^2 \cdot \mathrm{SS}_{xx}}}, \tag{4.2-16}$$

   in which n_a is the number of the replicates of a measured value y_i. All other symbols are explained above.
9. *Confidence interval of the predicted concentration*, $\mathrm{CI}(\hat{x})$
   The confidence interval CI of the predicted concentration $\mathrm{CI}(\hat{x})$ for the significance level P is calculated by (4.2-17):

$$\mathrm{CI}(\hat{x}) = \hat{x} \pm t(P, \mathrm{df}) \cdot s(\hat{x}). \tag{4.2-17}$$

As already mentioned above, the degrees of freedom is $\mathrm{df} = n_c - 2$ for a linear regression function. The upper and lower confidence limits of a regression line are the dotted lines in Fig. 4.2-2.

Challenge 4.2-1

In an analytical laboratory, a method is to be introduced for the quality control of technical *n*-hexane produced by the hydrogenation of benzene. The limit value L_0 of the residual content of benzene is specified as $L_0 = 0.03\%$ (v/v). In order to minimize the cost, simple and inexpensive photometry should be chosen.

Before one selects an analytical method suitable for the task, one must check whether the *sensitivity* of the method is sufficient. The sensitivity of the

(continued)

photometry is determined by the absorptivity, which is $\alpha \approx 250\ \text{L mol}^{-1}\ \text{cm}^{-1}$ at $\lambda = 254$ nm for the analyte benzene.

(a) Check if the sensitivity of the photometry is sufficient for this analytical problem, as follows: estimate which concentration has to be prepared for the absorbances $A_{\text{min}} = 0.1$ and $A_{\text{max}} = 1.5$ using a cuvette $l = 1$ cm and decide whether the critical concentration lies inside this absorbance range. Note that the limit values of the absorbance A result from Challenge 2.2.5-2.

With the answer to this question, the choice of the concentration range given in Table 4.2-1 is confirmed. The calibration standards are prepared as following: in five 100 mL volumetric flasks which are about half-filled with n-hexane (CRS), the volumes of benzene (CRS) given in Table 4.2-1 are injected *into* the n-hexane solution in order to avoid loss of the volatile analyte benzene. Then, the flasks are filled with n-hexane and closed. After each calibration solution has been filled in a closable cuvette, starting with standard 1, the absorbance A is measured at the wavelength $\lambda = 254$ nm. This procedure is repeated with every calibration solution. The measured values of the absorbance A_i are also given in Table 4.2-1.

The density of benzene is $\rho = 0.8765\ \text{g cm}^{-3}$ (at room temperature).

(b) Determine the following parameters for the calibration function:

- Intercept with its confidence interval
- Slope with its confidence interval
- Calibration error
- Analytical error
- Relative standard deviation of the analytical process

(c) Let us assume that the calibration function is valid. The following values A_i are obtained for a sample of a batch of n-hexane measured under the same conditions in triplicate:

(continued)

Table 4.2-1 Preparation of standard solutions for the calibration of benzene in n-hexane as well as the twofold determination of the absorbance of the standard solutions

Standard	V_{inj} in μL	Measured absorbance A	
		First determination	Second determination
1	7	0.1991	0.2008
2	14	0.3958	0.3992
3	21	0.6076	0.6012
4	28	0.7999	0.8016
5	35	1.0013	1.0095

V_{inj} is the injected volume of benzene (CRS) in 100 mL volumetric flasks which are about half-filled with n-hexane (CRS)

y_i	0.8304	0.8301	0.8309

Check if this batch fulfills the quality requirement stated in the quality document of the plant or, in other words, can the batch be released?

(d) According to the Lambert–Beer law $A_i = \alpha_\lambda \cdot c \cdot l$, the basis of photometric analysis, the slope is the absorptivity α_λ at the wavelength λ, which is the *sensitivity* of the photometric method. Estimate the absorptivity α_λ for benzene at the wavelength $\lambda = 254$ nm in the usual units L mol^{-1} cm^{-1} and m^2 mol^{-1}, respectively.

Solution to Challenge 4.2-1

(a) The critical limit is $L_0 = 0.03\%$ (v/v) or 0.03 mL benzene in 100 mL solution which is, taking into account the density of benzene, 0.2630 g L^{-1}. Thus the critical concentration is

$$c_{crit} = \frac{0.2630\,\text{g}}{78\,\text{g mol}^{-1} \cdot 1\,\text{L}} = 0.00337\,\text{mol L}^{-1} = 3.37\,\text{mmol L}^{-1}. \qquad (4.2\text{-}18)$$

Is the sensitivity for concentrations $c_{crit} \approx 3.5$ mmol L^{-1} sufficient for the reliable measurement of the absorbance in the range $A_{min} = 0.1$ to $A_{max} = 1.5$, which is the requirement of the photometry?

The concentrations for the lower and upper limits of the absorbance A_{min} and A_{max}, respectively, calculated according to the Lambert–Beer law with a cuvette of $l = 1$ cm are

$$c_{min} = \frac{0.1}{250\,\text{L mol}^{-1}\,\text{cm}^{-1} \cdot 1\,\text{cm}} = 4 \cdot 10^{-4}\,\text{mol L}^{-1} = 0.4\,\text{mmol L}^{-1} \qquad (4.2\text{-}19)$$

and

$$c_{max} = \frac{1.5}{250\,\text{L mol}^{-1}\,\text{cm}^{-1} \cdot 1\,\text{cm}} = 6 \cdot 10^{-3}\,\text{mol L}^{-1} = 6\,\text{mmol L}^{-1}. \qquad (4.2\text{-}20)$$

As (4.2-19) and (4.2-20) show, the critical concentration $c_{crit} \approx 3.5$ mmol L^{-1} lies within the range of reliable measurement of the absorbance.

The estimate of the concentration shows that the sensitivity of the photometry is sufficient for the determination of benzene in *n*-hexane in

(continued)

routine analysis and the chosen concentration range given in Table 4.2-1 should be appropriate: the absorbance lies in a range of low relative error (see Fig. 2.2.5-1) and the critical concentration lies within this range.

(b) Sometimes, as in the current calibration, the number of *independent* concentration standards n_c has to be reconceived.

As Fig. 4.2-3 shows, in contrast to a structureless absorption band (b), the absorbance of the analyte benzene (a) has to be determined by a vibrationally structured absorption band. However, measurements of the absorbance at a rising or decreasing edge with strong variation of the absorbance within the range of the measurement can be linked to a higher distribution of the measured absorbance, as in the maximum of a structureless absorption band such as (b) in Fig. 4.2-3. Therefore, and because the preparation of the standard solutions consists only of the simple pipetting of the analyte benzene with high quality CRS, the random error of the calibration is nearly all produced by the measurement of the absorbance A_i, i.e. $s_A \gg s_c$.

If this is valid, the twofold measurement of the absorbance of the same standard solution may be considered as *two* "independent" measurements. Therefore, the number of calibration standards is $n_c = 10$ and, thus, the degrees of freedom df = 8.

After the estimation of the number n_i of calibration standards, the concentrations of the standard solutions must be calculated by (4.2-21) and (4.2-22):

(continued)

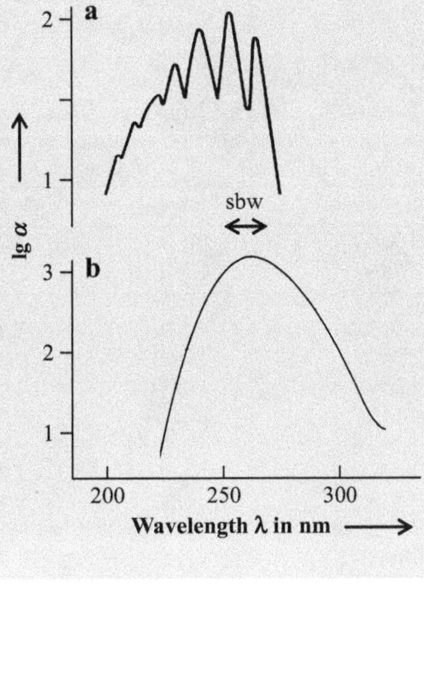

Fig. 4.2-3 The vibrationally structured absorption spectra of the analyte benzene (**a**) and a structureless absorption band obtained, for example, with phenol (**b**). The range marked with an *arrow* is roughly the spectral band width (sbw) for the measurement of the absorbance

$$c = \frac{m\ \mathrm{g}}{M\ \mathrm{g\,mol^{-1}} \cdot V_{\mathrm{sol}}\ \mathrm{L}} \tag{4.2-21}$$

$$c = \frac{\rho\ \mathrm{g\,mL^{-1}} \cdot V_{\mathrm{benzene}}\ \mathrm{mL}}{M\ \mathrm{g\,mol^{-1}} \cdot V_{\mathrm{sol}}\ \mathrm{L}} = \frac{0.8765\ \mathrm{g\,mL^{-1}} \cdot V_{\mathrm{benzene}}\ \mathrm{mL}}{78\ \mathrm{g\,mol^{-1}} \cdot 0.1\ \mathrm{L}}. \tag{4.2-22}$$

As an example, the calculation for standard solution 1 is given:

$$c = \frac{0.8765\ \mathrm{g\,mL^{-1}} \cdot 0.007\ \mathrm{mL}}{78\ \mathrm{g\,mol^{-1}} \cdot 0.1\ \mathrm{L}} = 0.0007866\ \mathrm{mol\,L^{-1}} = 0.7866\ \mathrm{mmol\,L^{-1}}.$$

The further concentrations are listed in Table 4.2-2. With the intermediate quantities given in Table 4.2-2, the calibration parameters may be calculated and they are summarized in Table 4.2-3.

The intermediate quantities of the calculation, intercept a_0, slope a_1, and the residuals e_i, are presented in Table 4.2-2.

(c) As discussed in Sect. 2.2.4, the quality criterion is given by

$$\hat{x}_{\mathrm{critical}} < L_0. \tag{4.2-23}$$

Thus, the critical value $\hat{x}$ calculated by

$$\hat{x}_{\mathrm{critical}} = \hat{x} + \mathrm{CI}_{\mathrm{upper}}(\hat{x}) = \hat{x} + t(\bar{P}_{\mathrm{one-sided}}, \mathrm{df}) \cdot s_{\hat{x}} \tag{4.2-24}$$

must be smaller than the limit value L_0 specified in the documents of AQA which in the given case is $L_0 = 0.03\%$ (v/v).

(continued)

Table 4.2-2 Intermediate quantities of the calculation of the calibration parameters intercept a_0, slope a_1 and the residuals e_i for the photometric determination of benzene in technical n-hexane

$c(x)$ in mmol $\mathrm{L^{-1}}$	$A(y)$	$(x_i-\bar{x})^2$	$(x_i-\bar{x})(y_i-\bar{y})$	$10^6(y_i-\hat{y}_i)^2$
0.7866	0.1991	2.475	0.633	0.11
0.7866	0.2008	2.475	0.631	4.10
1.5732	0.3958	0.619	0.162	19.00
1.5732	0.3992	0.619	0.159	0.97
2.3598	0.6076	0.000	0.000	36.00
2.3598	0.6012	0.000	0.000	0.16
3.1464	0.7999	0.619	0.156	9.70
3.1464	0.8016	0.619	0.157	2.00
3.9330	1.0013	2.475	0.629	9.80
3.9330	1.0095	2.475	0.642	26.00
$\bar{x} = 2.360$	$\bar{y} = 0.602$			
Sum		12.375 SS_{xx}	3.169 SS_{xy}	107.80

Table 4.2-3 Calibration parameters for the photometric determination of benzene in technical *n*-hexane

Parameter	Equation	Value
Intercept a_0	(4.2-4)	−0.00265
Slope a_1 in L mmol^{-1}	(4.2-5)	0.2561
Calibration error $s_{y.x}$	(4.2-7)	0.00367
Analytical error $s_{x.0}$ in mmol L^{-1}	(4.2-9)	0.01434
$s_r\%$	(4.2-10)	0.61
Standard deviation of the intercept s_{a_0}	(4.2-11)	0.00272
Confidence interval of the intercept CI(a_0)	(4.2-12)	±0.00628
Standard deviation of the slope s_{a_1} in L mmol^{-1}	(4.2-13)	0.00104
Confidence interval of the slope CI(a_1) in L mmol^{-1}	(4.2-14)	±0.00241
Number of calibration standards n_c		10
Degrees of freedom df		8
Student's *t*-factor $t(P = 95\%, \text{df} = 8)$		2.306

Note that in this case the upper confidence interval is calculated with the one-sided *t*-factor.

The predicted concentration $\hat{x}$ calculated by (4.2-15) with the mean value $\bar{\hat{y}} = 0.83047$ obtained by three replicates is $\hat{x} = 3.254$ mmol L^{-1}.

The one-sided confidence interval at the significance level $P = 95\%$ calculated according to (4.2-16) and (4.2-17) with $t(\bar{P}_{\text{one-sided}}, \text{df} = 8) = 1.860$ and the parameters given above is $\text{CI}(\hat{x}) = 0.0188$ mmol L^{-1}. The upper value is thus $\hat{x} + \text{CI}_{\text{one-sided}}(\hat{x}) = 3.272$ mmol L^{-1}.

Transformation into the units % (v/v) gives *c* in % (v/v) = 0.0291. The upper concentration of benzene in *n*-hexane is smaller than the limit value $L_0 = 0.03\%$ (v/v), and thus the batch fulfills the quality requirements and can be released.

(d) The absorptivity is the slope of the calibration function. The slope was determined using a cuvette with $l = 1$ cm, therefore $\alpha\,(\lambda = 254\,\text{nm}) = 256$ L mol^{-1} cm^{-1} which is within the range of the literature values.

Sometimes, the values of the absorptivity are given in the units m^2 mol^{-1}. The conversion for benzene gives

$$\alpha = \frac{256\,\text{L}}{\text{mol cm}} = \frac{256 \cdot 1{,}000\,\text{cm}^3}{\text{mol cm}} = \frac{25.6\,\text{m}^2}{\text{mol}}. \qquad (4.2\text{-}25)$$

Challenge 4.2-2

In an analytical laboratory the determination of iron in the range of 0–3 mg L^{-1} must be carried out in a routine using simple and inexpensive photometry. In contrast to the analysis of organic compounds, inorganic ions

(continued)

must be transferred in a complex with high absorptivity by addition of a reagent. However, the chemical equilibrium for the formation of the complex is a source of a relatively high random error, which means that the error of the measurement of the absorbance *A* will be much smaller than that of the preparation of the measurement solution. Therefore, in this case, the double measurement of the absorbance of the same sample does not comply with the requirement of two independent *determinations* because the main error (chemical equilibrium) is not included. The mean value is calculated from both measurement values and corresponds to *one* determination.

Another problem is the choice of an appropriate reagent for the iron complex. In the literature one finds, for example, ferrozine ($\alpha = 2{,}790\ m^2\ mol^{-1}$ at 562 nm) [2] as well as sulfosalicylic acid (SSA) with absorptivity $\alpha = 560\ m^2\ mol^{-1}$ at the maximum of the absorption band.

The standard solutions for the calibration are prepared as follows:

The volumes of a Fe standard solution V_{st} given in Table 4.2-4 are pipetted into eight 100 mL volumetric flasks. The concentration of the standard solution is 4 mg L^{-1} Fe. After filling with water the absorbances are measured in duplicate. The results are also listed in Table 4.2-4.

(a) Confirm that ferrozine is suitable as a reagent for the photometric determination of Fe in the given concentration range, but not sulfosalicylic acid.
For which analytical problems can a reagent with a considerably lower absorptivity be applied?

(b) Determine the following calibration parameters for the general calibration function $A_i = f(c_i$ in μmol $L^{-1})$:

- Intercept
- Slope
- Calibration error
- Analytical error
- Relative standard deviation (s_r%)

(continued)

Table 4.2-4 The volumes V_{st} for the preparation of the calibration standard solutions and the measured values of the absorbance *A*

Calibration standard	V_{st} in mL	Absorbance *A*	
		A_1	A_2
1	5	0.1056	0.1076
2	15	0.2951	0.2923
3	25	0.5103	0.5109
4	35	0.6933	0.6987
5	45	0.9075	0.9082
6	55	1.1002	1.0009
7	65	1.2899	1.2904
8	75	1.5089	1.5095

What are the calibration and the analytical functions?
Show the regression function.

(c) 100 g plant ash was repeatedly extracted. The total volume of the extract is 500 mL. The Fe content of the plant ash must be determined in ppm (w/w). The measurement solutions were prepared as follows: 75 mL of extract was pipetted into two 100 mL volumetric flask. After addition of the reagents, each absorbance A is measured twice, giving the following results:

Sample 1	$A_{11} = 0.7682$	$A_{12} = 0.7689$
Sample 2	$A_{21} = 0.7473$	$A_{22} = 0.7478$

Calculate the result.

(d) Propose some methods of minimizing the confidence interval.

Solution to Challenge 4.2-2

(a) The concentrations of the lowest and highest limit values in the units mol L^{-1} are $c_{\text{lowest}} = 5.376 \cdot 10^{-6}$ mol L^{-1} and $c_{\text{highest}} = 5.3763 \cdot 10^{-5}$ mol L^{-1}, calculated with $M_{\text{Fe}} = 55.8$ g mol^{-1}.

According to the Lambert–Beer law $A = \alpha \cdot c \cdot l$ the values of the absorbance obtained with the lowest and highest concentrations are $A_{\text{lowest}} = 0.030$ and $A_{\text{highest}} = 0.301$ for the iron complex with SSA ($\alpha = 5{,}600$ L mol^{-1} cm^{-1}) and $A_{\text{lowest}} = 0.150$ and $A_{\text{highest}} = 1.500$ for the iron complex with ferrozine ($\alpha = 27{,}900$ L mol^{-1} cm^{-1}), respectively, using the standard cuvette with $l = 1$ cm.

Thus, the absorbance is considerably smaller than 0.1 in the range of the lowest limit with the SSA complex, but for this small absorbance the relative error is very high (see Fig. 2.2.5-1). However, the absorbance values obtained by the ferrozine complex lie in the optimal range.

With about a tenfold higher concentration the samples have to be diluted, because the absorbances cannot be directly measured any more. In order to avoid the dilution stage, a reagent can be used which yields a complex with a lower absorptivity, i.e. lower sensitivity.

(b) The concentrations of the calibration standard solutions c_{cal} for 100 mL volumetric flasks calculated by (4.2-26) with the stock concentration $c_{\text{st}} = 4$ mg L^{-1} are given in Table 4.2-5.

$$c_{\text{cal}} = c_{\text{st}} \frac{V_{\text{st}}}{V_{\text{flask}}} = 71.68 \ \mu\text{mol L}^{-1} \frac{V_{\text{st}} \text{ mL}}{100 \text{ mL}} \tag{4.2-26}$$

(continued)

Table 4.2-5 Concentrations of the calibration standard solutions

Calibration standard	V_{st} in mL	c in μmol L^{-1}
1	5	3.58
2	15	10.75
3	25	17.92
4	35	25.09
5	45	32.26
6	55	39.43
7	65	46.59
8	75	53.76

Table 4.2-6 Intermediate quantities and results of the calibration parameters

c in μmol L^{-1}	$\bar{A}$	$(x_i - \bar{x})^2$	$(x_i - \bar{x})(y_i - \bar{y})$	$10^5(y_i - \hat{y}_i)^2$
3.58	0.1066	629.49	17.286	1.62
10.75	0.2937	321.17	8.994	4.73
17.92	0.5106	115.62	3.064	14.45
25.09	0.6960	12.85	0.357	0.03
32.26	0.9079	12.85	0.402	17.60
39.43	1.0506	115.62	2.742	176.69
46.59	1.2902	321.17	8.863	0.02
53.76	1.5092	629.49	17.904	42.49
Sum		2158.25 SS_{xx}	59.613 SS_{xy}	257.63
$\bar{x}$	28.674	$\bar{y}$	0.796	

The volumes of the stock solution V_{st} are given in Table 4.2-4.

The calibration parameters are best calculated by the corresponding Excel functions given above, but for the calculation of a_0, a_1, and $s_{y.x}$ according to (4.2-4), (4.2-5), and (4.2-7), respectively, the intermediate quantities are presented in Table 4.2-6.

As already mentioned above, the two measured values given for each standard are only *double measurements* of the absorbance from which the means is formed and *not double determinations* with double degrees of freedom. Therefore, the mean values $\bar{A}$ obtained by the double measurements of the absorbances must be used for the calculation of the calibration parameters.

The regression parameters obtained by Excel functions or calculations according to the formulae are summarized in Table 4.2-7 and the regression function is shown in Fig. 4.2-4.

The *calibration function* is

$$\hat{y} = 0.00357 + 0.02762 \text{ L } \mu\text{mol}^{-1} \cdot x \qquad (4.2\text{-}27)$$

and the *analytical function* for the predicted values $\hat{x}$ is

(continued)

Table 4.2-7 Calibration parameters of the photometric determination of iron using the reagent ferrozine

Parameter	Equation	Value
Intercept a_0	(4.2-4)	0.00357
Slope a_1 in L μmol^{-1}	(4.2-5)	0.02762
Calibration error $s_{y.x}$	(4.2-7)	0.02072
Analytical error $s_{x.0}$ in μmol L^{-1}	(4.2-9)	0.75020
$s_r\%$	(4.2-10)	2.62
Number of calibration standards n_c		8
Degrees of freedom df		6
$t(P = 95\%, \mathrm{df} = 6)$		2.447

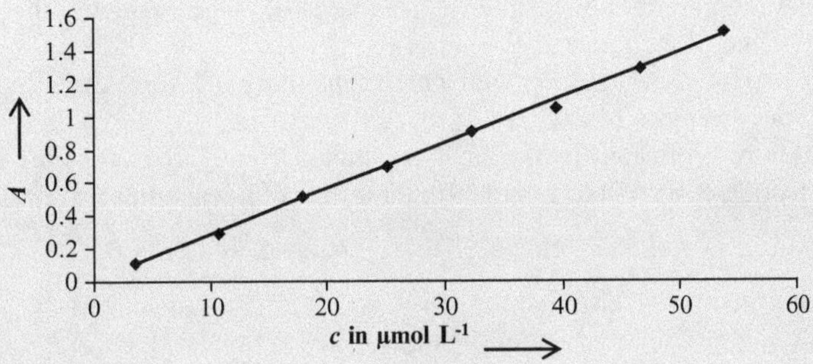

Fig. 4.2-4 Regression function of the spectrophotometric determination of iron using the reagent ferrozine

$$\hat{x} \text{ in } \mu\text{mol L}^{-1} = \frac{\hat{y} - 0.00357}{0.02762 \text{ L } \mu\text{mol}^{-1}}. \tag{4.2-28}$$

(c) Because of two replicates, the number of analyses $n_a = 2$ and the measured grand mean is $\hat{\bar{y}} = 0.75805$.

With the other parameters required for (4.2-15)–(4.2-17), the predicted concentration of the measured samples is $\hat{x} \pm \mathrm{CI}(\hat{x}) = 27.32 \pm 1.45$ μmol L^{-1}Fe. Note that in contrast to Challenge 4.2-1, the two-sided t-factor must be used because in this case $\pm\,\mathrm{CI}(\hat{x})$ must be calculated.

However, we must still consider the dilution of the sample of the extract. Because 75 mL of extract solution was diluted to 100 mL of measuring solution, the dilution factor is 1.333. Therefore, the concentration of iron in the extract is $c = 36.4 \pm 1.9$ μmol L^{-1} Fe, which equates to 1016 ± 54 μg Fe in the total extract of $V_{\text{extract}} = 500$ mL obtained from 100 g plant ash. Thus, the content of Fe in the plant ash is 10.2 ± 0.5 ppm (w/w).

(d) For the analytical problems discussed in Sect. 2.2-4, besides the predicted value $\hat{x}$ its confidence interval is also important, particularly whenever

(continued)

decisions must be made. If the confidence interval is too big then specified limit values can be easily crossed with the consequence that quality requirements are not fulfilled. Therefore, the question arises as to how one can minimize the confidence interval.

According to (4.2-16) and (4.2-17) the confidence interval is mainly determined by parameters of the *calibration* from which $s_{x.0}$ and $t(P, \mathrm{df})$ are directly proportional to $\mathrm{CI}(\hat{x})$, whereas from the *analysis* stage only the square root of the number of replicates n_a is considered in (4.2-16). This fact is important for the strategy of an analytical method. In order to minimize the confidence interval one should choose

(1) a high number of calibration standards n_c, which also diminishes the value of the t-factor.

(2) a method, if possible, with a high sensitivity a_1, which diminishes $s_{x.0}$ according to (4.2-9).

Many replicates in the analysis increase the time and cost of the analysis but have hardly any effect on the confidence interval.

Challenge 4.2-3

The determination of Cd by flame AAS (air/C_2H_2, $\lambda = 228.8$ nm) was carried out under various conditions with results presented in Tables 4.2-8 and 4.2-9. Evaluate the results with regard to an effective procedure for the determination of Cd in routine analysis giving a small confidence interval. In calibration procedure I (Table 4.2-8) the absorbance was determined without replicates whereas in calibration II (Table 4.2-9) each standard was determined by two replicates. For two samples, the predicted value $\hat{x}$ was determined with two and four replicates. The predicted value $\hat{x}$ with its confidence interval is

(continued)

Table 4.2-8 Concentrations c and measured values of the absorbance A for calibration I with $n_c = 8$ as well as the analysis results obtained with $n_a = 2$ and $n_a = 4$, respectively

	Calibration I		Analysis	
Standard	c in mg L^{-1}	A	Sample 1	Sample 2
1	2	0.2156	A	A
2	3	0.3244	0.5851	0.5863
3	4	0.4463	0.5872	0.5842
4	5	0.5409		0.5887
5	6	0.6474		0.5854
6	7	0.7538	Replicates	
7	8	0.8936	2	4
8	9	0.9706		

Table 4.2-9 Concentrations c and measured values of the absorbance A for calibration II with $n_c = 16$ as well as the analysis results obtained with $n_a = 2$ and $n_a = 4$, respectively

Calibration II				Analysis	
Standard	c in mg L^{-1}	A	A	Sample 1	Sample 2
1	2	0.2154	0.2168	A	A
2	3	0.3245	0.3243	0.5851	0.5863
3	4	0.4461	0.4465	0.5872	0.5842
4	5	0.5409	0.5409		0.5887
5	6	0.6475	0.6474		0.5854
6	7	0.7535	0.7541	Replicates	
7	8	0.8937	0.8935	2	4
8	9	0.9703	0.9709		

to be calculated for both samples using the parameters of both calibration functions.

Evaluate the result.

Solution to Challenge 4.2-3

After the detailed calculation of the calibration parameters was presented in two Challenges, we will now use the respective Excel functions for further calculations, but the important intermediate quantities will be given in order to understand and reproduce the calculations.

Table 4.2-10 gives the parameters of the calibrations obtained by the two procedures.

Note that because of the double degrees of freedom, the values for $s_{x.0}$ and $t(P, \mathrm{df})$ are smaller in calibration II.

The predicted values (i.e. the analytical results) $\hat{x}$ and their confidence intervals CI($\hat{x}$) calculated by (4.2-15)–(4.2-17) are given in Table 4.2-11.

As the results in the table show, the number of replicates n_a and the number of calibration standards do not have an influence on the predicted value $\hat{x}$ but the confidence interval is largely determined by n_c and n_a. The

(continued)

Table 4.2-10 Parameters for calibration procedures I and II

Parameters	Calibration I	Calibration II
a_0	−0.000693	−0.000392
a_1 in L mg^{-1}	0.10905	0.10902
$s_{y.x}$	0.011807	0.010923
$s_{x.0}$ in mg L^{-1}	0.10828	0.10021
$\bar{x}$ in mg L^{-1}	5.5	5.5
$s_r\%$	1.97	1.82
n_c	8	16
df	6	14
$t(P = 95\%, \mathrm{df})$	2.447	2.145

Table 4.2-11 Analytical results $\hat{x} \pm \Delta\hat{x}$ in mg L^{-1} obtained under different conditions

Parameter	Calibration I		Calibration II	
	Sample 1	Sample 2	Sample 1	Sample 2
df	6		14	
n_a	2	4	2	4
SS_{xx} in mg^2 L^{-2}	42		84	
$\hat{\bar{y}}$	0.58615	0.58615	0.58615	0.58615
$\hat{x}$ in mg L^{-1}	5.38	5.38	5.38	5.38
CI($\hat{x}$) in mg L^{-1}	0.21	0.16	0.16	0.12

same confidence interval is reached with half of the numbers n_a calculated by the parameters of calibration II with the higher number of the degrees of freedom in the calibration. For routine analysis where many analyses must be carried out, calibration parameters should be determined with a high number of calibration standards n_c and not with a high number of replicates n_a.

Challenge 4.2-4

In general one would like to illustrate the calibration function with confidence intervals graphically, which is possible using common commercial software. But if such software is not available, the graph must be constructed with an Excel spreadsheet.

Table 4.2-12 gives the calibration data set for the determination of quinine by fluorimetry. The relative intensity of the fluorescence was determined by two replicates.

Provide the graph of the calibration function with confidence intervals.

Solution to Challenge 4.2-4

The calibration parameters calculated by Excel functions are given in Table 4.2-13 together with the lower and upper confidence intervals required for the construction of the graph.

(continued)

Table 4.2-12 Calibration of quinine by fluorimetric analysis

Standard i	$c(x_i)$ in µg L^{-1}	Measured I_{rel} (y_{ij}) in counts	
		y_{1j}	y_{2j}
1	0.01	97	99
2	0.02	170	162
3	0.03	238	250
4	0.04	320	321
5	0.05	411	416
6	0.06	492	495

Table 4.2-13 Calculation of the confidence interval for the calibration function of the fluorimetric determination of quinine according to the data set given in Table 4.2-12

Standard	c in μg L^{-1}	$\bar{I}_{rel}$ in counts	Calibration parameters	
	x	$\bar{y}$	SS_{xx} in $μg^2\ L^{-2}$	0.00175
1	0.01	98.0	a_0 in counts	9.600
2	0.02	166.0	a_1 in counts L $μg^{-1}$	7,990
3	0.03	244.0	$s_{y.x}$ in counts	7.537
4	0.04	320.5	$t(P = 95\%, \text{df} = 4)$	2.776
5	0.05	413.5	$\bar{y}$ in counts	289.3
6	0.06	493.5	n_a	2

Examples for the calculation of the confidence interval CI($\hat{x}$)

$\hat{x}$	$\hat{y}$	CI($\hat{x}$)	$\text{CI}_{\text{lower}}(\hat{x})$ $\hat{y} - \text{CI}(\hat{x})$	$\text{CI}_{\text{upper}}(\hat{x})$ $\hat{y} + \text{CI}(\hat{x})$
0.000	9.60	24.46	−14.86	34.06
0.001	17.59	24.11	−6.52	41.70
0.002	25.58	23.76	1.82	49.34
0.003	33.57	23.41	10.16	56.98
0.004	41.56	23.07	18.49	64.63
0.005	49.55	22.74	26.81	72.29
...	...	...	...	...
0.066	536.94	23.07	513.87	560.01
0.067	544.93	23.41	521.52	568.34
0.068	552.92	23.76	529.16	576.68
0.069	560.91	24.11	536.80	585.02
0.070	568.90	24.46	544.44	593.36

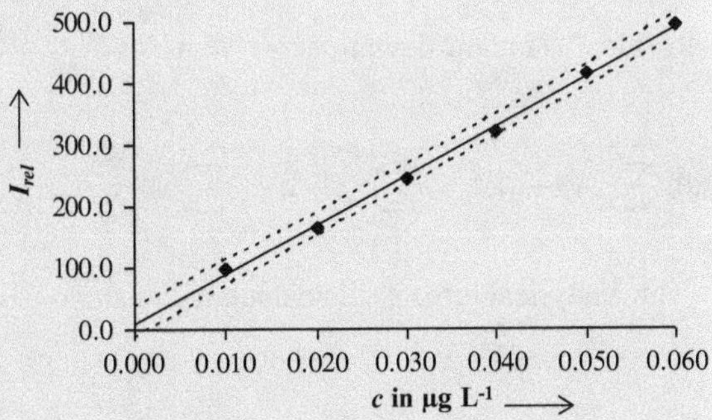

Fig. 4.2-5 Calibration function $\hat{y} = 9.60$ counts + 7,990 counts L $\mu\text{mol}^{-1} \cdot x$ with confidence intervals obtained by Excel functions

The confidence interval is calculated by

$$\text{CI}(\hat{y}) = \hat{y} \pm s_{y.x} \cdot t(P = 95\%, \text{df} = 4) \cdot \sqrt{\frac{1}{n_a} + \frac{1}{n_c} + \frac{(\bar{y} - \hat{\bar{y}})^2}{a_1^2 \cdot \text{SS}_{xx}}}. \tag{4.2-29}$$

The calibration function $\hat{y} = 9.60$ counts + 7,990 counts L $\mu\text{g}^{-1} x$ with confidence intervals is presented in Fig. 4.2-5.

4.3 Simplification of the Linear Calibration Function

The calibration function $\hat{y} = a_0 + a_1 \cdot x$ can be simplified if the intercept a_0 does not differ significantly from zero. This is the case if zero is included within the range of the confidence interval of the intercept $\mathrm{CI}(a_0)$.

The parameters of the simplified calibration function

$$\hat{y} = a_1' \cdot x \tag{4.3-1}$$

are calculated by (4.3-2)–(4.3-7) [3]:

$$\text{– Slope: } a_1' = \frac{\sum x_i \cdot y_i}{\sum x_i^2} \tag{4.3-2}$$

$$\text{– Variance of the slope: } s_{a_1}'^2 = \frac{s_{y.x}'^2}{\sum x_i^2} \tag{4.3-3}$$

$$\text{– Confidence interval of the slope: } \mathrm{CI}(a_1') = s_{a_1}' \cdot t(P, \mathrm{df}) \tag{4.3-4}$$

$$\text{– Degrees of freedom: } \mathrm{df} = n_c - 1 \tag{4.3-5}$$

$$\text{– Residual standard deviation: } s_{y.x}' = \sqrt{\frac{\sum (\hat{y}_i - y_i)^2}{\mathrm{df}}} \tag{4.3-6}$$

$$\text{with } \sum (\hat{y}_i - y_i)^2 = \sum y_i^2 - a'_1 \cdot \sum x_i \cdot y_i. \tag{4.3-7}$$

The analytical error $s_{x.0}'$ is calculated analogously to (4.2-8) and (4.2-9):

$$s_{x.0}' = \frac{s'_{y.x}}{a'_1} \tag{4.3-8}$$

$$s_r\% = \frac{s'_{x.o}}{\bar{x}} \times 100. \tag{4.3-9}$$

The predicted value $\hat{x}$ with its confidence interval is calculated by (4.3-10) and (4.3-11)

$$\hat{x} = \frac{\hat{\bar{y}}}{a_1'} \tag{4.3-10}$$

$$\mathrm{CI}(\hat{x}) = \hat{x} \pm s'_{x.0} \cdot t(P, \mathrm{df} = n_c - 1) \cdot \sqrt{\frac{1}{n_a} + \frac{1}{n_c} + \frac{(\bar{y} - \hat{\bar{y}})^2}{a_1'^2 \cdot \mathrm{SS}_{xx}}} \qquad (4.3\text{-}11)$$

in which n_c and n_a are the number of calibration standards and replicates in the analysis, respectively, SS_{xx} is the sum of squares of the x-values (calibration standards), $\bar{y}$ is the mean of the measured information values (response) and $\hat{\bar{y}}$ is the means of the repeated information values obtained by the sample; all other symbols are explained above.

Challenge 4.3-1
In a laboratory the determination of Zn in waste water for the range 0.5–5 mg L^{-1} must be carried out by flame AAS in routine analysis.

(a) The following calibration parameters are to be calculated with the data set given in Table 4.3-1: residual error, analytical error, $s_r\%$, and slope with its confidence interval. What is the calibration function?
(b) Draw the graph of the calibration function with the confidence intervals.
(c) Calculate the predicted value $\hat{x} \pm \mathrm{CI}(\hat{x})$ of a sample from the measured absorbance of two replicates $\hat{y}_1 = 0.9561$ and $\hat{y}_2 = 0.9610$.

Table 4.3-1 Calibration data set for the determination of Zn by flame AAS

Standard	c in mg L^{-1}	A
1	0.5	0.1727
2	1	0.3277
3	1.5	0.4650
4	2	0.6620
5	2.5	0.7617
6	3	0.9034
7	3.5	1.1082
8	4	1.3196
9	4.5	1.4148
10	5	1.6240

Solution to Challenge 4.3-1
(a) The parameters for the linear calibration function $\hat{y} = a_0 + a_1 \cdot x$ calculated by Excel are summarized in Table 4.3-2.

As Table 4.3-2 shows, the value zero is within the range of the confidence interval of the intercept $\mathrm{CI}(a_0)$, i.e. the intercept is not significantly different from 0. Therefore, the calibration function can be simplified into the form $\hat{y} = a'_1 \cdot x$.

(continued)

Table 4.3-2 Parameters of the calibration function and the lower and upper limits of the confidence interval $CI_{lower}(a_0)$ and $CI_{upper}(a_0)$, respectively

a_0	−0.00494
a_1 in L mg^{-1}	0.32031
df	8
SS_{xx} in mg^2 L^{-2}	20.63
s_{a_0}	0.02209
$t(P = 95\%, \text{df})$	2.306
$CI_{lower}(a_0)$	*−0.05587*
$CI_{upper}(a_0)$	*0.04598*

Table 4.3-3 Intermediate quantities for the calculation of the parameters of the simplified calibration function $\hat{y} = a_1' \cdot x$

Standard	$x_i y_i$	x_i^2	$\hat{y}_i$	$(y_i - \hat{y}_i)^2$
1	0.0864	0.25	0.1594	0.000176
2	0.3277	1	0.3189	0.000077
3	0.6974	2.25	0.4783	0.000179
4	1.3240	4	0.6378	0.000587
5	1.9041	6.25	0.7972	0.001266
6	2.7101	9	0.9567	0.002844
7	3.8786	12.25	1.1161	0.000064
8	5.2784	16	1.2756	0.001939
9	6.3667	20.25	1.4350	0.000409
10	8.1202	25	1.5945	0.000875
Sum	30.6936	96.25		0.008414
a_1' in L mg^{-1} (4.3-2)	0.3189	df (4.3-5)	9	
$s_{a_1'}$ in L mg^{-1} (4.3-3)	0.003117	$s_{y.x}'$ (4.3-6)	0.03058	
$s_{x.0}'$ in mg L^{-1} (4.3-8)	0.09588			

Intermediate quantities for the calculation of the parameters of the simplified calibration function $\hat{y} = a_1' \cdot x$ are given in Table 4.3-3. The calibration function is $\hat{y} = 0.3189 \text{ L mg}^{-1} \cdot x$.

(b) The graph for the calibration function with its confidence intervals is generated as described in Challenge 4.2-4. The data set of the confidence interval used for the Excel graph is calculated by

$$CI(\hat{x}) = \hat{y} \pm s_{y.x}' \cdot t(P = 95\%, \text{df} = 9) \cdot \sqrt{1 + \frac{1}{n_c} + \frac{(\bar{y} - \hat{\bar{y}})^2}{a_1'^2 \cdot SS_{xx}}} \quad (4.3\text{-}12)$$

with the values given in Table 4.3-3.

An extract of the data set is listed in Table 4.3-4 and the graph is presented in Fig. 4.3-1.

(c) The predicted value $\hat{x}$ and its confidence interval are calculated by (4.3-10) and (4.3-11), respectively.

(continued)

Table 4.3-4 Extract of intermediate quantities for the calculation of the lower and upper limit of the confidence interval

x_i c	$\hat{y}_i$	$s'(\hat{y}_i)$	$\mathrm{CI}_{\mathrm{lower}}(\hat{y}_i)$	$\mathrm{CI}_{\mathrm{upper}}(\hat{y}_i)$
0.5	0.15945	0.25153	−0.09208	0.41097
0.6	0.19134	0.24953	−0.05819	0.44086
0.7	0.22323	0.24760	−0.02438	0.47083
0.8	0.25512	0.24576	0.00936	0.50087
0.9	0.28700	0.24399	0.04302	0.53099
...	...	...	...	...
4.6	1.46691	0.24410	1.22281	1.71102
4.7	1.49880	0.24588	1.25293	1.74468
4.8	1.53069	0.24773	1.28296	1.77842
4.9	1.56258	0.24966	1.31292	1.81224
5	1.59447	0.25166	1.34281	1.84613

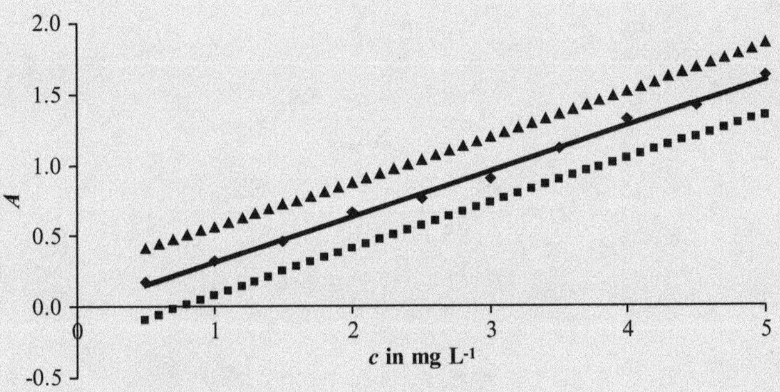

Fig. 4.3-1 Calibration function $\hat{y} = 0.3189$ L μmol$^{-1}\cdot x$ with the confidence intervals

With $\hat{\bar{y}} = 0.95855$, $n_a = 2$, and further parameters given in Table 4.3-3 the following result is obtained:

$$\hat{x} \pm \mathrm{CI}(\hat{x}) = 3.01 \pm 0.169 \text{ mg L}^{-1} \text{ Zn}. \tag{4.3-13}$$

4.4 Quadratic Regression Analysis

The pairs of values x, y given in Fig. 4.4-1 cannot be fitted by a linear regression, but a quadratic regression may be the better model

$$\hat{y} = a_0 + a_1 \cdot x + a_2 \cdot x^2. \tag{4.4-1}$$

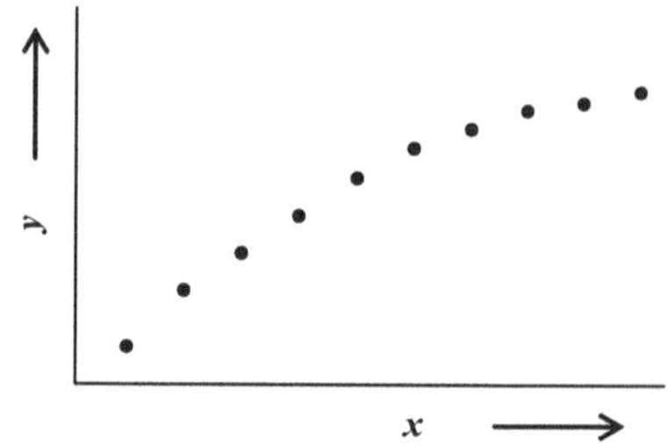

Fig. 4.4-1 Non-linear relationship between the x and y-values of a calibration experiment

For a quadratic regression model the calibration parameters are calculated by the following equations [4]:

- *Coefficients*, a_0, a_1, and a_2

$$a_0 = \frac{1}{n_c} \cdot \left(\sum y_i - a_1 \cdot \sum x_i - a_2 \cdot \sum x_i^2 \right), \tag{4.4-2}$$

$$a_1 = \frac{SS_{xy} - a_2 \cdot SS_{x^3}}{SS_{xx}}, \tag{4.4-3}$$

$$a_2 = \frac{SS_{xy} \cdot SS_{x^3} - SS_{x^2y} \cdot SS_{xx}}{(SS_{x^3})^2 - SS_{xx} \cdot SS_{x^4}}. \tag{4.4-4}$$

- *Sum of squares*, SS

$$SS_{x^3} = \sum x_i^3 - \frac{(\sum x_i) \cdot (\sum x_i^2)}{n_c}, \tag{4.4-5}$$

$$SS_{x^4} = \sum x_i^4 - \frac{(\sum x_i^2)^2}{n_c}, \tag{4.4-6}$$

$$SS_{x^2y} = \sum (x_i^2 \cdot y_i) - \frac{(\sum y_i) \cdot \sum x_i^2}{n_c}. \tag{4.4-7}$$

SS_{xx} and SS_{xy} are calculated by (4.1-3) and (4.1-5), respectively.

- *Residual standard deviation*, $s_{y.x}$

$$s_{y.x} = \sqrt{\frac{\sum (y_i - \hat{y})^2}{\text{df}}} \tag{4.4-8a}$$

with

$$\hat{y} = a_0 + a_1 \cdot x + a_2 \cdot x^2. \tag{4.4-8b}$$

– *Degrees of freedom* df for n_c calibration standards

$$\mathrm{df} = n_c - 3. \tag{4.4-9}$$

– *Predicted value* $\hat{x}$ (analytical result) with *negative* curvature

$$\hat{x} = -\frac{a_1}{2 \cdot a_2} - \sqrt{\left(\frac{a_1}{2 \cdot a_2}\right)^2 - \frac{a_0 - \hat{\bar{y}}}{a_2}} \tag{4.4-10}$$

with *positive* curvature

$$\hat{x} = -\frac{a_1}{2 \cdot a_2} + \sqrt{\left(\frac{a_1}{2 \cdot a_2}\right)^2 - \frac{a_0 - \hat{\bar{y}}}{a_2}}. \tag{4.4-11}$$

– *Confidence interval* $\mathrm{CI}(\hat{x})$ with n_a replicates

$$s(\hat{x}) = \frac{s_{y.x}}{(a_1 + 2 \cdot a_2 \cdot \hat{x})} \cdot \sqrt{\frac{1}{n_c} + \frac{1}{n_a} + \frac{A}{B}} \tag{4.4-12}$$

with

$$A = \left\{ (\hat{x} - \bar{x})^2 \cdot \mathrm{SS}_{x^4} + \left(\hat{x}^2 - \frac{\sum x_i^2}{n_c}\right)^2 \cdot \mathrm{SS}_{xx} - 2 \cdot (\hat{x} - \bar{x}) \cdot \left(\hat{x}^2 - \frac{\sum x_i^2}{n_c}\right) \cdot \mathrm{SS}_{x^3} \right\}$$

$$B = \mathrm{SS}_{x^4} \cdot \mathrm{SS}_{xx} - (\mathrm{SS}_{x^3})^2$$

$$\mathrm{CI}(\hat{x}) = \hat{x} \pm s(\hat{x}) \cdot t(P, \mathrm{df}). \tag{4.4-13}$$

– *Sensitivity*, Sens

$$\mathrm{Sens} = a_1 + 2 \cdot a_2 \cdot x. \tag{4.4-14}$$

Because, unlike linear regression, sensitivity is a function of x, it is usually given as the means of the range $\bar{x}$:

$$\mathrm{Sens}\,(\bar{x}) = a_1 + 2 \cdot a_2 \cdot \bar{x}. \tag{4.4-15}$$

– *Process standard deviation,* $s_{x.0}$

$$s_{x.0} = \frac{s_{y.x}}{\mathrm{Sens}}. \tag{4.4-16}$$

Table 4.4-1 Parameters for the quadratic regression function in Excel

Matrix position	Parameter	Matrix position	Parameter	Matrix position	Parameter
1, 1	a_2	1, 2	a_1	1, 3	a_0
2, 1	s_{a_2}	2, 2	s_{a_1}	2, 3	s_{a_0}
3, 1	r^2	3, 2	$s_{y.x}$		
4, 1	F-value	4, 2	df		
5, 1	SS_{yy}	5, 2	$\sum (y_i - \hat{y}_i)^2$		

– $s_r\%$ (*the error of the analytical procedure*)

$$s_r\% = \frac{s_{x.0}}{\bar{x}} \cdot 100. \qquad (4.4\text{-}17)$$

Using the Excel function

$$= \text{LINEST}(y\_\text{values}, [x, x^2\_\text{values}], 1, 1)$$

the regression parameters presented in Table 4.4-1 are given for a quadratic function. Note that the data input of a matrix must be made by using CTRL + ⇑ + Enter.

Challenge 4.4-1

The determination of organophosphorus pesticides in seepage water must be introduced in an analytical laboratory. After extraction with acetonitrile and clean-up by solid phase extraction (SPE), the analytes should be determined by gas chromatography using the highly specific flame photometric detector (FPD). However, this detector may have a non-linear response. In order to choose the correct regression model for the gas chromatograph software, the calibration function was tested with the pesticide malathion in methanolic solution. The results are given in Table 4.4-2.

(continued)

Table 4.4-2 Calibration data of the determination of malathion by GC-FPD

Standard	c in mg L^{-1}	Response y in mV
1	0.05	27
2	0.10	49
3	0.15	68
4	0.20	82
5	0.25	92
6	0.30	105
7	0.35	111
8	0.40	120
9	0.45	128
10	0.50	132

(a) The decision on the quadratic regression is to be made from the graphical representation of the function $\hat{y} = f(x)$. Note that if the quadratic regression is really the better model of the relationship between the x- and y-values, this will be tested with methods given in Sect. 5.3.
(b) Calculate the coefficients as well as further parameters of the quadratic calibration function.
(c) Calculate the analytical result as $\hat{x} \pm \mathrm{CI}(\hat{x})$ in mg L^{-1} for a malathion sample which was analyzed under the same conditions with two replicates $y_1 = 94.6$ mV and $y_2 = 94.1$ mV.

Solution to Challenge 4.4-1

(a) As the calibration function $y = f(x)$ in Fig. 4.4-2 shows, the linear regression model should be rejected and the x, y-values are better fitted by the quadratic regression model.
(b) The constants of the quadratic regression function are best obtained by using the Excel function = LINEST(y_values, [x, x^2_values], 1, 1) with the results given in Table 4.4-1. In order to recalculate the parameters of the quadratic calibration function according to (4.4-2)–(4.4-7), the intermediate quantities are given in Table 4.4-3.

Thus, the quadratic calibration function is

$$\hat{y} = 8.883\,\mathrm{mV} + 431.0\,\mathrm{mV\,L\,mg^{-1}} \cdot x - 374.2\,\mathrm{mVL^2\,mg^{-2}} \cdot x^2 \qquad (4.4\text{-}18)$$

(c) The calibration function shows a negative curvature, therefore (4.4-10) must be used for the calculation of the predicted value, which is $\hat{x} = 0.2545$ mg L^{-1} calculated with the mean measured value $\bar{y} = 94.35$ mV and the values a_0, a_1, and a_2 given in Table 4.4-3. The standard deviation of $\hat{x}$ is $s(\hat{x}) = 0.0077$ mg L^{-1} calculated by (4.4-12) with the values

(continued)

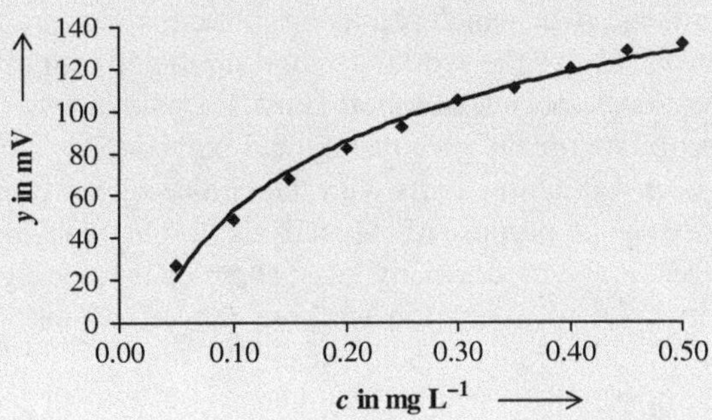

Fig. 4.4-2 Calibration function of the determination of malathion by GC-FPD

Table 4.4-3 Intermediate quantities and results of the calculation of the coefficients of the quadratic calibration function

$x_i \cdot y_i$	x_i^2	x_i^3	x_i^4	$x_i^2 \cdot y_i$	$(y_i - \hat{y})^2$
1.35	0.0025	0.00013	0.000006	0.068	6.2500
4.90	0.0100	0.00100	0.000100	0.490	0.5693
10.20	0.0225	0.00338	0.000506	1.530	8.2961
16.40	0.0400	0.00800	0.001600	3.280	3.5242
23.00	0.0625	0.01563	0.003906	5.750	1.5739
31.50	0.0900	0.02700	0.008100	9.450	0.2351
38.85	0.1225	0.04288	0.015006	13.598	8.4364
48.00	0.1600	0.06400	0.025600	19.200	2.0242
57.60	0.2025	0.09113	0.041006	25.920	0.8655
66.00	0.2500	0.12500	0.062500	33.000	1.3330
Sums					
297.8	0.9625	0.37813	0.158331	112.285	33.1076
$\sum x_i$	2.75	$\sum y_i$	914	n_c	10
Sums of squares					
SS_{xx}	0.20625	SS_{xy}	46.45	SS_{x^2y}	24.3125
SS_{x^3}	0.11344	SS_{x^4}	0.06569		
Results					
a_0	8.8833 in mV	a_1 in mV L mg^{-1}	431.0455	a_2 in mV L^2 mg^{-2}	−374.24
$s_{y,x}$	2.1748	s_r%	3.51	Sens($\bar{x}$) in mV L mg^{-1}	225.2

given in Table 4.4-3. Thus, the confidence interval calculated according to (4.4-13) is $CI(\hat{x}) = 0.018$ mg L^{-1} at the significance level $P = 95\%$.

The analytical result is 0.2545 ± 0.018 mg L^{-1} Zn.

4.5 Working Range and Calibration Standards

The working range is the difference between the highest and lowest values of the analyte in the sample. The working range of an analytical method is the concentration range over which results are obtained that are fit for a specific purpose. Note that outside of the working range predicted values $\hat{x}$ are not statistically certain. Therefore, special attention must be paid to the choice of the working range appropriate for the given analytical purpose.

Each validation starts with the choice of the provisional range, which is determined by the purpose of the analysis. In pharmaceutical analysis, required working ranges are given for many tests; some of them are given in Table 4.5-1.

In order to assess the working range and confirm its fitness for purpose, the concentration range should exceed the required limits by 10% or more. In general, the calibration standards should be evenly spaced across the range. To establish the suitability of the working range used for the validation of parameters such as

Table 4.5-1 Examples of the range of some tests for pharmaceuticals recommended by the International Conference on Harmonization (ICH) [5]

Test	Range
Assay of the API	80–120% of the documented concentration
Content uniformity	70–130% of the documented concentration
Release	±20% of the specified limit
Impurity	Specification up to 120% of the specified value

linearity, homogeneity of variances, limit of determination, etc., at least seven different concentration levels should be used in the process of method validation.

Another problem which should be addressed concerns linearity and sensitivity of the instrument response in relation to the working range studied. Thus, for example, non-linearity of the analytical method obtained by a statistical test can be caused by non-linearity of the analytical method (interfering compounds, an incomplete equilibrium, etc.) or by non-linearity of the instrument response or both. Checking the linearity of the instrument response across the required working range in an initial study can be done by analyzing standard solutions produced by CRS of appropriate concentrations. If the linearity of the response values obtained is confirmed (see Sect. 5.3), then any non-linearity observed in analyzing the calibration solutions has its origin in the calibration stage.

Sensitivity is mainly a problem in trace analysis. If the sensitivity is too small for an analytical problem, a change to another method with a higher sensitivity is required; for example, a change from UV detection in HPLC to fluorescence detection, fluorescent markers can be added to non-fluorescent substances, the use of an electron capture detector (ECD) instead of a flame ionization detector (FID) for GC determination of chlorinated hydrocarbon compounds, or graphite furnace atomization instead of flame atomization in the AAS.

If the working range checked by the tests given in the following chapters is appropriate, the provisional range is fixed as the actual working range of the analytical method.

Some practical advice for the preparation of calibration solutions:

- The preparation of standard solutions must be made by chemical reference materials (CRM) or chemical reference substances (CRS), i.e. substances which are characterized by a certificate. This also holds for the solvents used.
- In order to achieve the required independence of the calibration standards, each solution must be prepared separately. Because weighing is more precise than measuring volumes, weighing should be preferred over volume dosage.
- The preparing of calibration standards cannot always be made by separate weighings of the CRS because the amounts required are too small; for example, preparation of a stock solution in the sub-microgram range. However, further dilutions required for the preparation of the calibration standards have to be made starting with the stock solution and not by successive dilution steps.
- If possible, the same syringe should be used for the preparation of solution rows.
- The calibration solutions c in mol L^{-1} are prepared using (4.5-1a):

$$c \text{ in mol L}^{-1} = \frac{m_a \text{ in g}}{M_a \text{ in g mol}^{-1} \cdot V_{sol} \text{ in L}} \quad (4.5\text{-}1a)$$

$$= \frac{m_a \text{ in mg}}{M_a \text{ in g mol}^{-1} \cdot V_{sol} \text{ in mL}} \quad (4.5\text{-}1b)$$

in which m_a is the mass and M_a the molar mass of the analyte or CRS, respectively, and V_{sol} is the volume of the volumetric flask.

– The concentration c_2 obtained by the dilution of a solution 1 with concentration c_1 is given by (4.5-2):

$$c_2 = c_1 \cdot \frac{V_1}{V_2} \quad (4.5\text{-}2)$$

in which V_1 and V_2 are the volumes of the solutions 1 and 2, respectively.

Challenge 4.5-1

The *assay* of an API and the *impurity* of the byproduct X must be tested for tablets within the scope of quality control.

The following data are known in advance and will be given:

- The weight of a tablet is 200 mg and the content of the API is 10% (w/w).
- According to EUROPHARM the limit of the content of the impurity X should be 0.1% (w/w) of the API.
- According to the test regulation, ten tablets are to be dissolved per 100 mL eluent (methanol/water, v/v = 50/50).
- The CRS is certified to contain 95% (w/w) API and 4% (w/w) impurity X.

Instructions for the preparation of the calibration standard solutions are to be established for testing

(a) The *assay* of the API.
(b) The *impurity* X in tablets.
The volume of the calibration standards should be $V_{CS} = 100$ mL.

Solution to Challenge 4.5-1

(a) Calibration standards for the *assay* of the API:

According to the regulatory guidelines given in Table 4.5-1 the range for the assay test is established at 80–120% of the documented concentration.

(continued)

Table 4.5-2 Range of the calibration standards for the determination of the assay

Standard	1	2	3	4	5
mg CRS in 100 mL eluent	168.4	189.5	210.5	231.6	252.6

With a content of 20 mg API in one tablet, the concentration of the sample is 200 mg API in 100 mL eluent for ten tablets. Therefore, the required range is from 160 to 240 mg API in 100 mL eluent, which gives, taking in account the API content of the CRS, 168.4 up to 252.6 mg CRS in 100 mL eluent.

The five standard solutions given in Table 4.5-2 should be prepared.

If each calibration standard is determined by two replicates, the degrees of freedom df = 8.

Note that as discussed in Sect. 4.1, two HPLC-replicates from the same vial whose solution was prepared by weighing a certain amount of the CRS gives two independent determinations because the error is caused by the chromatographic stage and not by the preparation of the standard solution.

(b) Calibration standards for the *impurity* X in the tablets:

According to EUROPHARM the limit of the content of the impurity X should be 0.1% (w/w) of the API. Therefore, the limit of the content of X in one tablet with 20 mg API is $20 \cdot 0.001 = 0.02$ mg X. Because ten tablets are used for the preparation of each calibration standard, 0.2 mg X may be contained in each solution. According to the regulatory guidelines (Table 4.2-1), the range for the testing of impurities is set up to 120% of the specificied value. Therefore, the range is from 0.2 to 0.24 mg X in 100 mL eluent.

According to the CRS certificate, the content of the impurity X is 4% (w/w), and the following amounts of CRS are used in 100 mL eluent to prepare the lower and the upper limit concentrations, respectively:

$$x_{\mathrm{CRS}}(\mathrm{mg}) = \frac{0.2\,\mathrm{mg}}{4\% \cdot 0.01} = 5.0\,\mathrm{mg} \tag{4.5-3}$$

and

$$x_{CRS}(\mathrm{mg}) = \frac{0.24\,\mathrm{mg}}{4\% \cdot 0.01} = 6.0\,\mathrm{mg}. \tag{4.5-4}$$

The five standard solutions given in Table 4.5-3 should be prepared.

If each calibration standard is determined by two replicates the degrees of freedom df = 8 as discussed above.

Table 4.5-3 Range of the calibration standards for the determination of the impurity X

Standard	1	2	3	4	5
mg CRS in100 mL eluent	5.0	5.25	5.55	5.75	6.0

Challenge 4.5-2

In a laboratory the determination of BTXE (benzene, toluene, xylenes, and ethylbenzene) in solid samples must be carried out. The concentration range of the analytes may be between 10 and 50 ppm (w/w), and some alkanes may be also present, but they do not necessarily have to be determined. In order to choose the calibration range, the following questions must be answered:

(a) Which pretreatment of the samples is to be applied?

Let us assume that the extraction of each 1 g sample will be extracted with 10 mL methanol (CRS). From each clear extract, 4.0 mL is pipetted by a syringe which is equipped with a filter with pore size 0.45 μm into a 5 mL volumetric flask, which is then filled with methanol.

Instructions for the preparation of the calibration standards are to be compiled.

Considering the high price of high-quality solvent, only 10 mL volumetric flasks should be used.

(b) Which analytical method should be used?

GC analysis with FID using the internal standard method should be the best analytical procedure for this purpose. Let us assume that *n*-octane is not present in the samples, so this alkane can be used as an internal standard because it is also a hydrocarbon compound and its boiling point is in the area of those of the analytes.

(c) Why should the FID be used as detector?

To answer this question, the detector parameters *sensitivity* and *linearity* must be considered. Let us assume the split may be 10 : 1.

Solution to Challenge 4.5-2

(a) Calculation of the limit concentrations of the solutions obtained according to the proposed pretreatment:

Lower limit: 10 ppm (w/w) means 10 μg analyte per 1 g solid sample. Assuming quantitative extraction with the proposed 10 mL of solvent, the concentration of the extract is $c_{\text{extract}} = 1\ \mu\text{g mL}^{-1}$ analyte. Because 4 mL of this solution is diluted to 5 mL, the concentration of the lower limit is $c_{\text{low}} = 0.8\ \mu\text{g mL}^{-1}$ CRS.

(continued)

Table 4.5-4 Proposed working range for the GC determination of BTXE according to the extraction method

Standard	1	2	3	4	5	6	7
c in µg mL^{-1} CRS	0.6	1.2	1.8	2.4	3.0	3.6	4.2

Table 4.5-5 Preparation of the solutions St_0 for all BTXE analytes for the 10 mL volumetric flask

	B	T	*o*-X	*m*-X	*p*-X	EB
ρ in g mL^{-1}	0.877	0.862	0.876	0.860	0.857	0.867
$V(St_0)$ in µL	68	70	68	70	70	69

B – benzene, T– toluene; *o*-X – *o*-xylene, *m*-X – *m*-xylene, *p*-X – *p*-xylene, EB – ethylbenzene

Upper limit: The upper concentration is five times higher than c_{low}, therefore $c_{up} = 4$ µg mL^{-1} CRS.

The proposed working range given in Table 4.5-4 extends beyond the required concentration range by −25% and +5%, respectively.

In order to assess the working range and confirm its fitness for purpose, calibration standard solutions must be produced.

Because all analytes are liquids, the calibration solutions have to be made by defined volumes, but, taking into account the density, µg mL^{-1} means volumes in the nanoliter range which cannot be achieved. Therefore, the calibration solutions must be produced by appropriate dilutions of stock solutions.

Let us assume two dilution steps with each dilution 1:100 (i.e. 0.1 mL into a 10 mL volumetric flask) to give a solution from which the calibration standards are made.

According to (4.5-2) for the lower standard concentration given in Table 4.5-4 ($c_1 = 0.6\ \mu g\,mL^{-1}$) the concentration of stock solution $c(St_1)$ is given by (4.5-5):

$$c(St_1) = 0.6\frac{\mu g}{mL}\cdot\frac{10\,mL}{0.1\,mL} = 60\ \mu g\,mL^{-1}. \qquad (4.5\text{-}5)$$

The stock solution St_1 is obtained if the stock solution St_0 $c(St_0) = 6,000\ \mu g\,mL^{-1}$ is diluted $1:100$ (0.1 mL into 100 mL). Thus, only the stock solution St_0 is to be prepared from the analytes; all other solutions are made by dilutions.

Preparation of the *stock solution* St_0:

For each solution, 60 mg of the analyte must be added to the 10 mL volumetric flask. Taking into account the density of the analyte, the volumes given in Table 4.5-5 must be used for preparation of the stock solutions St_0 for the six analytes. Because of the high volatility of the

(continued)

Table 4.5-6 Preparation of the calibration solution by dilution of the stock solution St_1 in 10 mL volumetric flasks

Calibration standard CS	1	2	3	4	5	6	7
Solution St_1 in mL	0.1	0.2	0.3	0.4	0.5	0.6	0.7

analytes, each volumetric flask is half-filled with the solvent methanol and the analytes are pipetted *into* the solvent.

Preparation of *stock solution* St_1 for all BTXE analytes:

100 μL of stock solution St_0 is pipetted in a 10 mL volumetric flask and then filled with methanol.

Preparation of the solution of the *internal standard* (*n*-octane) $St_{0,IS}$:

A solution of 5 μL *n*-octane in 10 mL methanol is diluted 1:100 (100 μL in 10 mL). The concentration of the solution $St_{0,IS} = 0.005\ \mu L\ mL^{-1}$.

Preparation of the *calibration solutions* (CS):

The volumes of each analyte given in Table 4.5-6 are added to 10 mL volumetric flasks. After addition of 100 μL internal standard (solution $St_{0,IS}$), the flask is filled up.

Only 210 mL methanol (CRS) are used with the given procedure for the preparation of the seven calibration standards. If each calibration standard is measured by two replicates (two injections per vial) the degrees of freedom df = 12. Last but not least, the same 100-μL syringe can be used for the preparation of all the stock solutions St_0, and all other solutions are prepared with the same 1,000-μL syringe.

(b) As discussed in Sect. 4.2, in general, a linear calibration function is required for calculation of analytical results. In order to decide if any non-linearity is caused by the sample (interfering compounds, association equilibrium, or other effects), the linearity of the instrument response has to be given. But the FID with a linearity of 10^7 always shows a linear response.

The sensitivity of the FID is documented by its detection limit which is about 10 pg.

Let us estimate the amount of analyte to be transferred onto the GC column by the condition given above for the smallest concentration of $c = 0.6\ \mu g\ mL^{-1} (= 0.6\ ng\ \mu L^{-1})$. The usual injection volume for the split injection technique is 1 μL. Thus, with the split ratio 10:1 about 60 pg is transferred onto the GC column, which can be easily detected.

Challenge 4.5-3

Seven calibration standards should be prepared for the determination by HS-GC of benzene in waste water in the range 5–15 mg L^{-1}.

(continued)

Give instructions for the preparation of the calibration standard solutions. The degrees of freedom should be df = 5.

Solution to Challenge 4.5-3
In general, calibration solutions are prepared according to (4.5-1a) and (4.5-2): from a stock solution (c_{st} = constant) various volumes (V = variable) are pipetted into volumetric flasks which are then filled with the solvent. However this procedure cannot be applied to the preparation of water-insoluble organic analytes, such as benzene.

The preparation of calibration standard solutions for water-insoluble compounds is carried out as follows:

- A stock solution of the analyte in a modifier is produced for every calibration standard. The modifier must be soluble in water and it enables the analyte to be dissolved in water. Acetone is frequently used.
- For every calibration standard, volumetric flasks of a large volume (25 mL or more) are filled with water. In this case we will take 25 mL volumetric flasks.
- The same but low volume (V = constant) of the stock solutions of various concentrations (c_{st} = variable) is stirred into the water with very fast stirring using a magnetic stirrer. The volumetric flasks are then closed and the stirring is continued for 15 min or so. In this case we will use 25 μL of stock solution.

Seven calibration standard solutions are needed for df = 5. A stock solution St_0 is prepared using 13.0 mL benzene (CRS) (m = 11.4 g) in 100 mL acetone. The stock solution St_0 with $c_{st,0} = 114\ g\,L^{-1}$ benzene is used for the preparation of the seven stock solutions St_{add}. The volumes V (St_0) given in Table 4.5-7 are pipetted into 5 mL volumetric flasks which are then filled with methanol. The concentrations of the calibration solutions c_{cs} prepared by stirring in each of 25 μL stock solutions St_{add} in 25 mL water according to the steps given above are presented in Table 4.5-7.

The required lower and upper limits are extended by −8.5% and +22%.

Table 4.5-7 Preparation of the calibration standard solutions for the determination of benzene in waste water by HS-GC analysis

Standard	$V(St_0)$ in mL for 5 mL flask	$c(St_{add})$ in $g\,L^{-1}$	c_{CS} in $mg\,L^{-1}$
1	0.2	4.56	4.56
2	0.3	6.84	6.84
3	0.4	9.12	9.12
4	0.5	11.40	11.40
5	0.6	13.68	13.68
6	0.7	15.96	15.96
7	0.8	18.24	18.24

References

1. DIN ISO 8466–1 (1990) Water quality – calibration and evaluation of analytical methods and estimation of performance characteristics, part 1: statistical evaluation of the linear calibration function. Beuth, Berlin
2. Stookey LL (1970) Ferrocene – a new spectrophotometric reagent for iron. Anal Chem 42:780
3. Doerffel K (1990) Statistik in der analytischen Chemie 5. Aufl. Deutscher Verlag für Grundstoffindustrie, Leipzig
4. DIN ISO 8466-1 (2001) Water quality – calibration and evaluation of analytical methods and estimation of performance characteristics, part 2: calibration strategy for non-linear second-order calibration functions. Beuth, Berlin
5. ICH Harmonised Tripartite Guideline (2005) Validation of analytical procedures: text and methodology Q2(R1). http://www.ICH.org

Chapter 5
Validation of Method Performance

5.1 General Remarks

In the area of AQA, only validated methods should be used to solve the analytical problems. Analytical methods need to be validated or revalidated:

1. Before their introduction into routine use
2. Whenever the conditions change for which the method has been validated (for example, samples with a changed matrix or in a different concentration range)
3. Whenever the method is changed (for example, changing the determination of nitrite in waste water by photometry into the ion chromatographic method, or substitution of the determination of organic compounds according to the headspace GC extraction method)

The procedures of method validation are mandated by regulatory agencies. Guidelines with the required validation parameters for pharmaceutical and the environmental analysis are:

1. The US FDA CGMP requirements in section 211.165 (e)
2. ISO/IEC 17025, Sect. 4.4
3. The validation procedure of ICH
4. Method development and validation for the Resource Conservation and Recovery Act (RCRA) by the US EPA (Environmental Protection Agency)

Method validation given in ISO/IEC 17025 "is the confirmation by examination and the provision of objective evidence that the particular requirements for a specific intended use are fulfilled" [1]. According to the definition, the method validation is characterized by three requirements:

- The need for *experimental* investigations.
- The objective *evidence* which is provided by performance parameters.
- The *specific intended use*, which means that a method is valid only for the purpose for it has been validated. If, for example, the determination of benzene in waste water has been validated by HS-GC for the working range 5–15 ppm (w/w), then the same method cannot be applied either for concentrations outside this working range or for other aromatics in the same range.

M. Reichenbächer and J.W. Einax, *Challenges in Analytical Quality Assurance*, DOI 10.1007/978-3-642-16595-5_5, © Springer-Verlag Berlin Heidelberg 2011

Method validation assumes that an appropriate analytical method has been *selected* (for example, titration, photometry, AAS, chromatographic methods) and that the method has been *developed* (see Chap. 6).

The selection of an appropriate analytical method and the method validation procedure are determined by the type of samples which have to be analyzed in routine analysis. Some aspects are:

- The type of analytes and their concentration range which must be detected
- The type of sample matrices
- The interfering substances expected
- Whether qualitative or quantitative information is needed
- The required robustness of the method
- Last but not least, the cost per analysis

The extensive procedure of method validation starts with the development of an operating procedure or a validation master plan, goes through the definition of the performance parameters and acceptance criteria, the determination of the validation experiments, the performance of the validation experiments, the development of SOPs (Standard Operation Procedures) for executing the method in routine analysis, and ends with the documentation of the validation experiments and results in the validation report if the method is fit-for-purpose. From these steps, the main *performance parameters* (also called *validation parameters*) which are required by the regulatory agencies given above and their evaluation are discussed in the following sections.

5.2 Precision

The definition of precision given in ISO 3524-2 (2006) [2] is the "closeness of agreement between independent test results obtained under stipulated conditions". The precision of a set of results of measurements is quantified as a standard deviation obtained from replicate measurements of a sample which is representative in terms of the matrix and analyte concentration. As discussed in Sect. 3.2, three requirements have to be fulfilled for the calculation of the standard deviation from a data set: normal distribution, no outliers, and no trend. The precision of an analytical procedure is expressed as the variance s^2, standard deviation s, or the relative standard deviation $s_r\%$, sometimes also called coefficient of variation (CV) of a series of measurements.

In instrumental analysis the distribution of measurement values may be caused by two sources:

1. The instrument precision itself
2. The analytical procedure

In order to estimate the precision of the analytical method, knowledge of the instrument precision is necessary.

Precision of the instrument (measurement precision, system precision). In general, the instrument precision is given by the s_r% value calculated by the measured response values with a CRS compound or a stable, homogeneous sample. The number of replicates may be ten, but six replicates are required by ICH [Q2(R1)] [3] in pharmaceutical analysis.

Errors in instrumental precision are mostly caused by injection techniques. For example, because of the small injection volume of 1 μL or less combined with split injection techniques, the injection precision of GC analysis may be no better than 2%. In general, these sources of error are inherently smaller for HPLC and photometry, and the limit of injection precision is 1%.

Challenge 5.2-1

(a) In order to test the assay of the drug ergocalciferol (vitamin D_2) by HPLC analysis according to EUROPHARM, the injection precision has to be checked by six replicates of a test solution prepared by dissolving 0.5 g ergocalciferol (CRS) in 2.0 mL toluene without heating and then making up the eluent to 10.0 mL. The limit value of s_r% evaluated by the peak areas obtained from the chromatograms has to be no greater than 1.0%. The test was carried out using an autosampler with was equipped with two different syringes. The peak areas in counts obtained by the chromatograms are given in Table 5.2-1.
Check if the old syringe 1 is appropriate for testing the injection precision or if the new syringe 2 has to be used.

(b) In an analytical laboratory the photometric determination of nitrite-N by DIN EN 26777 (1993) [4] must be introduced. According to the analytical procedure, NO_2^- is transferred into an azo dye with $\lambda_{max} = 540$ nm, which is the wavelength for the measurement of the absorbance *A*. But this dye cannot be used for checking the instrument precision, i.e. the precision of the absorbance measurement, because it is generated by a chemical equilibrium. Therefore, in order to test the instrument precision a *stable* dye which absorbs in the visible range has to be used, which may be methylene blue with $\lambda_{max} = 665$ nm. 0.48 mg methylene blue ($M = 319.98$ g mol^{-1}) was dissolved in ethanol/water

(continued)

Table 5.2-1 Peak areas *A* obtained from the HPLC chromatograms for the determination of the injection precision using two different syringes

Replicate	1	2	3	4	5	6
Syringe 1						
A in counts	125,401	127,997	125,397	126,578	127,834	124,675
Syringe 2						
A in counts	128,321	128,298	128,732	128,395	128,201	128,163

Table 5.2-2 Measurement values of the absorbance A obtained by a methylene blue solution used as a test substance for the photometric determination of nitrite-N

Replicate	A
1	1.0223
2	1.0219
3	1.0222
4	1.0222
5	1.0220
6	1.0225
7	1.0219
8	1.0223
9	1.0221
10	1.0222

80% (v/v) and the absorbance was measured with ten replicates using a 1 cm cuvette. The results are listed in Table 5.2-2.
Calculate $s_r\%$ of the measurements of the absorbance A.

Solution to Challenge 5.2-1
There is no a hint of a trend in both data sets, but the tests for normal distribution and outliers must be carried out:

(a) The David test calculated by (3.2.1-1) gives 2.401 and 2.785 for syringe 1 and 2, respectively. These values lie within the limit values $q_{r,\text{lower}} = 2.28$ and $q_{r,\text{upper}} = 3.012$. Thus, both data set sets can be regarded as normally distributed.
As Table 5.2-3 shows, no outlier is detected in the data set of syringe 1 at the significance level $P = 95\%$. After rejection of peak area 128,732 as an outlier in the data set of syringe 2, the injection precision with the new syringe is $s_r\% = 0.07$, calculated with the standard deviation $s = 93.626$ counts and the mean value $\bar{x} = 128,275.6$ counts. This value of syringe 2 fulfills the regulatory requirement, whereas the relative standard deviation of the old syringe 1, $s_r\% = 1.1$, exceeds the required limit value 1%.

(b) The statistical tests for normal distribution and outliers are made as described above. The data set is normally distributed: the test value is $\hat{q}_r = 2.963$ which lies between the lower (2.67) and the upper (3.685) critical values of the David table at the significance level $P = 95\%$ and $n = 10$.
On inspection of the data set of the measured values of absorbance A, the value 1.0225 is suspected to be an outlier, and is checked by the Dixon test. Remember that for ten replicates the test value has to be calculated according to (3.2.3-1):

$$\hat{Q} = \frac{|x_1^* - x_2|}{|x_1 - x_{n-1}|}.$$

(continued)

Table 5.2-3 Results of the check for outliers by the Dixon test according to (3.2.3-1) for both HPLC syringes

	Syringe 1	Syringe 2
$x^* = x_{max}$	127,997	128,732
x_2	127,834	128,395
$x_n = x_{min}$	124,675	128,163
$\hat{Q}_{max}$	0.0491	0.592
$x^* = x_{min}$	124,675	128,163
x_2	125,397	128,201
$x_n = x_{max}$	127,997	128,732
$\hat{Q}_{min}$	0.217	0.067
$Q(P = 95\%, n = 6)$		0.560

Neither x_{max} nor x_{min} are confirmed to be an outlier. The test values $\hat{Q}_{max} = 0.1667$ and $\hat{Q}_{min} = 0$ do not exceed the critical value $Q(P = 95\%, n = 10) = 0.477$.
The precision of the absorbance measurements in the visible range is $s_r\% = 0.02$ calculated with $s = 0.000202$ and $\bar{x} = 1.0222$.

Precision of the Analytical Procedure. The precision of the analytical procedure may be considered under three headings [3, 5]:

1. Repeatability
2. Intermediate precision
3. Reproducibility

Repeatability. Repeatability expresses the precision under the same operation conditions, which "include the same measurement procedure, same operators, same measuring system, same operation conditions and same locations, and replicate measurement on the same or similar objects over a short time" [5]. Precision under repeatability conditions is also termed as "within-batch" or "intra-assay" precision. Repeatability reflects the differences between replicate measurements obtained in a single batch of analysis.

Repeatability is expressed quantitatively by the repeatability standard deviation s_r which is the deviation obtained from a series of n measurements under repeatability conditions, as well as the repeatability interval r which is also called the repeatability limit. The repeatability limit is calculated by (5.2-1)

$$r = t(P, \mathrm{df}) \cdot \sqrt{2} \cdot s_r \tag{5.2-1}$$

in which $t(P, \mathrm{df})$ is the quantile of the two-tailed t-distribution. It is the confidence interval representing the maximum permitted difference between two results obtained under repeatability conditions.

The repeatability standard deviation can be estimated by simple replication studies, which involve making repeated measurements on a suitable sample under the same conditions. The precision is expressed as the relative standard deviation $s_r\%$ given in (2.2-5a). The number of replicates should be at least six because otherwise the confidence interval becomes too wide.

Intermediate Precision. Intermediate precision is obtained from "condition of measurement, out of a set of conditions that includes the same measurement procedure, same location, and replicate measurements on the same or similar objects over an extent period of time, but may include other conditions involving changes" [5]. The changes can include new calibrations, operators, and measurement systems. The intermediate precision is also known as "within-laboratory reproducibility".

Reproducibility. Reproducibility is also defined in [5]. It is the precision obtained from "conditions of measurement, out of a set of conditions that includes different locations, operators, measuring systems, and replicate measurements on the same or similar objects". Reproducibility is expressed quantitatively by the reproducibility standard deviation s_R which is the experimental standard deviation obtained from a series of measurements under reproducibility conditions. The number of measurements n should be sufficiently large to estimate a representative standard deviation. The reproducibility interval R or reproducibility limit is a confidence interval representing the maximum permitted difference between two single measured results under reproducibility conditions. It is calculated by (5.2-2)

$$R = t(P, \mathrm{df}) \cdot \sqrt{2} \cdot s_R \tag{5.2-2}$$

in which $t(P, \mathrm{df})$ is the quantile of the two-tailed Student's t-distribution. The degrees of freedom df relate to the number of replicates by which s_R has been established. As mentioned above, s_R is estimated by a large number of replicates, therefore the Student's t-value is approximately 2 and, according (5.2-2), $R = 2.8 \cdot s_R$.

In order to estimate the intermediate precision or the reproducibility, a nested (or hierarchical) design can be used. If, for example, the intermediate precision of an analytical procedure has to be studied using various sets of equipments or carried out by different analysts, portions of the same bulk are analyzed by replicates under repeatability conditions using different equipment or performed by different analysts. Another example concerns interlaboratory studies with regard to method validation or production of certified parameters of chemical reference materials (CRM). The results from this type of study are calculated by one-way ANOVA (see Sect. 3.6).

Limit Values of the Precision. The precision of the analytical procedure which is acceptable is determined by the complexity of the method, the matrix, and the concentration.

In EUROPHARM [6–8], the regulatory documents of pharmaceutical analysis, limit values of the precision of analytical procedures are established. Thus, the limit value of the relative standard deviation $s_{r_{max},r}\%$ is given by

$$s_{r_{max},r}\% = \frac{K \cdot B \cdot \sqrt{n}}{t(P = 90\%, \mathrm{df})} \cdot 100 \qquad (5.2\text{-}3)$$

in which the constant K = 0.349, B is the difference between the upper limit value of the assay and 100% given in each special monograph, n is the number of replicated injections of a appropriate reference solution, and df is the degrees of freedom calculated by df $= n - 1$. The limit values calculated by (5.2-3) are valid only for the determination of the assay but not for impurities or related substances whose content is much smaller. The limit values of the related substances and other byproducts are determined by their content. The following values are common:

$s_r\% \leqslant 5$ for content of 1 to 10% (w/w) and
$s_r\% \leqslant 10$ for content of 0.1 to 1% (w/w).

However, in environmental analysis much smaller values of the content are possible. The limit values of the precision were established by Horowitz [9, 10] from the results of a very large number of interlaboratory trials. The limit value of the precision under reproducibility conditions $s_{r_{max},R}\%$ is given by (5.2-4)

$$s_{r_{max},R}\% = 2^{(1-0.5 \log c)} \qquad (5.2\text{-}4)$$

in which c is the concentration of the analyte in the sample (as a decimal).

The corresponding precision under repeatability conditions is calculated by (5.2-5):

$$s_{r_{max},r}\% = 0.67 \cdot s_{r_{max},R}\%. \qquad (5.2\text{-}5)$$

Challenge 5.2-2

(a) Calculate the limit values of the precision under repeatability and reproducibility conditions for the analyte concentration given in Table 5.2-4, and complete the table.

(b) During preliminary investigations of the analysis of dioxin in water solutions, a test analysis was carried out by the recovery of the content of a stock solution ($c_{st} = 0.45$ nmol L^{-1} 2,3,7,8-TCDD). The mean values obtained by six replicates are given in Table 5.2-5.

1. Check whether the precision of the analytical procedure is satisfactory.
2. Present the analytical result as $\bar{x} \pm \Delta\bar{x}$ nmol L^{-1} 2,3,7,8-TCDD.

(continued)

Table 5.2-4 Limit values of the precision for various analyte concentrations

Relative amount	$s_{r_{max},R}\%$	$s_{r_{max},r}\%$
10%		
5%		
1%		
0.1%		
100 ppm		
10 ppm		
1 ppm		
100 ppb		
10 ppb		
1 ppb		

Table 5.2-5 Analytical results for the determination of 2,3,7,8-TCDD in a water sample

Replicate	1	2	3	4	5	6
$\bar{x}$ in ppt (w/w) [parts per trillion]	125	80	115	203	90	165

3. Estimate whether the experimentally determined content of the stock solution used is within the confidence interval of the analytical result.

   The sum formula of 2,3,7,8-TCDD is $C_{12}H_4O_2Cl_4$. The value $\rho = 1$ g cm^{-3} can be used for the density of the stock solution.

Solution to the Challenge 5.2-2

(a) The values for $s_{r_{max},R}\%$ and $s_{r_{max},r}\%$ calculated according to (5.2-4) and (5.2-5), respectively, are summarized in Table 5.2-6.

(continued)

Table 5.2-6 Limit values of the precision for various analyte concentrations

Relative amount	$s_{r_{max},R}\%$	$s_{r_{max},r}\%$
10%	2.8	1.9
5%	3.1	2.1
1%	4.0	2.7
0.1%	5.7	3.8
100 ppm	8.0	5.4
10 ppm	11.3	7.6
1 ppm	16.0	10.7
100 ppb	22.6	15.2
10 ppb	32.0	21.4
1 ppb	45.3	30.3

(b) 1. The value of the precision is $s_r\% = 36.0$ calculated by (2.2-5a) with mean value $\bar{x} = 129.7$ ppt and standard deviation $s = 46.7$ ppt. The limit value of $s_{r_{max},r}\%$ calculated by (5.2-5) for the range 130 ppt is $s_{r_{max},r}\% = 41.2$. Thus, the experimentally determined relative standard deviation under repeatability conditions does not exceed the limit value calculated by (5.2-5). The value of the precision, although high, can be accepted and the analytical result may be calculated.
2. With $n = 6$ and $t(P = 95\%, \mathrm{df} = 5) = 2.571$, the confidence interval calculated by (2.3-1) is 130 ± 49 ppt (w/w). Conversion into the units $\mathrm{nmol\,L^{-1}}$ gives 0.40 ± 0.15 nmol $\mathrm{L^{-1}}$ 2,3,7,8 - TCDD.
3. The stock solution $c_{at} = 0.45\ \mathrm{nmol\,L^{-1}}$ is within the range of the analytical result which is 0.25–0.55 nmol $\mathrm{L^{-1}}$.

Challenge 5.2-3
Let us assume that the content of three drugs is fixed in EUROPHARM, measured in relation to the CSR standard:

Drug I: $c = 98.5$–102% (w/w)
Drug II: $c = 98.0$–102.5% (w/w)
Drug III: $c = 98$–103% (w/w)

Calculate the limit of the repeatability standard deviations $s_{r_{max},r}\%$ for 3, 4, 5, and 6 replicates.

Solution to Challenge 5.2-3
The values of the repeatability standard deviation $s_{r_{max},r}\%$ calculated by (5.2-3) are listed in Table 5.2-7.

Challenge 5.2-4
In order to certify the Cd content of a soil sample, an interlaboratory trial was organized between seven laboratories. Each laboratory had to determine the Cd content in a portion of the same homogenous bulk. The results are listed in Table 5.2-8.

(a) Check the homogeneity of variances of the seven groups at the significance level $P = 95\%$
(b) Estimate the certificate of the soil sample as $\bar{\bar{x}} \pm \Delta\bar{\bar{x}}$ ppm (w/w) Cd, also with $P = 95\%$.

(continued)

Table 5.2-7 Repeatability standard deviations $s_{r_{max},r}\%$ for various values of $B\%$ (w/w) and different numbers of replicated injections

Drug	$B\%$ (w/w)	Number of replicate injections n			
		3	4	5	6
		Degrees of freedom df			
		2	3	4	5
		$t(P = 90\%, \mathrm{df})$			
		2.920	2.353	2.132	2.015
		$s_{r_{max},r}\%$			
I	2.0	0.41	0.59	0.73	0.85
II	2.5	0.52	0.74	0.92	1.06
III	3.0	0.62	0.89	1.10	1.27

Table 5.2-8 Cd content in ppm (w/w) of a soil sample determined by seven laboratories with five replicates

Replicates	Laboratory						
	A	B	C	D	E	F	G
1	45.09	45.20	45.37	45.23	45.40	45.63	44.92
2	45.19	45.27	45.45	45.26	45.41	45.65	44.95
3	45.22	45.30	45.48	45.31	45.45	45.73	44.93
4	45.25	45.40	45.60	45.39	45.61	45.85	45.18
5	45.31	45.75	45.62	45.44	45.60	45.86	45.17

Table 5.2-9 Analytical results of the determination of Cd in ppm (w/w) using the CRM with the parameters determined by the data set of Table 5.2-8

Replicate	Laboratory H	Laboratory I
1	45.61	45.17
2	45.63	44.83
3	45.73	44.95
4	45.85	44.83
5	45.84	45.18
6	45.96	45.18
7	45.73	45.00
8	45.54	44.98
9	45.63	44.99
10	45.78	45.10
11	45.76	45.12
12	45.81	45.03

(c) Estimate the mean values of laboratories F and G with respect to the confidence interval determined by the interlaboratory trial.

(d) The same certified reference material was used to test the performance of laboratories H and I with regard to the determination of Cd in soil samples. The results are given in Table 5.2-9.
Check whether for laboratories H and I

1. The requirement of the *precision* of the analytical procedure is achieved.
2. The analytical results are *true*.

(continued)

(e) The method of determination of Cd in soil samples was used in laboratory K. The results obtained were $x_1 = 40.45$ ppm (w/w) and $x_2 = 40.86$ ppm (w/w) in a sample determined by two replicates. Is the calculation of the mean value $\bar{x}$ ppm (w/w) permitted?

Solution to Challenge 5.2-4

(a) As Table 5.2-10 shows, the test values $\hat{q}_r$ of laboratories E, F, and G lie outside the limit values at the significance level $P = 95\%$, which means that a normal distribution is not present. However, this result should be ignored because of the small data quantities in each data set.
All seven data sets are free of outliers according to the Dixon test at the significance level $P = 95\%$.
The data sets of equal size can be checked for the homogeneity of the group variances by the Cochran test. As Table 5.2-11 shows, the test values calculated by the intermediates given in Table 5.2-11 does not exceed the critical value. The group variances can be regarded as homogeneous at the significance level $P = 95\%$.
The alternative check, the Bartlett test, yields the same result. The test value calculated according to (3.4-2), $\hat{\chi}^2 = 2.303\,[28 \cdot \log 0.0161 - (-52.941)] = 6.209$, does not exceed the critical value $\chi^2(P = 99\%, \mathrm{df} = 6) = 16.812$.

(b) The grand mean value $\bar{\bar{x}}$, the mean of the laboratory mean values, is $\bar{\bar{x}} = 45.38$ ppm (w/w) Cd.
The equation which one has to use for the calculation of its confidence interval is determined by the relation of the variances between

(continued)

Table 5.2-10 Results of the check for normal distribution and outliers according to the David and Dixon tests, respectively

Laboratory	A	B	C	D	E	F	G
David test according to (3.2.1-1)							
$\hat{q}_r$	2.704	2.536	2.380	2.389	2.037	2.128	1.957
Limit values at $P = 95\%$, $n = 5$				Lower	2.15	Upper	2.753
Dixon test according to (3.2.3-1)							
$x^* = x_{\max}$	45.31	45.75	45.62	45.44	45.61	45.86	45.18
x_2	45.25	45.4	45.6	45.39	45.6	45.85	45.17
$x_n = x_{\min}$	45.09	45.2	45.37	45.23	45.4	45.63	44.92
$\hat{Q}_{\max}$	0.273	0.636	0.080	0.238	0.048	0.043	0.038
$x^* = x_{\min}$	45.09	45.2	45.37	45.23	45.4	45.63	44.92
x_2	45.19	45.27	45.45	45.26	45.41	45.65	45.93
$\hat{Q}_{\min}$	0.455	0.127	0.320	0.143	0.048	0.087	0.038
$Q(P = 95\%, n = 5)$				0.642			

Table 5.2-11 Intermediate quantities and result of the Cochran test of homogeneity of the group variances calculated by (4.4-1)

Laboratory	s_i^2
A	0.0066
B	0.0470
C	0.0110
D	0.0077
E	0.0106
F	0.0117
G	0.0176
$\sum s_i^2$	0.1124
s_{max}^2	0.0470
$\hat{C}$	0.4185
$C(P = 95\%, k = 7, \mathrm{df} = 4)$	0.4307

(s_{bw}^2) and within (s_{in}^2) the laboratories' values, which is checked by an F-test:

$$\hat{F} = \frac{s_{bw}^2}{s_{in}^2}. \tag{5.2-6}$$

The variances s_{bw}^2 and s_{in}^2 are calculated by the one-way ANOVA procedure (see Sect. 3.6).

If the test value $\hat{F}$ is larger than the table value $F(P, \mathrm{df}_1 = \mathrm{df}_{bw}, \mathrm{df}_2 = \mathrm{df}_{in})$, then the variance between laboratories is significantly larger than the variance within laboratories, and the confidence interval is calculated by (6.5.2-7):

$$\Delta\bar{\bar{x}} = \frac{s_{bw} \cdot t(P, \mathrm{df}_{bw})}{\sqrt{n}}. \tag{5.2-7}$$

The number of degrees of freedom between the k laboratories is $\mathrm{df}_{bw} = k - 1$, and n is the total number of measured values obtained with the k laboratories and n_j replicates $n = k \cdot n_j$.

If the variances between and within laboratories are homogeneous, which is the case if the test value $\hat{F}$ calculated by (5.2-6) is smaller than the critical value $F(P, \mathrm{df}_{bw}, \mathrm{df}_{in})$, the confidence interval is calculated by (5.2-8):

$$\Delta\bar{\bar{x}} = \frac{s_{tot} \cdot t(P, \mathrm{df}_{tot})}{\sqrt{n}}. \tag{5.2-8}$$

The total standard deviation s_{tot} is calculated by (5.2-9):

$$s_{tot} = \sqrt{\frac{\mathrm{SS}_{tot}}{\mathrm{df}_{tot}}}. \tag{5.2-9}$$

(continued)

The total degrees of freedom df_{tot} is given by (5.2-10):

$$df_{tot} = n - 1. \tag{5.2-10}$$

The total sum of squares SS_{tot} is:

$$SS_{tot} = SS_{bw} + SS_{in}. \tag{5.2-11}$$

Estimation of the variances s_{bw}^2 and s_{in}^2 by one-way ANOVA:
The intermediate quantities and results of one-way ANOVA calculated according to the computational scheme given in Table 3.6-2 are summarized in Table 5.2-12.
According to the results presented in Table 5.2-12, the test value $\hat{F}$ exceeds the quantiles of the F-distribution $F(P = 95\%, df_{bw} = 6, df_{in} = 28) = 2.445$. This means that the variances between the laboratory results are significantly greater than the variances within the laboratories at the significance level $P = 95\%$. The confidence interval must be calculated by (5.2-7) which gives $\Delta\bar{\bar{x}} = 0.21$ ppm (w/w) Cd calculated with $s_{bw} = 0.5118$, $n = 35$, and $t(P = 95\%, df_{bw} = 6) = 2.447$.
The certificate of the soil sample is $\bar{\bar{x}} \pm \Delta\bar{\bar{x}} = 45.38 \pm 0.21$ ppm (w/w) Cd.

(c) As discussed in Sect. 2.2.3, the true value is within the range 45.17–45.59 ppm (w/w) Cd but this range does not include the mean values of the laboratories F ($\bar{x}_F = 45.74$ ppm (w/w)and G ($\bar{x}_G = 45.03$ ppm (w/w)). The question arises whether the results of the laboratories F and G have to be rejected as outliers.
However, the test values of both laboratories calculated by the Dixon test

(continued)

Table 5.2-12 Intermediate quantities and results of one-way ANOVA

Laboratory						
A	B	C	D	E	F	G
SS_i						
0.0265	0.1881	0.0441	0.0309	0.0425	0.0467	0.0706
$\sum SS_i = SS_{in}$		0.4495				
$\bar{x}_i$						
45.21	45.38	45.50	45.33	45.49	45.74	45.03
$\bar{\bar{x}}$	45.38	n	35	n_i	5	
$n_i(\bar{x}_i - \bar{\bar{x}})^2$						
0.1494	0.0000	0.0710	0.0173	0.0596	0.6449	0.6296
$\sum n_i(\bar{x}_i - \bar{\bar{x}})^2 = SS_{bw}$			1.5718	k	7	
df_{in}	28	s_{in}^2	0.0161			
df_{bw}	6	s_{bw}^2	0.2620	$\hat{F}$	16.319	

$$\hat{Q}_{\mathrm{F}} = \frac{|45.74 - 45.50|}{|45.74 - 45.03|} = 0.336$$

$$\hat{Q}_{\mathrm{G}} = \frac{|45.03 - 45.21|}{|45.03 - 45.74|} = 0.255$$

do not exceed the critical value $Q(P = 95\%, n = 7) = 0.507$. Thus, there is no cause to reject the mean values obtained by laboratories F and G.

(d) The precision of laboratories H and I (s_{Lab}) is checked by the F-test

$$\hat{F} = \frac{s_{\mathrm{Lab}}^2}{s_{\mathrm{in}}^2}. \tag{5.2-12}$$

The variance within the laboratories s_{in}^2 is obtained by the interlaboratory trial given above.

The trueness is checked by the t-test

$$\hat{t} = \frac{|\bar{x}_{\mathrm{Lab}} - \bar{\bar{x}}|}{s_{\mathrm{Lab}}} \cdot \sqrt{n_{\mathrm{Lab}}}, \tag{5.2-13}$$

in which $\bar{x}_{\mathrm{Lab}}$ is the mean values obtained by n_{Lab} replicates in laboratories H and I, respectively, and $\bar{\bar{x}}$ is the grand mean value obtained by the interlaboratory trial, i.e. the mean value of the certificate.

The test values are obtained by the Excel functions explained in the previous chapter, giving the following results:

Both data sets are *normally distributed*, as checked by the David test (3.2.1-1).

Test values: $\hat{q}_{r,\mathrm{H}} = 3.50$, $\hat{q}_{\mathrm{r,I}} = 2.82$

Critical values:

$$q_{r,\mathrm{lower}}(P = 95\%, n = 12) = 2.80,$$

$$q_{r,\mathrm{upper}}(P = 95\%, n = 12) = 3.91.$$

Outlier test by Dixon according to (3.2.3-1):

For $n = 12$, test values must be calculated by (5.2-14)

$$\hat{Q} = \left| \frac{x_1^* - x_3}{x_1^* - x_{n-1}} \right|. \tag{5.2-14}$$

(continued)

For example, the test value for the maximum value of the data set of laboratory H is

$$\hat{Q}_{x_{\max},\mathrm{H}} = \left|\frac{45.96 - 45.84}{45.96 - 45.61}\right| = 0.343.$$

The other test values are

$$\hat{Q}_{x_{\min},\mathrm{H}} = 0.290,$$

$$\hat{Q}_{x_{\max},\mathrm{I}} = 0.029,$$

$$\hat{Q}_{x_{\min},\mathrm{I}} = 0.343.$$

The critical value with $Q(P = 95\%, n = 12) = 0.546$ is larger than all four test values, and therefore both data sets can be regarded as outlier-free.

Estimation of the *precision* of the laboratories s_{Lab}:

The test values $\hat{F}$ are $\hat{F}_{\mathrm{H}} = 0.899$ and $\hat{F}_{\mathrm{I}} = 0.957$. The critical value $F(P = 95\%, \mathrm{df}_1 = \mathrm{df}_{\mathrm{Lab}} = 11, \mathrm{df}_2 = \mathrm{df}_{\mathrm{in}} = 28) = 2.151$ is larger than the test values, which means that the required precision in the analytical procedure is achieved in both laboratories.

Test for *trueness*:

The check for trueness is performed by the t-test according to (5.2-13). The test values are $\hat{t}_{\mathrm{H}} = 10.359$ and $\hat{t}_{\mathrm{I}} = 9.782$ calculated with the following parameters:

$\bar{\bar{x}} = 45.38$ ppm (w/w), $\bar{x}_{\mathrm{H}} = 45.74$ ppm (w/w), $\bar{x}_{\mathrm{I}} = 45.03$ ppm (w/w), $s_{\mathrm{H}} = 0.120$ ppm (w/w), $s_{\mathrm{I}} = 0.124$ ppm (w/w), and $n = 12$

Both test values are larger than the critical value $t(P = 95\%, \mathrm{df}_{\mathrm{Lab}} = 11) = 2.201$, and thus the analytical results of both laboratories are false!

Note that the mean values obtained in laboratories H and I are the same as in laboratories F and G, whose results could not be rejected as outliers from the data sets of the interlaboratory trial, but the same values obtained by using the CRM yield a false result.

(e) The confidence interval for the difference between two results obtained under repeatability conditions is calculated by (5.2-1). With $s_{\mathrm{r}} = \sqrt{s_{\mathrm{in}}^2} = 0.1267$ and $t(P = 95\%, \mathrm{df}_{\mathrm{in}} = 28) = 2.048$ the repeatability interval is $r = 0.37$ ppm (w/w) Cd, but the difference between the two measured values is $\Delta = 0.41$ ppm (w/w) Cd. This means that calculation of the mean values is not permitted.

5.3 Linearity of Calibration Lines

5.3.1 *General Remarks*

As discussed in Sect. 4.2, the objective of regression analysis is to use the mathematical expression relating response to concentration to predict concentrations of unknown samples. In general, *linear* regression is used to establish a relationship between the x and y variables. But the question is whether the *linear* regression function is really the best mathematical model for this relationship. Therefore, validation of the regression model is necessary to verify that the chosen model adequately describes the relationship between the two variables x and y. This means one has to verify that the best model is a straight line or whether the data are better described by a curve.

Remember that according to (4.2-15) the coefficients of the linear regression a_0 and a_1 are the basis of the calculation of the predicted values $\hat{x}$ of unknown samples. But if these constants are not valid, the analytical results are false. It is clear that checking the linearity is an important validation parameter which is included in all the regulatory requirements given above.

In practice the correlation coefficient r_{xy} (see Sect. 4.1) and the coefficient of determination r_{xy}^2 are frequently used in order to verify the linearity of the regression model, but this is incorrect. The correlation coefficient as a measure of the linear relationship cannot be applied for calibration. Concentrations (or contents) as x-values are commonly defined and, thus, fixed in advance in analytical practice. Consequently, these values are not random variables. However, there are various procedures for testing the linearity, which are given below.

5.3.2 *Quality Coefficient*

A suitability check for linearity with homoscedastic measurements is the estimation of the *quality coefficient* (QC) [11] which is calculated by (5.3.2-1):

$$\mathrm{QC} = 100 \cdot \sqrt{\frac{\sum \left(\frac{y_i - \hat{y}_i}{\bar{y}}\right)^2}{\mathrm{df}}}. \qquad (5.3.2\text{-}1)$$

Each residual $(y_i - \hat{y}_i)$ is related to the mean of all observations $\bar{y}$. The degrees of freedom are $\mathrm{df} = n - 2$ as proposed in [11].

If a target value for the quality coefficient QC has been specified (for example, obtained from previous experiments), the suitability of the linearity can be checked.

Challenge 5.3.2-1

Let us assume from previous experiments that the target value of the quality coefficient has been specified as 1%.

(a) Calculate the QC value for calibration of the photometric determination of benzene in *n*-hexane according to Challenge 4.2-1 and check whether the linearity is valid.
(b) Calculate the QC value of the data set for the determination of malathion by GC-FPD given in Table 4.4-2 and evaluate the result.

Solution to Challenge 5.3.2-1

(a) The intermediate quantities for the calculation of the QC value are given in Table 5.3.2-1 calculated with the parameters $a_0 = -0.00265$, $a_1 = 0.2561\ \mathrm{L\,mmol^{-1}}$, $\bar{y} = 0.6016$ obtained by Table 4.2.3.
The quality coefficient is QC = 0.61% calculated by (5.3.2-1) with df = 8. The QC value is smaller than the target value of 1%, and thus linearity can be assumed.
(b) The intermediate quantities presented in Table 5.3.2-2 are calculated using $a_0 = 29.467$ mV, $a_1 = 225.212\ \mathrm{mV\,L\,mg^{-1}}$, and $\bar{y} = 91.4$ mV. The quality coefficient is QC = 8.61% calculated by (5.3.2-1) using df = 8. The QC value is greater than the target value which means linearity cannot be assumed.

Table 5.3.2-1 Intermediate quantities for the calculation of the quality coefficient QC for the photometric determination of benzene

n_i	x_i	$y_i\ (A_i)$	$\hat{y}_i$	$e_i = (y_i - \hat{y}_i)$	$10^5 \frac{(y_i-\hat{y}_i)^2}{\bar{y}}$
1	0.787	0.1991	0.1980	0.00033	0.0250
2	0.787	0.2008	0.1988	0.00203	1.1064
3	1.573	0.3958	0.4002	−0.00439	5.4661
4	1.573	0.3992	0.4002	−0.00099	0.3034
5	2.360	0.6076	0.6016	0.00600	9.6287
6	2.360	0.6012	0.6016	−0.00040	0.0682
7	3.146	0.7999	0.8030	−0.00312	2.9107
8	3.146	0.8016	0.8030	−0.00142	0.6601
9	3.933	1.0013	1.0044	−0.00313	2.9991
10	3.933	1.0095	1.0044	0.00507	6.6487
$\bar{y}$		0.6016		$\sum \left(\frac{y_i-\hat{y}_i}{\bar{y}}\right)^2$	29.8164

The concentrations x_i are given in mmol $\mathrm{L^{-1}}$ and y_i are the measured values of the absorbance A_i.

Table 5.3.2-2 Intermediate quantities for the calculation of the quality coefficient QC for the determination of malathion by GC-FPD

n_i	x_i	y_i	$\hat{y}_i$	$e_i = (y_i - \hat{y}_i)$	$\left(\frac{y_i - \hat{y}_i}{\bar{y}}\right)^2$
1	0.050	27	40.7273	−13.7273	0.0226
2	0.100	49	51.9879	−2.9879	0.0011
3	0.150	68	63.2485	4.7515	0.0027
4	0.200	82	74.5091	7.4909	0.0067
5	0.250	92	85.7697	6.2303	0.0046
6	0.300	105	97.0303	7.9697	0.0076
7	0.350	111	108.2909	2.7091	0.0009
8	0.400	120	119.5515	0.4485	0.00002
9	0.450	128	130.8121	−2.8121	0.0009
10	0.500	132	142.0727	−10.0727	0.0121
$\bar{y}$		91.4		$\sum\left(\frac{y_i - \hat{y}_i}{\bar{y}}\right)^2$	0.0593

The concentrations x_i are given in mg L^{-1} and the measured y-values in mV.

5.3.3 Visual Examinations

Sometimes a *visual inspection* of the calibration line $y = f(x)$ can already give information as to whether linearity should be rejected. For example, the calibration function presented in Fig. 4.4-2 shows clearly that the relationship between the

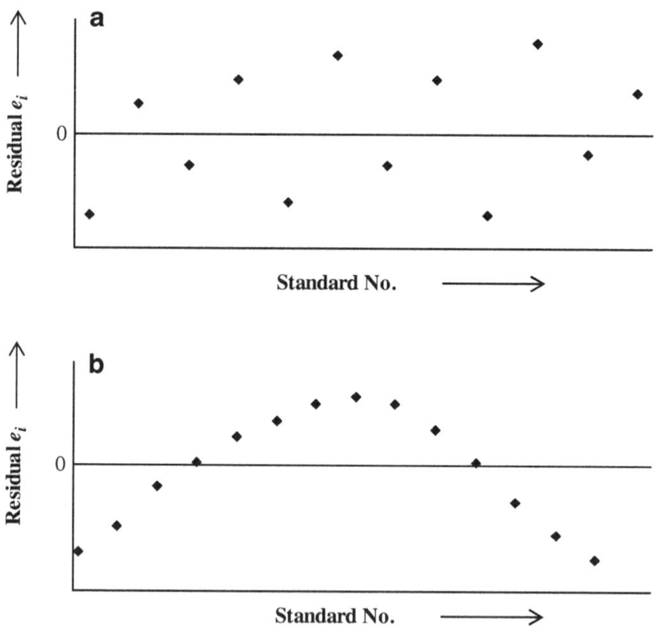

Fig. 5.3.3-1 Examples of residual plots

x and y values is better fitted by a *non-linear* function. However, this simple check does not usually give unequivocal information.

A better result can be obtained by *residual analysis*. The residuals e_i calculated by (4.2-3) plotted against the x_i-values or the standard numbers can provide valuable information concerning the goodness of fit of the mathematical model. In Fig. 5.3.3-1 two possible patterns of residual plots are given. The residuals in Fig. 5.3.3-1a are randomly distributed within a horizontal band with equal (or approximately equal) numbers of negative and positive residuals. This means there is a good fit between the data and the linear regression model. But the U-shaped residual plot in Fig. 5.3.3-1b illustrates a residual plot typical of when the calibration line is fitted by a non-linear regression model.

Challenge 5.3.3-1

(a) In Challenge 4.2-1 the regression coefficients of the photometric determination of benzene in *n*-hexane were calculated by establishing a linear regression model.
Check by visual examination whether the assumed linearity is valid.

(b) The calibration function of the determination of malathion by GC-FDP presented in Fig. 4.4-2 of Challenge 4.4-1 shows that the data set is best fitted by a curve.
Check whether the quadratic regression function can be also confirmed by examination of the residual plot.

Solution to Challenge 5.3.3-1

(a) The calibration function of the photometric determination of benzene in *n*-hexane presented in Fig. 5.3.3-2 shows that the linear regression function may be valid, which is also confirmed by the pattern of the residual plot in Fig. 5.3.3-3 obtained with the residuals e_i from Table 5.3.2-1 (fifth column). The residuals are distributed randomly around zero with a pattern which is similar to Fig. 5.3.3-1a.

(continued)

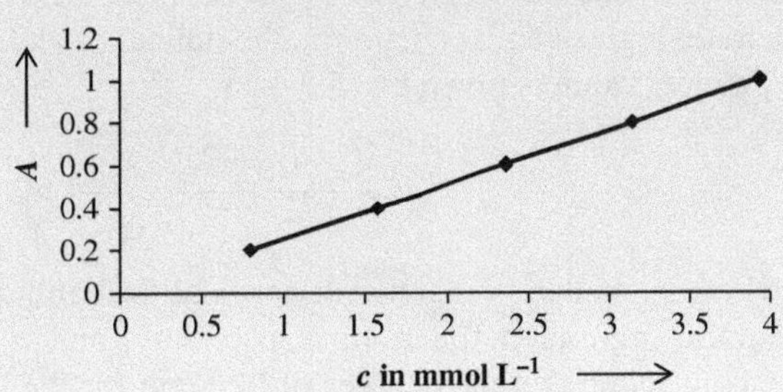

Fig. 5.3.3-2 Calibration function of the photometric determination of benzene in *n*-hexane derived from the data set given in Table 4.2-1

Fig. 5.3.3-3 Plot of the residuals e_i from Table 5.3.2-1

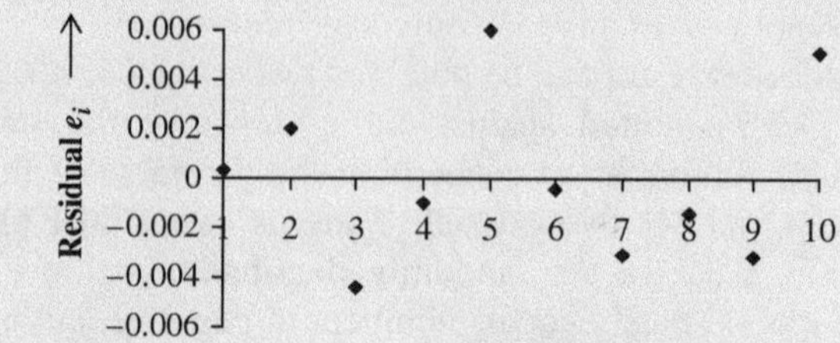

Fig. 5.3.3-4 Plot of the residuals e_i from Table 5.3.2-2

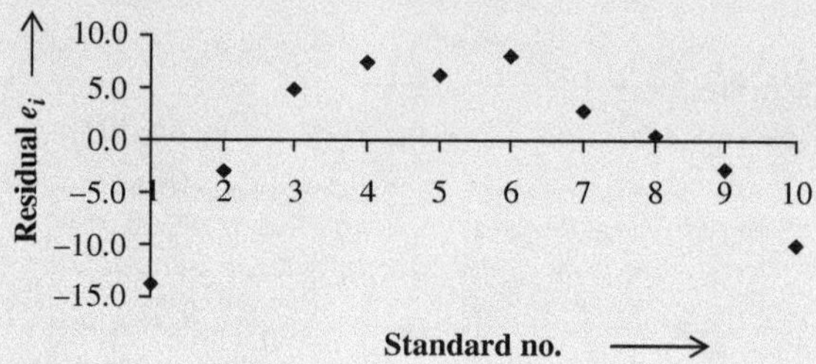

(b) The residual plot of the values e_i of Table 5.3.2-2 is presented in Fig. 5.3.3-4. The residual plot is similar to the pattern of a quadratic regression function illustrated in Fig. 5.3.3-1b. The non-linearity is confirmed.

5.3.4 Mandel Test

Because visual tests do not usually deliver an unequivocal result, it is necessary in addition to apply a mathematical linearity test, which is also included in the most common software. Such a test was proposed by Mandel and is recommended in a DIN [12]. According to Mandel the residual error is calculated for a quadratic regression function $s_{y.x,2}$. An F-test is then used to decide whether or not the quadratic regression is a better mathematical model than the linear regression.

The test value is given by (5.3.4-1)

$$\hat{F} = \frac{s_{y.x}^2 \cdot (n-2) - s_{y.x,2}^2 \cdot (n-3)}{s_{y.x,2}^2}, \qquad (5.3.4\text{-}1)$$

in which $s_{y.x}^2$ and $s_{y.x,2}^2$ are the variances of the calibration error of the linear and the quadratic regression function, respectively, and n is the number of calibration standards. Note that the degrees of freedom for the linear and quadratic regression function are $df_1 = n - 2$ and $df_2 = n - 3$, respectively.

The calculated test value $\hat{F}$ is compared with the critical value $F(P, \mathrm{df}_1 = 1, \mathrm{df}_2 = n - 3)$. If $\hat{F}$ does not exceed the critical F-value at the statistical significance P (in this case usually $P = 99\%$) the quadratic regression model does not provide a better description of the relationship between the x and y variables.

The residual standard deviation for the linear regression $s_{y.x}$ is calculated by (4.2-6) or (4.2-7), and for calculation of the quadratic regression $s_{y,x,2}$ (4.4-8) is used.

Challenge 5.3.4-1

(a) Check whether the linearity tested by the visual examination of the calibration function and by the residual pattern in Challenge 5.3.3-1a can be confirmed by the Mandel test.
(b) Check whether a non-linear regression function for the data set given in Table 4.4-1 obtained by visual examinations of Figs. 4.4-2 and 5.3.3-4 can be confirmed by the Mandel test.

Solution to Challenge 5.3.4-1

The calculation of the residual error of the linear regression $s_{y.x}$ can be realized by the Excel function = STEYX(y, x): $s_{y.x} = 0.003671$.

The residual error of the quadratic linear regression $s_{y.x,2}$ must be calculated according to (4.4-8):

$$s_{y.x,2} = \sqrt{\frac{(y_i - \hat{y}_i)^2}{n - 3}} \tag{5.3.4-2}$$

with $\hat{y}_i = a_0 + a_1 \cdot x_i + a_2 \cdot x_i^2$ or by the Excel function (see Table 4.4-1).

(a) The residual standard deviation for the linear regression function is $s_{y.x} = 0.003671$ calculated according to (4.2-6). The intermediate quantities for the calculation of the residual standard deviation of the quadratic regression function and the result are listed in Table 5.3.4-1.
Comparison of the calculated $\hat{F}$ value with the critical $\hat{F}$ value shows that the quadratic regression is not the better model and, hence, the results of the tests given above are confirmed.
(b) The residual standard deviation for the linear regression function is $s_{y.x} = 7.8684$ calculated according to (4.2-6). The intermediate quantities for the calculation of the residual standard deviation of the quadratic regression function and the result of the Mandel test are given in Table 5.3.4-2.

(continued)

Table 5.3.4-1 Intermediate quantities and the result of the Mandel test

$\sum x_i$	23.598	$\sum y_i$	6.016
$\sum x_i^2$	68.062	$\sum x_i y_i$	17.365
$\sum x_i^2 y_i$	55.905	$\sum x_i^3$	219.017
$\sum x_i^4$	749.608		
SS_{xx} (4.1-3)	12.375	SS_{xx} (4.1-5)	3.169
SS_{x^3} (4.4-5)	58.405	SS_{x^2y} (4.4-7)	14.959
SS_{x^4} (4.4-6)	286.367		
a_0 (4.4-2)	−0.00082	a_1 (4.4-3)	0.25407
a_2 (4.4-4)	0.000421	$s_{y.x,2}$ (4.4-8)	0.003889
$\sum (y_i - \hat{y}_i)^2$	0.0001059	n	10
Test result			
$\hat{F}$ (5.3.3-1)			0.126
$F(P = 99\%, \mathrm{df}_1 = 1, \mathrm{df}_2 = 8)$			12.246

Table 5.3.4-2 Intermediate quantities and the result of the Mandel test

$\sum x_i$	2.75	$\sum y_i$	914
$\sum x_i^2$	0.9625	$\sum x_i y_i$	297.800
$\sum x_i^2 y_i$	112.285	$\sum x_i^3$	0.3781
$\sum x_i^4$	0.1583		
SS_{xx} (5.6-3)	0.206	SS_{xx} (5.6-5)	46.450
SS_{x^3} (5.4-5)	0.1134	SS_{x^2y} (5.4-7)	24.313
SS_{x^4} (5.4-6)	0.06569		
a_0 (5.4-2)	8.883	a_1 (5.4-3)	431.045
a_2 (5.4-4)	−374.242	$s_{y.x,2}$ (5.4-8)	2.1748
$\sum (y_i - \hat{y}_i)^2$	33.108	n	10
Test result			
$\hat{F}$ (5.3.3-1)			97.722
$F(P = 99\%, \mathrm{df}_1 = 1, \mathrm{df}_2 = 8)$			12.246

The test value $\hat{F} = 97.722$ is much greater than the critical $\hat{F}$ value, and thus the quadratic regression better describes the relationship between the x and y values. The non-linearity obtained by the visual test as well as by the quality coefficient is confirmed by the Mandel test.

5.3.5 The Lack-of-Fit Test by ANOVA

Analysis of variance (ANOVA) can be applied in order to verify whether the model chosen is the correct one. For this test *replicate measurements* are needed but, in practice, this procedure is frequently used anyway.

The total sum of squares SS_{tot}

$$SS_{tot} = \sum_{i}^{k} \sum_{j}^{n_i} (y_{ij} - \bar{\bar{y}})^2 \tag{5.3.5-1}$$

is composed of the following sums of squares [11]:

$$SS_{tot} = SS_{PE} + SS_{LOF} + SS_{Reg}. \tag{5.3.5-2}$$

SS_{PE} is the *pure error sum of squares*, a component which measures the pure experimental error. It is calculated by

$$SS_{PE} = \sum_{i}^{k} \sum_{j}^{n_i} (y_{ij} - \bar{y}_i)^2. \tag{5.3.5-3}$$

SS_{LOF} is the *sum of squares due to lack-of-fit* which measures the variation of the group means $\bar{y}_i$ about the regression line. It is calculated by

$$SS_{LOF} = \sum_{i}^{k} n_i \cdot (\bar{y}_i - \hat{y}_i)^2. \tag{5.3.5-4}$$

SS_{Reg} is the *sum of squares due to regression*, which is calculated by

$$SS_{Reg} = \sum_{i}^{k} n_i \cdot (\hat{y}_i - \bar{\bar{y}})^2. \tag{5.3.5-5}$$

SS_R is the *residual sum of squares* which is the sum of SS_{PE} and SS_{LOF}

$$SS_R = SS_{PE} + SS_{LOF}, \tag{5.3.5-6}$$

where:

k is the number of calibration levels, i.e. different x-values
n_i is the number of replicate measurements made at x_i
y_{ij} is one of the n_i replicate measurements at x_i
$\sum_{i=1}^{k} n_i = n$ is the total number of all measurements, including all replicates
$\bar{\bar{y}}$ is the grand mean, i.e. the mean of all observations
$\bar{y}_i$ is the mean value of the replicates y_{ij} at x_i
$\hat{y}_i$ is the value of y_i at x_i estimated by the regression function. All replicates at x_i have the same estimated value $\hat{y}_i$

The mean squares MS are obtained by dividing the sums of squares SS by their corresponding degrees of freedom df:

Table 5.3.5-1 ANOVA scheme for the linearity test of the regression model with replicate measurements

Source of variation	SS	df	MS (5.3.5-7)	$\hat{F}$
Regression	SS_{Reg} (5.3.5-5)	1	MS_{Reg}	
Residual	SS_R (5.3.5-6)	$n-2$	MS_R	
Lack-of-fit	SS_{LOF} (5.3.5-4)	$k-2$	MS_{LOF}	$\frac{MS_{LOF}}{MS_{PE}}$ (5.3.5-8)
Pure error	SS_{PE} (5.3.5-3)	$n-k$	MS_{PE}	
Total	SS_{tot} (5.3.5-1)	$n-1$		

$$MS = \frac{SS}{df}. \tag{5.3.5-7}$$

The ANOVA scheme is given in Table 5.3.5-1.

The mean square MS_{PE} is an estimate of σ^2, the pure error of the measurement, and MS_{LOF} is an estimate of σ^2 if it is chosen as the correct one. It estimates $\sigma^2 + (\text{bias})^2$ if the model is not adequate.

The test value $\hat{F}$ calculated by (5.3.5-8) is compared with the one-sided F-distribution at the significance level P and the degrees of freedom $df_1 = (k-2)$, $df_2 = (n-k)$. If the $\hat{F}$ – values of the lack-of-fit test is greater than the critical value $F_{one-sided}(P, df_1, df_2)$, one concludes that the model chosen is inadequate, because the variation of the group means along the line cannot be explained in terms of pure experimental uncertainty. If the test value $\hat{F}$ does not exceed the critical F value, the model is justified.

Challenge 5.3.5-1

The validation of the determination of Zn by flame AAS in waste water was verified at six levels with three replicates. The results are listed in Table 5.3.5-2.

(a) Check if the linear regression model is valid and show the calibration line.

(b) Check the linearity of regression if the observation $y_{61} = 0.805$ (expressed by the value in italics in Table 5.3.5-2) is substituted by the value $y_{61} = 0.960$.

Table 5.3.5-2 Determination of Zn by flame AAS

Level	1	2	3	4	5	6
Concentration c in mg L^{-1}						
x_i	1	2	3	4	5	6
Absorbance A						
y_{i1}	0.040	0.260	0.422	0.605	0.754	*0.805*
y_{i2}	0.055	0.261	0.409	0.612	0.725	0.778
y_{i3}	0.041	0.271	0.420	0.601	0.728	0.785

Solution for the Challenge 5.3.5-1

(a) The intermediate quantities for the determination of the required sums of squares SS_{PE} and SS_{LOF} according to (5.3.5-3) and (5.3.5-4) with the regression coefficients $a_0 = -0.056178$ and $a_1 = 0.1521143\ \mathrm{L\,mg^{-1}}$ are listed in Table 5.3.5-3.

According to the ANOVA scheme given in Table 5.3.5-1, the mean squares MS_{PE}, MS_{LOF} and the test value $\hat{F}$ are calculated to give the following results:

Level $k = 6$, number of the observations $n = 18$, degrees of freedom of the pure error $df_{PE} = 12$, degrees of freedom of the lack-of-fit $df_{LOF} = 4$, mean square of the pure error $MS_{PE} = 0.00010633$, mean square of lack of fit $MS_{LOF} = 0.0086061$. The test value is $\hat{F} = 80.935$. The one-sided critical value is $F_{\text{one-sided}}(\bar{P} = 95\%, df_{LOF} = 4, df_{PE} = 12) = 3.259$ which is much smaller than the test value $\hat{F}$. Thus, the linearity of the regression function must be rejected.

As Fig. 5.3.5-1 shows, the relationship between the x-and y-values can be better described by a quadratic calibration curve.

(b) The results of ANOVA using the observation $y_{61} = 0.960$ are listed in Table 5.3.5-4. The new regression coefficients are $a_0 = -0.073400$ and

(continued)

Table 5.3.5-3 Intermediate quantities and results for the calculation of the sums of squares SS_{PE} and SS_{LOF} (c in mg L^{-1})

Level	$c\ x_i$	A y_{ij}	$\bar{y}_i$	$\hat{y}_i$	$3 \cdot (\bar{y}_i - \hat{y}_1)^2$	$\sum_i^k \sum_j^{n_i} (y_{ij} - \bar{y}_1)^2$
1	1	0.040				0.0000284
		0.055				0.0000934
		0.041	0.0453	0.0959	0.00768	0.0000188
2	2	0.260				0.0000160
		0.261				0.0000090
		0.271	0.2640	0.2481	0.00076	0.0000490
3	3	0.422				0.0000250
		0.409				0.0000640
		0.420	0.4170	0.4002	0.00085	0.0000090
4	4	0.605				0.0000010
		0.612				0.0000360
		0.601	0.6060	0.5523	0.00866	0.0000250
5	5	0.754				0.0003361
		0.725				0.0001138
		0.728	0.7357	0.7044	0.00293	0.0000588
6	6	*0.805*				0.0002454
		0.788				0.0001284
		0.785	0.7893	0.8565	0.01354	0.0000188
Sum					0.034424 SS_{LOF}	0.0012760 SS_{PE}

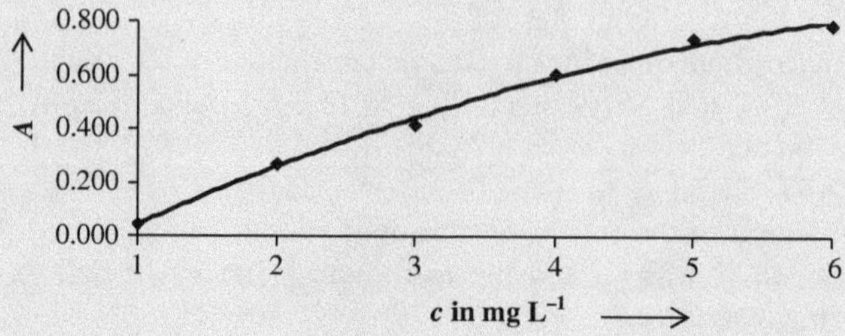

Fig. 5.3.5-1 Calibration curve for the determination of Zn by flame AAS obtained by the data set given in Table 5.3.5-2 with the observation value $y_{6.1} = 0.805$

Table 5.3.5-4 Intermediate quantities and results for the calculation of the sums of squares SS_{PE} and SS_{LOF} (c in mg L^{-1})

Level	$c(x_i)$	$A(y_{ij})$	$\bar{y}_i$	$\hat{y}_i$	$3 \cdot (\bar{y}_i - \hat{y}_1)^2$	$\sum_i^k \sum_j^{n_i} (y_{ij} - \bar{y}_1)^2$
1	1	0.040				0.0000284
		0.055				0.0000934
		0.041	0.0453	0.0861	0.004985	0.0000188
2	2	0.260				0.0000160
		0.261				0.0000090
		0.271	0.2640	0.2456	0.001017	0.0000490
3	3	0.422				0.0000250
		0.409				0.0000640
		0.420	0.4170	0.4051	0.000426	0.0000090
4	4	0.605				0.0000010
		0.612				0.0000360
		0.601	0.6060	0.5646	0.005147	0.0000250
5	5	0.754				0.0003361
		0.725				0.0001138
		0.728	0.7357	0.7241	0.000403	0.0000588
6	6	*0.960*				0.0141610
		0.788				0.0039690
		0.785	0.8410	0.8836	0.005437	0.0031360
Sum					0.017414	0.0221493
					(= SS_{LOF})	(= SS_{PE})
MS_{LOF}					0.0043535	
MS_{PE}						0.0018458

$a_1 = 0.159495$ L mg^{-1}. The test value is $\hat{F} = 2.359$, and therefore it is smaller than the critical value which is the same as in Challenge 5.3.5-1a. The increase in measurement error, the denominator of (5.3.5-8), reduces the test value $\hat{F}$ and the linearity may be valid, which can also be seen by the calibration line in Fig. 5.3.5-2.

(continued)

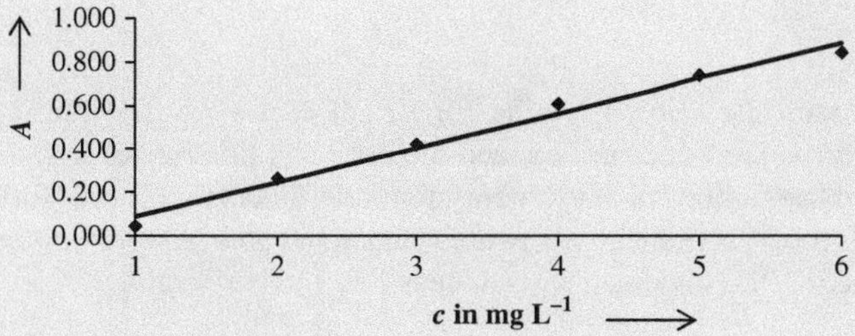

Fig. 5.3.5-2 Calibration line for the determination of Zn by flame AAS obtained by the data set given in Table 5.3.5-2 with the observation value $y_{6.1} = 0.960$

However the observation $y_{61} = 0.960$ must still be checked as to whether it has to be rejected as an outlier. The test value of the Dixon test is $\hat{Q} = 0.962$ and the critical value is $Q(P = 95\%, n = 3) = 0.941$. Therefore, it must be removed from the data set. After rejecting the outlier value, the test value is $\hat{F} = 98.878$ which exceeds the critical value. The linearity is not confirmed.

5.3.6 Test of the Significance of the Quadratic Regression Coefficient $\mathbf{a_2}$

The linearity is confirmed if the quadratic regression coefficient a_2 of the equation given in (4.4-1)

$$y = a_0 + a_1 \cdot x + a_2 \cdot x^2$$

is not significant. But if a_2 is significantly different from zero, a polynomial regression may better describe the relationship between the x and y values, i.e. the non-linearity should be tested.

The hypothesis that the quadratic term is zero or not
$H_0 : a_2 = 0$
$H_1 : a_2 \neq 0$
can be checked by two methods:

1. Check whether zero is included in the coefficient interval of a_2
   The coefficient interval is calculated by

$$\mathrm{CI}(a_2) = a_2 \pm t(P, \mathrm{df} = n - 3) \cdot s_{a_2}. \qquad (5.3.6\text{-}1)$$

2. Using a t-test.
   The absolute test value $\hat{t}$

$$\hat{t} = \left| \frac{a_2}{s_{a_2}} \right| \tag{5.3.6-2}$$

is compared with the critical t-value for $t(P, \mathrm{df} = n - 3)$. The null hypothesis is valid if the test value $\hat{t}$ does not exceed the critical value $t(P, \mathrm{df} = n - 3)$.
The standard deviation of the regression coefficient s_{a_2} used for these tests is found in most statistical software packages and also in Excel with the function = LINEST(y_values, [x, x^2_values], 1, 1); see Sect. 4.4.

Challenge 5.3.6-1
Check whether the relationship between the x and y values in

(a) Table 5.3.2-1
(b) Table 5.3.2-2

can be better described by a second-degree equation or not, i.e. can the linearity tested with the previous methods be confirmed by using the test of the significance of the quadratic regression coefficient a_2?

Solution to Challenge 5.3.6-1
Table 5.3.6-1 presents the *LINEST-data matrix* obtained by the Excel function using the data set of Table 5.3.2-1, and in Table 5.3.6-2 gives the respective values with the data set of Table 5.3.2-2.

(a) 1. Coefficient interval of a_2
The coefficient interval CI(a_2) calculated by (5.3.6-1) with $t(P = 95\%, \mathrm{df} = 7) = 2.365$ is $CI(a_2) = -0.000421 \pm (2.365 \cdot 0.001188) = -0.000421 \pm 0.002809$. Zero is included in the range of CI(a_2) (from −0.00239 to 0.00323), and therefore the regression
(continued)

Table 5.3.6-1 Regression parameters with their standard deviation for the data set given in Table 5.3.2-1 obtained by the Excel function = LINEST(y_values, [x, x^2_values], 1, 1)

y_i	x_i	x_i^2	Excel output data matrix		
0.1991	0.7866	0.61937	a_2	a_1	a_0
0.2008	0.7866	0.61937	0.000421	0.254068	−0.000820
0.3958	1.5732	2.47433	s_{a_2}	s_{a_1}	s_{a_0}
0.3992	1.5732	2.47433	0.001188	0.005715	0.005899
0.6076	2.3598	5.56960	df	7	
0.6012	2.3598	5.56960			
0.7999	3.1464	9.89732			
0.8016	3.1464	9.89732			
1.0013	3.9330	15.46849			
1.0095	3.9330	15.46849			

Table 5.3.6-2 Regression parameters with their standard deviation for the data set given in Table 5.3.2-2 obtained by the Excel function = LINEST(y_values, [x, x^2_values], 1, 1)

y_i	x_i	x_i^2	Excel output data matrix		
27	0.05	0.0025	a_2	a_1	a_0
49	0.10	0.0100	−374.242	431.045	8.883
68	0.15	0.0225	s_{a_0}	s_{a_1}	s_{a_2}
82	0.20	0.0400	37.858	21.365	2.558
92	0.25	0.0625	df	7	
105	0.30	0.0900			
111	0.35	0.1225			
120	0.40	0.1600			
128	0.45	0.2025			
132	0.50	0.2500			

constant a_2 cannot be statistically distinguished by zero. The null hypothesis is valid; i.e. linearity is confirmed.

2. t-test
   The test value calculated by (5.3.6-2)

$$\hat{t} = \left|\frac{-0.000421}{0.001188}\right| = 0.354$$

   does not exceed the critical t-value $t(P = 95\%, \mathrm{df} = 7) = 2.365$. Thus, the null hypothesis is valid, and linearity of the regression function is confirmed.

(b) The test values are calculated as described above for the values of Table 5.3.6-2. The following results are obtained:

1. Coefficient interval of a_2
   The range of the confidence interval (from −463.76 to −284.72) does not include zero, i.e. the quadratic term is not zero. Consequently non-linearity is demonstrated.
   The check of significance of the quadratic regression coefficient a_2 reveals the same results as obtained by the other test procedures.
2. t-test
   The test value $\hat{t} = 9.885$ exceeds the critical value $t(P = 95\%, \mathrm{df} = 7) = 2.365$ which means that the null hypothesis has to be rejected and the alternative hypothesis $\mathrm{H}_1 : a_2 \neq 0$ is valid.

There are two possibilities if non-linearity is significantly detected:

– Reducing the working range or, if this is not possible
– A quadratic calibration function must be used

Note that significant non-linearity does not imply that the data are correctly fitted by a second-degree model. But there are no general rules for the solution of a non-linear calibration function; each problem requires an individual solution.

5.4 Test for Outliers in the Linear Regression Function

Although testing for outliers in the regression function is not explicitly required by all the regulatory agencies given in Sect. 5.1, the linear regression function used for the determination of analytical results in routine analysis should be checked for outliers in the course of method validation. In general, the regression function is obtained by using chemical reference materials certified for the calibration. Under such nearly "ideal" conditions observations with a large distance from the regression line should not occur. However, if this is the case, the cause must be sought and may be, for example, an single mistake in the automated sampling or a human mistake in the preparation of the calibration standards, or others. Whatever the cause, an observation which is inconsistent with the rest of the data set will affect both the slope and the intercept of the calibration line, resulting in false analytical results. Therefore, the absence of such observations in the calibration data, called *outliers in the linear regression line*, is an indispensable requirement and should be checked early in the method validation procedure. If the test result is positive the causes must be sought and removed and the calibration procedure should be completely repeated. Note that a test for outliers is included in most of the software for method validation. Therefore, we will present two statistical tests for outliers.

An outlier is an observation which lies outside the confidence interval of the linear regression function $y = a_0 + a_1 x$ (see the point x_{OL} in Fig. 5.4-1).

Such a value shows an unusually high or low residual (see residual x in Fig. 5.4-2). However, a statistical test is necessary to decide whether this suspicious value is in fact an outlier, because visual inspection is usually not sufficient.

Several diagnostics have been proposed for the identification of regression outliers, but for linear regression two tests are usually applied [13].

1. The *F-test*

   First, the x and y values detected as suspicious outliers in the residual plot are removed from the data set and the calibration error is again calculated. Then, the test value $\hat{F}$ is estimated by (4.4-1):

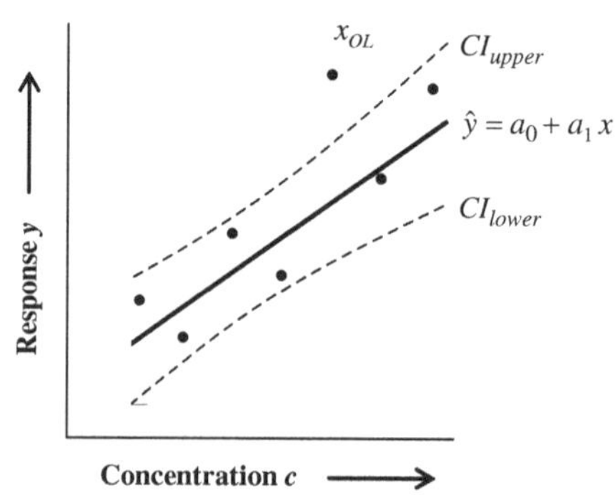

Fig. 5.4-1 Calibration function $\hat{y} = a_0 + a_1 x$ with the lower and upper limits of the two-sided confidence intervals CI_{lower} and CI_{upper}, respectively, as well as the outlier in the regression line x_{OL}

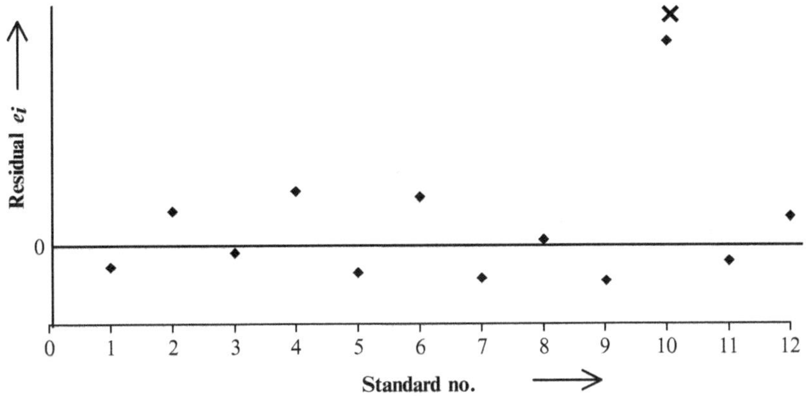

Fig. 5.4-2 Residual plot for the calibration function $\hat{y} = a_0 + a_1 \cdot x$ with an unusually high value of the residual (*multiplication sign*) which is suspected to be an outlier

$$\hat{F} = \frac{s_{y.x}^2 \cdot \mathrm{df} - s_{y.x,\mathrm{OL}}^2 \cdot \mathrm{df}_{\mathrm{OL}}}{s_{y.x,\mathrm{OL}}^2}, \tag{5.4-1}$$

where $s_{y.x}$, $s_{y.x,\mathrm{OL}}$ are the calibration errors calculated using the whole data set and the data set in which the x and y values suspected as an outlier are removed, respectively. The df and $\mathrm{df}_{\mathrm{OL}}$ are the degrees of freedom for these data sets.

Then, the test value $\hat{F}$ is compared with the critical F-value $F(P = 99\%, \mathrm{df}_1 = 1, \mathrm{df}_2 = n - 3)$, where n_{OL} is the number of standards without the removed x_{OL}, y_{OL}-values. If the test value $\hat{F}$ is greater than the critical F-value, the suspicious y-value is in fact an outlier and the calibration has to be repeated. If $\hat{F}$ is smaller than the critical F value, the suspicious y-value is not an outlier at the significance level P and the x- and y- values have to be included in the calibration data set.

2. The *t-test*

After removing the x_{OL}, y_{OL}-value suspected of being an outlier from the data set, the prediction interval $\mathrm{PI}(\hat{y}_{\mathrm{OL}})$ is recalculated according to Eqs. (5.4-2) and (5.4-3) for the concentration x_{OL}

$$\mathrm{PI}(\hat{y}_{\mathrm{OL}}) = \hat{y}_{\mathrm{OL}} \pm t(P, \mathrm{df}_{\mathrm{OL}}) \cdot s_{y.x,\mathrm{OL}} \cdot \sqrt{1 + \frac{1}{n_{\mathrm{OL}}} + \frac{(x_{\mathrm{OL}} - \bar{x}_{\mathrm{OL}})^2}{\mathrm{SS}_{xx,\mathrm{OL}}}} \tag{5.4-2}$$

with

$$\hat{y}_{\mathrm{OL}} = a_{0,\mathrm{OL}} + a_{1,\mathrm{OL}}\, x_{\mathrm{OL}}. \tag{5.4-3}$$

$s_{y.x,\mathrm{OL}}$ is the residual error, $\mathrm{SS}_{xx,\mathrm{OL}}$ is the sum of squares and $t(P,\mathrm{df}_{\mathrm{OL}})$ is the t-value for the degrees of freedom $\mathrm{df}_{\mathrm{OL}}$ at the significance level P. The index $_{\mathrm{OL}}$ means that all parameters are calculated without the x_{OL}, y_{OL}-values.

Finally, if the $\hat{y}_{\mathrm{OL}}$ value lies outside the prediction interval, it must be regarded as an outlier at the significance level $P = 99\%$, but if the $\hat{y}_{\mathrm{OL}}$ value lies inside the limits of the prediction interval, then the x_{OL}, y_{OL}-values must be included in the calibration data set.

Challenge 5.4-1

In the course of the validation of a HPLC method for the determination of an API in tablets in routine analysis, two calibration data sets were compiled which are presented in Tables 5.4-1 and 5.4-2.

(a) Check whether the linearity of the regression function is valid for each calibration set.

(b) Use the *F*-test and the *t*-test to determine whether each calibration data set is free of outliers.

Table 5.4-1 Calibration data set **I** based on the peak areas obtained from the HPLC measurements of the API

Standard	*c* in g L^{-1}	A in counts
1	3.750	7,367
2	5.625	11,652
3	7.500	15,953
4	9.375	19,605
5	11.250	23,937
6	13.125	27,551
7	15.000	31,599
8	16.875	36,005
9	18.750	40,010
10	20.625	45,096

Table 5.4-2 Calibration data set II based on HPLC measurements of the API

Standard	*c* in g L^{-1}	*A* in counts
1	3.750	7,370
2	5.625	11,648
3	7.500	15,980
4	9.375	19,615
5	11.250	23,935
6	13.125	27,448
7	15.000	31,167
8	16.875	35,012
9	18.750	40,088
10	20.625	44,580

Solution to Challenge 5.4-1

Note that the regression parameters are calculated by the Excel function LINEST.

(continued)

Table 5.4-3 Regression parameters for calibration sets I and II obtained by the Excel function LINEST

Calibration data set I			
Linear regression			
a_0 in counts	−828.558	a_1 in counts L g^{-1}	2,191.266
$s_{y.x}$ in counts	374.873	df	8
Quadratic regression			
a_0 in counts	18.867	a_1 in counts L g^{-1}	2,018.458
a_2 in counts L^2 g^{-2}	7.090		
$s_{y.x,2}$ in counts	337.266	df	7
Calibration data set II			
Linear regression			
a_0 in counts	−603.236	a_1 in counts L g^{-1}	2,156.926
$s_{y.x}$ in counts	478.857	df	8
Quadratic regression			
a_0 in counts	208.385	a_1 in counts L g^{-1}	1,991.419
a_2 in counts L^2 g^{-2}	6.790		
$s_{y.x,2}$ in counts	468.059	df	7

(a) *Linearity test*
The regression parameters for calibration data sets I and II are listed in Table 5.4-3. The residuals calculated by (4.2-3) are given in Table 5.4-4 for both calibration sets and are presented as plots in Figs. 5.4-3 and 5.4-4. In both plots the residuals are randomly distributed around zero which means the linearity may be valid for both regression functions. This is confirmed by the Mandel test. The test values calculated by (5.3.4-1) with the data given in Table 5.4-3 are $\hat{F} = 2.884$ and $\hat{F} = 1.373$ for the calibration data sets I and II, respectively. Neither test value exceeds the critical F-value $F(P = 99\%, \text{df}_1 = 1, \text{df}_2 = 7) = 12.246$, which means the linearity of the proposed regression function is valid.

(b) *Outlier F-test*
According to Figs. 5.4-3 and 5.4-4 and the residual data sets in Table 5.4-4, the greatest residual belongs to calibration level 10 and 8 in calibration data set I and II, respectively. After rejection of these x and y values from the data sets the residual standard deviation $s_{y.x,\text{OL}}$ calculated by Excel function $=$ LINEST(y_values, [x, x^2_values], 1, 1) is $s_{y.x,\text{OL}}(\text{I}) = 210.705$ and $s_{y.x,\text{OL}}(\text{II}) = 393.62$ for data set I and II, respectively. The test values $\hat{F}$ calculated by (4.4-1) are $\hat{F}(\text{I}) = 18.323$ and $\hat{F}(\text{II}) = 4.772$ for the calibration data sets I and II, respectively. The test value $\hat{F}(\text{I})$ obtained with calibration data set I is greater than the critical values of the F distribution for the significance level $P = 99\%$ which is $F(P = 99\%,$

(continued)

Table 5.4-4 Residuals e_i for the regression function obtained for calibration data sets I and II

n_i	x_i	y_i	$\hat{y}_i$	$e_i = y_i - \hat{y}_i$
Calibration data set I				
1	3.750	7,367	7,388.7	−21.7
2	5.625	11,652	11,497.3	154.7
3	7.500	15,953	15,605.9	347.1
4	9.375	19,605	19,714.6	−109.6
5	11.250	23,937	23,823.2	113.8
6	13.125	27,551	27,931.8	380.8
7	15.000	31,599	32,040.4	−441.4
8	16.875	36,005	36,149.1	−144.1
9	18.750	40,010	40,257.7	−247.7
10	20.625	45,096	44,366.3	729.7
Calibration data set II				
n_i	x_i	y_i	$\hat{y}_i$	$e_i = y_i - \hat{y}_i$
1	3.750	7,370	7,485.2	−115.2
2	5.625	11,648	11,529.5	118.5
3	7.500	15,980	15,573.7	406.3
4	9.375	19,615	19,617.9	−2.9
5	11.250	23,935	23,662.2	272.8
6	13.125	27,448	27,706.4	−258.4
7	15.000	31,167	31,750.7	−583.7
8	16.875	35,012	35,794.9	*−782.9*
9	18.750	40,088	39,839.1	248.9
10	20.625	44,580	43,883.4	696.6

The greatest absolute value of the residuals is given in italics.

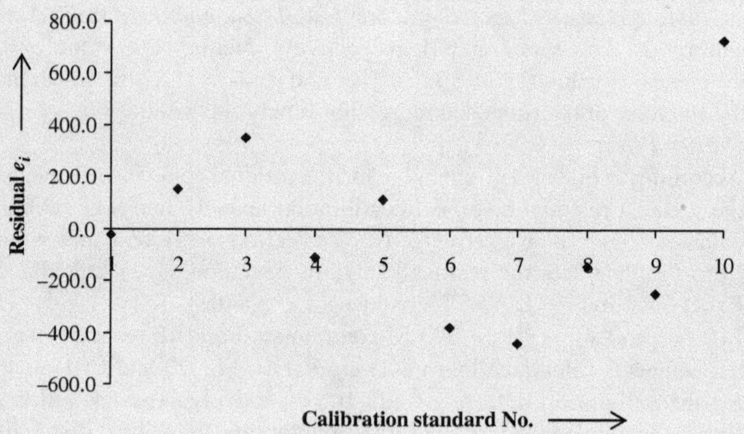

Fig. 5.4-3 Residual plot for calibration data set I

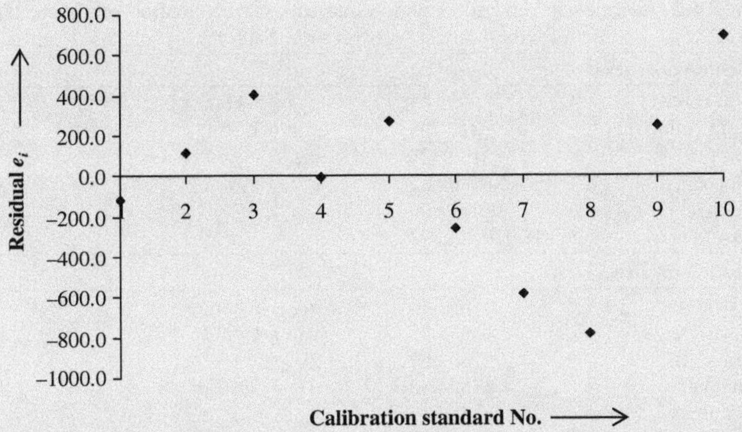

Fig. 5.4-4 Residual plot for calibration data set II

$df_1 = 1, df_2 = 7) = 12.246$, which means that observation $y_{10} = 45,096$ is statistically confirmed as an outlier.

The test value calculated for data set II obtained by a repeated calibration is $\hat{F} = 4.772$ which does not exceed the critical value; this is the same as for calibration I. The observation $y_8 = 35,012$ is not an outlier, and therefore the x and y values must be included in the data set.

Outlier t-test

After removing the x_{OL},y_{OL}-values from the calibration data sets, i.e. level 10 from the data set I and level 8 from the data set II, the prediction interval $PI(\hat{y}_{OL})$ is recalculated according to (5.4-2) and (5.4-3). The intermediate quantities and the results are summarized in Table 5.4-5. The regression parameters are obtained by respective Excel functions.

As the results given in Table 5.4-5 show, the test value $y_{10} = 45,096$ counts in calibration data set I lies outside the limits of the prediction interval. Thus, the measured value $y_{10} = 45,096$ counts must be regarded as an outlier. The whole calibration must be repeated, resulting in calibration data set II given in Table 5.4-2. The measured value of calibration standard 8 $y_8 = 35,012$ counts lies inside the limits of the prediction interval, which means that calibration level 8 is not identified as an outlier and the values of calibration level 8 must be included in the calibration data set. The result of the outlier F-test is confirmed.

The linearity of the regression function of the calibration data set II was confirmed and the data set is free of outliers, and thus calibration data set II is appropriate for further method validation tests.

Table 5.4-5 Intermediate quantities and results for the calculation of the prediction interval $PI(\hat{y}_{OL})$ with the calibration data sets given in Tables 5.4-1 and 5.4-2

Calibration data set I			
$a_{0,OL}$ in counts	−544.8	$a_{1,OL}$ in counts L g^{-1}	2,158.84
$s_{y.x,OL}$ in counts	210.705	$\bar{x}$ in g L^{-1}	11.25
SS_{xx} in g^2 L^{-2}	210.9375	df_{OL}	7
$\hat{y}_{OL}$ in counts	43,981.2	$t(P = 99\%, df)$	3.499
y_{10} in counts	45,096		
$PI(\hat{y}_{OL})$ in g L^{-1}	43,070–44,893		
Calibration data set II			
$a_{0,OL}$ in counts	−693.46	$a_{1,OL}$ in counts L g^{-1}	2,172.28
$s_{y.x,OL}$ in counts	394.75	$\bar{x}$ in g L^{-1}	11.67
SS_{xx} in g^2 L^{-2}	265.662	df_{OL}	7
$\hat{y}_{OL}$ in counts	35,9662	$t(P = 99\%, df)$	3.499
y_{10} in counts	35,012		
$PI(\hat{y}_{OL})$ in g L^{-1}	34,440–37,483		

Challenge 5.4-2

According to the British Standard BS 6748 [14], the content of Cd in ceramics is determined by the flame AAS method after extraction with 4% (v/v) acetic acid. Using a standard solution with the certified content $c_{st} = 500 \pm 0.5$ mg L^{-1}, five calibration solutions are prepared in the following manner: the volumes of the standard solution given in Table 5.4-6 are pipetted into 100 mL volumetric flasks and the flasks are filled up with distilled water. The absorbance A_i of these calibration solutions is then measured in triplicate. The experimental results are given in Table 5.4-6.

(a) Is the acceptance of the linearity of the regression function justified?
(b) Determine whether the calibration set is free of outliers at the significance level $P = 95\%$ using the F- and t- tests.

Note that the uncertainty given for the standard solution was neglected; this is a problem discussed in Chap. 10.

Table 5.4-6 Preparation of the calibration levels and the measured absorbance A_i by the flame AAS method (V_{st} = volume of the standard solution)

Calibration level	1	2	3	4	5
V_{st} in μL	20	60	100	140	180
A_1	0.028	0.084	0.134	0.180	0.215
A_2	0.027	0.083	0.132	0.181	0.231
A_3	0.059	0.081	0.133	0.183	0.216

Solution to Challenge 5.4-2

Preparation of the calibration solutions:

According to the formula for the dilution of solutions given in (4.5-2), the concentration of the calibration solution CL with the concentration of the standard solution $c_1 = c_{st} = 500 \text{ mg L}^{-1}$ and the volumetric flask $V_2 = 100$ mL, the concentration of the calibration levels $c_2 = c_{CL}$ is calculated by (5.4-4):

$$c_{CL} \text{ in mg L}^{-1} = \frac{500 \text{ mg L}^{-1} \cdot V_{st} \,\mu\text{L}}{100 \text{ mL} \cdot 1{,}000}. \quad (5.4\text{-}4)$$

For example, with $V_{st} = 20$ μL the concentration of calibration level 1 is 0.1 mg L^{-1} and so on.

Note that "triplicates" refers to three measurements of the absorbance and not to three determinations. Thus, the mean values of the measured absorbance $\bar{A}$ are used for the calculation of the regression parameters and the degrees of freedom df $= n - 2 = 3$, where n is number of calibration levels.

Choice of an appropriate method for the linearity test:

The Mandel test cannot be applied for checking the linearity because it requires least seven calibration levels; therefore, the check of the quadratic regression coefficient a_2 is applied.

The hypothesis that the quadratic term is zero or not

H_0: $a_2 = 0$

H_1: $a_2 \neq 0$

can be tested by means of the confidence interval for a_2 or by means of the t-test (see Sect. 5.3.6).

The concentrations calculated by (5.4-4), the mean values of the measured absorbance $\bar{A}$, linear and quadratic regression parameters obtained by Excel functions, the residuals $e_i = (y_i - \hat{y}_i)$, and the intermediates and results for checking the linearity are listed in Table 5.4-7.

(a) The residuals shown in Fig. 5.4-5 are statistically distributed around zero, and therefore the linearity of the regression function may be valid.
As the results in Table 5.4-7 show the null hypothesis H_0 is valid, and thus the linearity of the regression function is confirmed.

(b) According to Table 5.4-7 and the residual plots presented in Fig. 5.4-5, the largest value of the residuals is observation number 4, which must be checked as to whether it is an outlier or not.
The intermediate quantities are obtained by the Excel function after removing the x_4, y_4-values.

(continued)

Table 5.4-7 Concentrations of the five calibration levels, mean values of the measured absorbance $\bar{A}$, regression parameters, residuals $e_i = (y_i - \hat{y}_i)$, and intermediates and results for checking the linearity for the flame AAS analysis of Cd

Linear regression parameters					
Level	1	2	3	4	5
c_i in mg L^{-1}	0.1	0.3	0.5	0.7	0.9
$\bar{A} = \bar{y}$	0.0380	0.0827	0.1330	0.1813	0.2207
a_0	0.01513	a_1 in L mg^{-1}	0.2320	$s_{y.x}$	0.00332
$\hat{y}_i$	0.03833	0.08473	0.13113	0.17753	0.22393
e_i	−0.00033	−0.0021	0.0019	0.0038	−0.0033
Linearity check by testing the quadratic regression coefficient a_2					
Quadratic regression parameters					
a_2 in L^2 mg^{-2}	−0.02262	s_{a_2} in L^2 mg^{-2}	0.02192	df	2
Results of the t-test according to (5.3.6-2)					
$\hat{t}$	1.032	$t(P = 95\%, \text{df} = 2)$			4.303
Results of the test of PI(a_2) according to (5.3.6-1)					
PI(a_2)	0.09430	range from	−0.1169 to 0.0717		

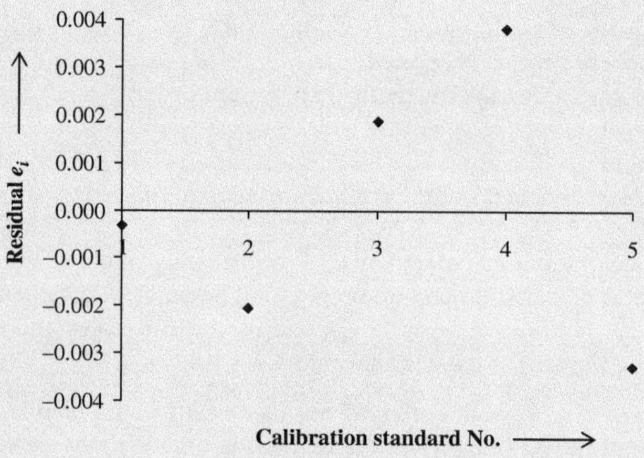

Fig. 5.4-5 Plot of residuals for the calibration of Cd determination by the flame AAS method

(c) Outlier F-test

The test value calculated by (5.4-1) is $\hat{F} = 3.341$, calculated with $s_{y.x,\text{OL}} = 0.002485$, $\text{df}_{\text{OL}} = 2$ and further data given in Table 5.4-7. The test value is smaller than the critical value $F(P = 99\%, \text{df}_1 = 1, \text{df}_2 = 2) = 98.50$, which means that calibration level 4 is not an outlier.

(continued)

(d) Outlier t-test:
After removing the x_{OL}, y_{OL}-values of calibration level 4 the predicted interval is:

$$\mathrm{PI}(\hat{y}_{OL}) = 0.1759 \pm 9.925 \cdot 0.00248 \cdot \sqrt{1 + \frac{1}{4} + \frac{(0.7 - 0.45)^2}{0.35}}$$
$$= 0.1759 \pm 0.2946 \qquad (5.4\text{-}5)$$

The experimental information value $y_4 = 0.1833$ lies inside the limits of the prediction interval 0.1464–0.2054 at the significance level $P = 99\%$. Both test methods give the same result. The values of calibration level 4 must be included in the data set. The calibration parameters are appropriate for further method validation steps.

5.5 Homogeneity of Variances

Like tests for outliers in the calibration line, the test for homogeneity of variances in the calibration line is not a validation parameter required in the regulatory guidelines of ICH (Q2A) or FDA but is a requirement given, for example, in the DIN ISO Guide for water analysis [15]. Therefore, we will consider the test for homogeneity of variances as a validation parameter.

Remember that one of the conditions for calibration is the homogeneity of the observations y_i (see Sect. 4.1). Inhomogeneity of variances does not only diminish the *precision* but it can also influence the *trueness* of the results caused by changing of slope. As Fig 5.5-1 shows, the variances of the information values increase with the concentration. But if the increase in the variance is significant at a chosen probability P, this must be checked by a statistical test.

In order to test the homogeneity of variances recommended by DIN ISO [15], the homogeneity of variances is checked by the variances obtained by ten replicates only at the lower and upper end of the calibration standards x_1 and x_n, respectively. The F-test is carried out after checking both data sets for normal distribution and outliers. The test value $\hat{F}$ is calculated by (5.5-1)

$$\hat{F} = \frac{s_1^2}{s_2^2} \qquad (5.5\text{-}1)$$

with the condition $s_1^2 > s_2^2$, see (3.3-1) in chap. 3.3.

The hypothesis that the variances differ significantly or not $H_0 : s_1^2 = s_2^2$ $H_1 : s_1^2 > s_2^2$ is checked by comparison of the test value $\hat{F}$ with the tabulated *one-sided* F-value for $\mathrm{df}_1 = n_1 - 1$ and $\mathrm{df}_2 = n_2 - 1$ degrees of freedom at the

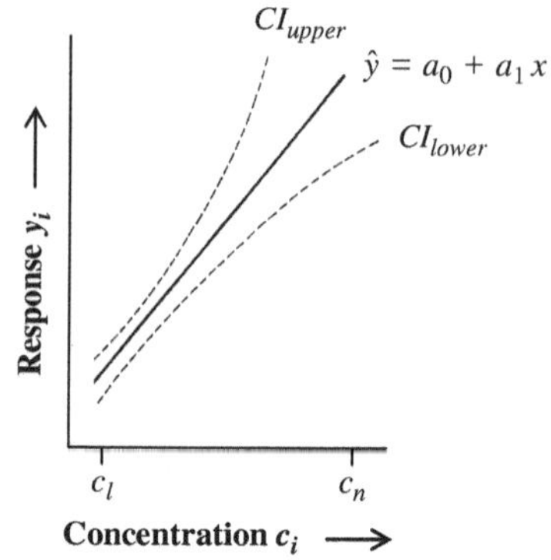

Fig. 5.5-1 Calibration line with its upper and lower confidence intervals

chosen significance level P, for example $P = 99\%$ as recommended in DIN ISO [15]. The indices 1 and 2 refer to data sets 1 and 2, respectively. Note that the denominator s_1^2 of (5.5-1) is not necessarily the variance obtained by the replicates with the lower concentration x_1; it is the larger of the two variances.

If the test value $\hat{F}$ does not exceed the critical F-value, the null hypothesis $s_1^2 = s_2^2$ is valid, which means that the homogeneity of variances, checked at the lower and upper ends of the calibration line, is confirmed, and one assumes that the variances *between* the limit values x_1 and x_n are also homogenous.

The homogeneity of variances is not always essential for analytical purposes, but if the predicted observations are to be used for the evaluation of limit values, i.e. if the confidence interval of the analytical results is necessary, then the homogeneity of variances must be checked.

What one can do if the check confirms inhomogeneity of variances? One possibility may be the shortening of the working range, if this is possible, i.e. if the analytical purpose is still fulfilled. Another possibility is the use of weighted regression which is described in the next Chapter.

Challenge 5.5-1

Control of limit values of Cd in waste water should be carried out by flame AAS (air/C_2H_2, $\lambda = 228.8$ nm). Control of limit values requires not only knowledge of the means of the samples but also of their confidence interval. Therefore, the homogeneity of variances has to be tested in the course of method validation. The assumed threshold value $L_0 = 4.5$ mg L^{-1} Cd and the working range 2, 3, 4, 5, 6, and 7 mg L^{-1} Cd is chosen.

In order to check the homogeneity of variances at the lowest ($x_1 = 2$ mg L^{-1}) and the highest concentration level ($x_n = 7$ mg L^{-1}), ten replicate measurements of each were carried out. The measured values of the absorbance are given in Table 5.5-1.

(continued)

Table 5.5-1 Measured mean values of the absorbance A for the test of homogeneity of variances obtained by flame AAS at the lowest (level 1), the second highest (level 5), and the highest calibration level (level 6), obtained by ten replicates each

Replicate	Level 1	Level 5	Level 6
1	0.2154	0.6152	0.7500
2	0.2165	0.6175	0.7541
3	0.2197	0.6148	0.7593
4	0.2166	0.6145	0.7519
5	0.2158	0.6161	0.7581
6	0.2164	0.6187	0.7525
7	0.2149	0.6137	0.7594
8	0.2177	0.6155	0.7509
9	0.2163	0.6165	0.7610
10	0.2159	0.6109	0.7519

If the check confirms inhomogeneity of variances the working range should be shortened, i.e. the highest concentration standard should be $c = 6\ \mathrm{mg\,L^{-1}}$. In order to check whether the shortening of the working range will result in homogeneity of variances the second highest concentration level $x_{n-1} = 6\ \mathrm{mg\ L^{-1}}$ Cd is also tested, and the results are also summarized in Table 5.5-1.

(a) Check the homogeneity of variance for the whole and shortened working range.
(b) According to the results obtained in part a. shortening of the working range is necessary. This is allowed because the homogeneity of variances are given and the analytical purpose can be fulfilled with the assumed limit value $L_0 = 4.5\ \mathrm{mg\,L^{-1}}$ Cd. In routine analysis two replicates will be carried out.

Table 5.5-2 lists the calibration data for the determination of Cd by flame AAS for the shortened working range.
The following values of the absorbance were measured for a sample: $A_1 = 0.4495$ and $A_2 = 0.4498$.
Check whether the limit value is exceeded or not.

Table 5.5-2 Calibration data for the determination of Cd by flame AAS obtained for the shortened working range

Level	1	2	3	4	5
x_i (c_i) in mg $\mathrm{L^{-1}}$	2	3	4	5	6
y_i (A_i)	0.2168	0.3241	0.4468	0.5422	0.6159

Solution to Challenge 5.5-1

(a) Remember that for the calculation of standard deviations the data set must be normally distributed, it must be free of outliers, and the data set may show no trend. Note there are no hints of a trend, and therefore tests are not required.

The intermediate quantities and the results of the check for normal distribution by the David test and that for outliers by the Dixon test are summarized in Tables 5.5-3 and 5.5-4, respectively.

As Table 5.5-3 shows, the test values $\hat{q}_r$ lie between the critical values, and thus the data sets are normally distributed at the significance level $P = 95\%$.

After ranking the data sets in ascending order for checking the lowest observation or in descending order for checking the highest observation, the test values $\hat{Q}$ are calculated according to (4.2-3) with $n = 10$ observations:

$$\hat{Q} = \left| \frac{x_1^* - x_2}{x_1^* - x_{n-1}} \right|.$$

However, in practice, the values x_2 and x_{n-1} are obtained with the non-ranked data set by the Excel functions = LARGE(data, 2) and = SMALL (data, 2), respectively. The intermediate quantities and results of the Dixon outlier test (see Sect. 3.2.3) are given in Table 5.5-4.

(continued)

Table 5.5-3 Intermediate quantities and results of the David test for normal distribution (see Sect. 3.2.1)

Level	1	5	6
x_{min}	0.2149	0.6109	0.7500
x_{max}	0.2197	0.6187	0.7610
$\hat{q}_r$	3.561	3.640	2.681
$q_{r,lower}(P = 95\%, n = 10)$			2.67
$q_{r,upper}(P = 95\%, n = 10)$			3.685

Table 5.5-4 Intermediate quantities and results of Dixon's outlier test on the calibration levels 1, 5, and 6

Level	Check for x_{min}			Check for x_{max}		
	1	5	6	1	5	6
x_1	0.2149	0.6109	0.7500	0.2197	0.6187	0.7610
x_2	0.2154	0.6137	0.7509	0.2177	0.6175	0.7594
x_{n-1}	0.2177	0.6195	0.7594	0.2154	0.6137	0.7509
$\hat{Q}$ (4.2-3)	0.1786	0.4242	0.0957	0.4651	0.2400	0.1584
$Q(P = 95\%, n = 10)$						0.477

According to the results given in Table 5.5-4, all data sets are free of outliers. Because the data are also normally distributed, the test for homogeneity of variances can be carried out.

Let us start with the test for homogeneity of variances at the whole working range.

The standard deviations obtained for calibration levels 1 and 6 are $s_1 = 0.00135$ and $s_6 = 0.00410$, respectively. The test value is $\hat{F}_{1/6} = 9.261$ calculated by (5.5-1) with $s_1 = 0.00410$ and $s_2 = 0.00135$. The critical value of the one-sided F-distribution is $F(P = 95\%, \text{df}_1 = \text{df}_2 = 9) = 3.179$. The critical value is smaller than the test value which means that the homogeneity of variances in the working range with the six calibration levels is not confirmed. Therefore, a shortening of the working range should be checked as long as the following questions are satisfied:

1. Is the analytical purpose still fulfilled by the shortened working range?
2. Is homogeneity of variances present in the shortened range from 2 to 6 mg L^{-1} Cd?

Although the working range should markedly exceed the limit value, shortening of the range is possible because the limit value $L_0 = 4.5\ \text{mg}\,\text{L}^{-1}$ Cd is still inside the shortened calibration range with the highest calibration standard $x_5 = 6.0\ \text{mg}\,\text{L}^{-1}$ Cd.

The test of homogeneity of variances obtained by the data of levels 1 and 5 given in Table 5.5-1 follows the same procedure as described above.

The test value calculated with the standard deviations $s_1 = 0.00135$ as the denominator s_2 in (5.5-1) and $s_5 = 0.00214$ as the numerator s_1 is $\hat{F}_{1/5} = 2.527$. This value is smaller than the critical F-value which is 3.179 as given above. Thus, the variances are homogeneous in the shortened working range which is confirmed to be valid for the analytical purpose.

(b) According to (4.2-24), the critical value $x_{\text{crit}} = \hat{x} + \text{CI}_{\text{one-sided}}(\hat{x})$ may not exceed the declared limit value L_0. The predicted value $\hat{x}$ is calculated by (4.2-15) and the one-sided confidence interval by (5.5-2):

$$\text{CI}(\hat{x}) = \frac{s_{y.x}}{a_1} \cdot t_{\text{one-sided}}(\bar{P}, \text{df} = n-2) \cdot \sqrt{\frac{1}{n_a} + \frac{1}{n_c} + \frac{(\hat{\bar{y}} - \bar{y})^2}{a_1^2 \cdot \text{SS}_{xx}}}. \quad (5.5\text{-}2)$$

The parameters are obtained by the calibration data given in Table 5.5-2 using the respective Excel functions: intercept $a_0 = 0.02264$, slope $a_1 = 0.10163\ \text{L}\,\text{mg}^{-1}$, residual standard deviation $s_{y.x} = 0.01642$, $t_{\text{one-sided}}$ $(\bar{P} = 95\%, \text{df} = 3) = 2.353$, number of the replicates in routine analysis $n_a = 2$, number of the calibration standards $n_c = 5$, mean value of the measured values $\bar{y} = 0.4292$, the mean value obtained by two replicates $\hat{\bar{y}} = 0.44965$, and sum of squares of the x-values $\text{SS}_{xx} = 10$. The predicted

(continued)

value is $\hat{x} = 4.20$ mg L^{-1} Cd and $\mathrm{CI}_{\text{one-sided}}(\hat{x}) = 0.32$ mg L^{-1} Cd. Thus, the critical value is $x_{\text{crit}} = 4.52$ mg L^{-1}which exceeds the declared limit value $L_0 = 4.5$ mg L^{-1} Cd.

5.6 Weighted Linear Least Squares Regression

If the measured values are not homogeneous, the least squares procedure described in Sect. 4.5.2 cannot be used. Except for the shortening of the working range proposed and verified in Sect. 5.5, the problem of the inhomogeneity of variances (heteroscedasticity) can be solved by a transformation or by a weighted least squares procedure, which is described in this section.

In weighted linear least squares regression, the issue of heteroscedasticity is overcome by introducing weighting factors that are, e. g., inversely proportional to the variance [11]:

$$w_i = \frac{1}{s_{y_i}^2}. \tag{5.6-1}$$

The variances must be obtained experimentally from replicate measurements performed across the whole working range.

The weighted slope $a_{1,w}$ and the weighted intercept $a_{0,w}$ are calculated from [11] by (5.6-2) and (5.6-3), respectively:

$$a_{1,w} = \frac{\sum w_i(x_i - \bar{x}_w) \cdot (y_i - \bar{y}_w)}{\sum w_i(x_i - \bar{x}_w)^2}, \tag{5.6-2}$$

$$a_{0,w} = \bar{y}_w - a_{1,w} \cdot \bar{x}_w. \tag{5.6-3}$$

The weighted mean values are calculated as follows:

$$\bar{x}_w = \frac{\sum w_i \cdot x_i}{\sum w_i}, \tag{5.6-4}$$

$$\bar{y}_w = \frac{\sum w_i \cdot y_i}{\sum w_i}. \tag{5.6-5}$$

Using the weighted regression parameters the predicted $\hat{x}$ value is calculated by (5.6-6):

$$\hat{x} = \frac{\hat{y} - a_{0,w}}{a_{1,w}} \tag{5.6-6}$$

and its confidence interval by (5.6-7)

$$\mathrm{CI}(\hat{x}) = \frac{s_{y.x,w}}{a_{1,w}} \cdot t(P, \mathrm{df}) \cdot \sqrt{\frac{1}{w_s \cdot n_s} + \frac{1}{\sum w_i} + \frac{\left(\hat{\bar{y}}_s - \bar{y}_w\right)^2 \cdot \sum w_i}{a_{1,w}^2 \cdot \left(\sum w_i \cdot \sum w_i \cdot x_i^2 - \left(\sum w_i \cdot x_i\right)^2\right)}}. \quad (5.6\text{-}7)$$

The weighted residual standard deviation is calculated by (5.6-8):

$$s_{y.x,w} = \sqrt{\frac{\sum w_i (y_i - \hat{y}_i)^2}{\mathrm{df}}} \quad (5.6\text{-}8)$$

where

- n_s is the number of calibration standards and replicates, respectively, for the sample with the mean response value $\hat{\bar{y}}_s$
- $s_{y.x,w}$ is the residual standard deviation
- $t(P, \mathrm{df})$ is the t-factor at the statistical significance level P with df degrees of freedom for the calibration levels
- w_s is the weighting factor of the sample calculated according to (5.6-1).

Challenge 5.6-1

The content of polyasparagine acid (PAA) in cooling water in the range 20–90 mg L^{-1} can be determined by fluorimetry as described in [16]. Because the standard deviation increases with the concentration, the large working range means that the variance at the highest calibration standard will be much higher than that at the lowest calibration standard. Because the highest concentration is nine times greater than the lowest, homogeneity of variances cannot be expected, and therefore weighted least squares regression should be applied. In order to estimate the weighting factor, five replicates are measured at each of the eight calibration levels. The results are listed in Table 5.6-1.

In the course of the method validation the following tasks must be done:

(a) Check the linearity of the calibration function.
(b) Inspect the calibration line with its confidence intervals for the presence of outliers.
(c) Confirm the inhomogeneity of variances using an appropriate test.
(d) Determine the parameters of the unweighted and weighted least squares regression and evaluate the results.

(continued)

Table 5.6-1 Calibration data for the determination of PAA by fluorometry

Replicate	1	2	3	4	5
c_i in mg L^{-1}	Fluorescence intensity I in counts				
20	41	42	41	40	40
30	59	57	60	59	61
40	80	78	82	79	83
50	98	100	95	103	97
60	121	126	122	117	120
70	142	137	144	141	146
80	158	152	160	161	154
90	178	172	185	177	180

The excitation wavelength was $\lambda_{ex} = 336$ nm and the emission was measured at the emission wavelength $\lambda_{em} = 411$ nm. The response y_i represents the fluorescence intensity (I) in counts

(e) For two water samples the following response, in arbitrary units, was obtained with three replicates:

Sample 1	44	42.5	44	I in counts
Sample 2	174	176	173	I in counts

Predict the concentration with the correct confidence interval for both samples.
Compare and estimate the results obtained by unweighted and weighted regression analysis.

Solution to Challenge 5.6-1

(a) Tests for linearity are described in Sect. 5.4. Let us test the significance of the quadratic regression coefficient a_2.
The hypothesis that the quadratic term is zero or not
$H_0 : a_2 = 0$
$H_1 : a_2 \neq 0$
can be checked by means of a t-test. The regression of the second degree polynomial and its standard deviation obtained by Excel function LINEST are:

$$a_2 = -0.000929 \text{ counts } L^2\,mg^{-2}$$

$$s_{a_2} = 0.001369 \text{ counts } L^2\,g^{-2}.$$

The test value calculated by (5.3.6-2)

(continued)

$$\hat{t} = \left|\frac{a_2}{s_{a_2}}\right| = \left|\frac{-0.000929}{0.001369}\right| = 0.678 \tag{5.6-9}$$

does not exceed the quantiles of the t-distribution for df $= n - 3$ degrees of freedom at the chosen significance level P, $t(P = 95\%, \text{df} = 5) = 2.571$. Therefore, the null hypothesis H_0: $a_2 = 0$ is valid.

The hypothesis test may also be carried out by means of the confidence interval for the quadratic regression coefficient a_2, which is calculated by (5.3.6-1) with the 95% significance interval:

$$\text{CI}(a_2) = -0.000929 \pm 2.571 \cdot 0.001369 = -0.000929 \pm 0.003520.$$

Zero is included in the range of $\text{CI}(a_2)$, -0.00444 to $+\,0.00259$, and thus the null hypothesis is valid.

(b) As Fig. 5.6-1 shows, all measured mean values y_i lie inside the upper and lower confidence intervals, and thus no outliers are present in the calibration line.

(c) According to the F-test of the variances at the lower and upper working range, homogeneity of variances is not present. The test value using the variances $s_8^2 = 22.30$ and $s_1^2 = 0.7$ obtained by the measured value of the upper and the lower working range, respectively, is

$$\hat{F} = \frac{s_1^2}{s_2^2} = \frac{22.30}{0.7} = 31.857, \tag{5.6-10}$$

which is larger than the critical value $F_{\text{one-sided}}(P = 99\%, \text{df}_1 = \text{df}_2 = 4) = 15.977$. The null hypothesis $H_0 : s_1^2 = s_8^2$ has to be rejected and the alternative hypothesis $H_1 : s_1^2 \neq s_8^2$ is valid. Because of the heteroscedasticity of variances, the weighted least squares procedure must be applied.

(d) The design of the calibration, measuring *all* calibration standards by replicates, enables the calculation of the weighting factors required for the weighted least squares procedure. Table 5.6-2 and its continuation

(continued)

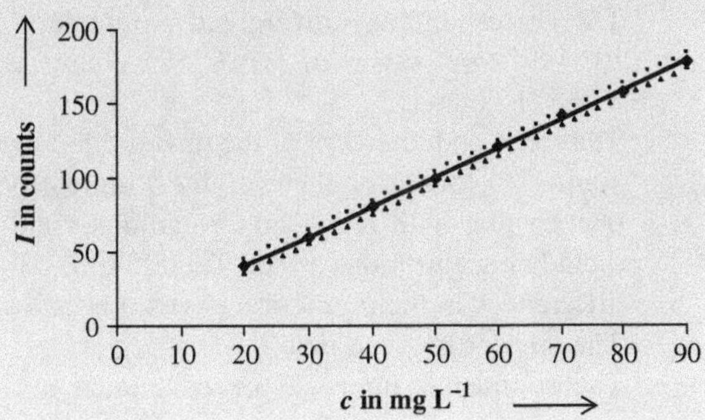

Fig. 5.6-1 Calibration line of the function $\bar{y}_i = f(x_i)$ calculated by the unweighted least squares regression procedure with upper and lower confidence intervals

Table 5.6-2 Data for the computation of the weighted regression line (continuation in Table 5.6-3)

Level	x_i	$\bar{y}_i$	s_i	w_i	$w_i{\cdot}x_i$	$w_i \cdot \bar{y}_i$
1	20	40.8	0.837	1.429	28.571	58.286
2	30	59.2	1.483	0.455	13.636	26.909
3	40	80.4	2.074	0.233	9.302	18.698
4	50	98.6	3.050	0.108	5.376	10.602
5	60	121.2	3.271	0.093	5.607	11.327
6	70	142.0	3.391	0.087	6.087	12.348
7	80	157.0	3.873	0.067	5.333	10.467
8	90	178.4	4.722	0.045	4.036	8.000
Sum				2.515	77.950	156.636

Table 5.6-3 Data for the computation of the weighted regression line (continuation of Table 5.6-2)

Level	$x_i - \bar{x}_w$	$\bar{y}_i - \bar{y}_w$	$w_i(x_i - \bar{x}_w)^2$	$w_i(x_i - \bar{x}_i)(\bar{y}_i - \bar{y}_w)$
1	−10.993	−21.478	172.622	337.277
2	−0.993	−3.078	0.448	1.388
3	9.007	18.122	18.869	37.962
4	19.007	36.322	38.848	74.236
5	29.007	58.922	78.639	159.737
6	39.007	79.722	132.312	270.414
7	49.007	94.722	160.116	309.473
8	59.007	116.122	156.138	307.268
Sum			757.990	1497.757

Table 5.6-3 presents the intermediate quantities for the computation of the weighted regression line according to (5.6-1)–(5.6-3) with $\bar{x}_w = 30.9925$ and $\bar{y}_w = 62.278$.

The weighted regression equation is:

$$\hat{y} = 1.03776 \text{ counts} + 1.97596 \text{ counts L mg}^{-1} \cdot \text{x}. \qquad (5.6\text{-}11)$$

The corresponding unweighted parameters obtained by Excel function LINEST are slope $a_1 = 1.97571$ counts L mg^{-1} and intercept $a_0 = 1.03571$ counts.

The slope and intercept of the unweighted regression parameters are very similar to those for the weighted regression (5.6-11), with the consequence that both regression equations yield similar results for the predicted concentrations $\hat{x}$ (see Table 5.6-5). However, are there significant differences in the prediction errors, i.e. in the confidence interval CI($\hat{x}$)? The answer is given below.

(e) The confidence intervals are calculated by (4.2-17) and (5.6-7) for the unweighted and weighted regression equations, respectively.

(continued)

Table 5.6-4 Intermediate quantities for the calculation of the confidence interval for the weighted regression line

Level	w_i	$\hat{y}_i$	$w_i(y_i - \hat{y}_i)^2$	$w_i \cdot x_i^2$
1	1.42857	40.5569	0.0844	571.4286
2	0.45455	60.3165	0.5666	409.0909
3	0.23256	80.0761	0.0244	372.0930
4	0.10753	99.8357	0.1642	268.8172
5	0.09346	119.5952	0.2407	336.4486
6	0.08696	139.3548	0.6084	426.0870
7	0.06667	159.1144	0.2980	426.6667
8	0.04484	178.8740	0.0101	363.2287
Sum	2.5151		1.9968	3173.8606

Table 5.6-4 gives the intermediate quantities for the calculation of the confidence interval using the data given in Tables 5.6-2 and 5.6-3.
The weighted residual standard deviation calculated by (5.6-8) is $s_{y.x,w} = 0.57689$ counts and the critical t-value is $t(P = 95\%, \mathrm{df} = 6) = 2.447$. With these data, intermediate quantities given in Tables 5.6-2 – 5.6-4 as well as the individual data of both samples
Sample 1:

$\bar{y}_{s,1} = 43.5$	$n_s = 3$	$w_{s,1} = 1.3333$

Sample 2:

$\bar{y}_{s,1} = 174.333$	$n_s = 3$	$w_{s,2} = 0.42857$

the confidence intervals for the weighted regression give the following results:
Sample 1: $\mathrm{CI}(\hat{x}_{1,w}) = 0.626\ \mathrm{mg\,L^{-1}}$
Sample 2: $\mathrm{CI}(\hat{x}_{2,w}) = 1.663\ \mathrm{mg\,L^{-1}}$
The respective unweighted confidence intervals are calculated by (4.2-17) and (4.2-17) using intermediate quantities which are obtained by Excel functions. The results of both regression models are summarized in Table 5.6-5, which shows that the unweighted and weighted regression analyses yield similar predicted concentrations $\hat{x}$ but give very different uncertainties for the predicted results, i.e. confidence intervals $\mathrm{CI}(\hat{x})$. In

(continued)

Table 5.6-5 Comparison of the results obtained by unweighted and weighted least squares regression analysis

Sample	Regression model	$\hat{x}$ in mg L^{-1}	$\mathrm{CI}(\hat{x})$ in mg L^{-1}
1	Unweighted	21.490	1.786
	Weighted(correct data)	21.489	0.626
2	Unweighted	87.703	1.771
	Weighted(correct data)	87.702	1.663

the weighted regression analysis the confidence interval increases with the concentration, and this reflects the heteroscedasticity.

Generalization of the results

The correct confidence interval is important if the uncertainty of the predicted concentration has to be known, for example when monitoring limit values.

Let us assume that the limit of an analytical parameter is 23.0 mg L^{-1}. The limit is exceeded with the unweighted result for sample 1 ($\hat{x} = 21.49\ \text{mg}\,\text{L}^{-1} + 1.79\ \text{mg}\,\text{L}^{-1} - 23.28\ \text{mg}\,\text{L}^{-1}$), but thc truc valuc obtained using weighted regression analysis is smaller than the limit value ($\hat{x} = 21.49\ \text{mg}\,\text{L}^{-1} + 0.63\ \text{mg}\,\text{L}^{-1} = 22.11\ \text{mg}\,\text{L}^{-1}$), and thus the limit is not exceeded.

On the other hand, a limit value of, for example, 90 mg L^{-1} is not exceeded neither with the unweighted ($\hat{x} = 87.70\ \text{mg}\,\text{L}^{-1} + 1.77\ \text{mg}\,\text{L}^{-1} = 89.47\ \text{mg}\,\text{L}^{-1}$), nor the weighted results ($\hat{x} = 87.70\ mg\,\text{L}^{-1} + 1.66\ \text{mg}\,\text{L}^{-1} = 89.36\ \text{mg}\,\text{L}^{-1}$).

5.7 Tests for Trueness

Trueness is a validation parameter which is explicitly mentioned in all the regulatory guidelines. Note that some regulations use the term "accuracy" but this can be misleading because accuracy includes both *trueness* and *precision* [17].

Measurement trueness is defined as the "closeness of agreement between the average of an infinte number of replicate measured quantity values and a reference quantity value" [5]. In pharmaceutical analysis trueness is usually reported as percent recovery by assay, using a proposed analytical procedure.

Before we turn to tests for trueness, we will look at the systematical errors which have an influence on the analytical results.

The analytical procedure can be influenced by

1. Previous steps such as extraction or others,
2. Effects caused by the matrix such as interferences

Both influences result in constant and/or proportional systematic errors.

5.7.1 Systematic Errors in the Least Squares Regression Procedure

Constant systematic errors In constant systematic errors, the deviation is independent of the concentration of the analytical component; therefore, the calibration line is shifted sideways as shown in Fig. 5.7.1-1.

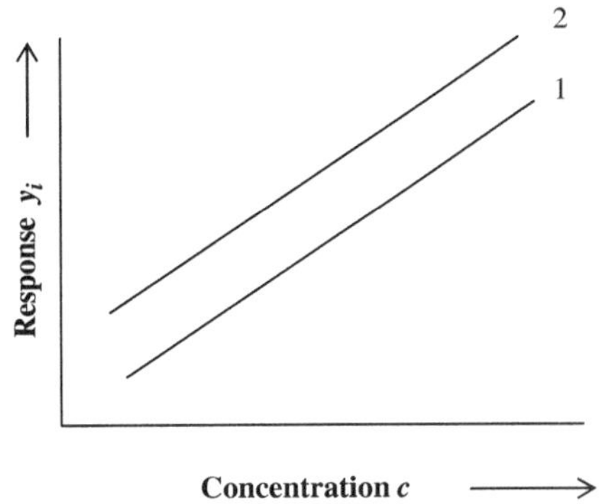

Fig. 5.7.1-1 Representation of a parallel-shifted calibration line (2) versus the calibration line (1), caused by a constant systematic error

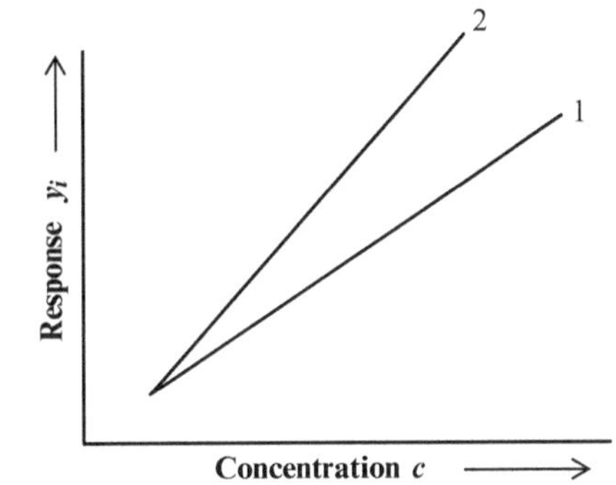

Fig. 5.7.1-2 Representation of a proportional systematic deviation (2) versus the error-free calibration line (1)

The origin of this additive shift could be co-registering of a matrix component because of lack of specificity of the analytical procedure, as described in Sect. 6.2.

Proportional systematic error In proportional systematic errors, the amount of the deviation is a function of the concentration. This leads to a change of the slope, as shown in calibration line 2 in Fig. 5.7.1-2.

This multiplicative deviation can originate in stages of sample preparation, such as extraction or matrix effects.

Of course, both systematic deviations can be present simultaneously.

In contrast to random errors, systematic errors must be avoided or eliminated if their origins become known, because they lead to false analytical results. There are various procedures for checking the presence of systematic errors which will be described in the following sections.

5.7.2 Mean Value t*-Test*

The mean value $\bar{x}$ obtained by n replicates of a sample is compared with the "true" value μ of a certified reference material (CRM) or substance (CRS) by means of a t-test (see Sect. 3.5). The test value calculated by (3.5-5) is compared with the quantiles of the two-sided t-distribution. The null hypothesis $H_0 : \bar{x} = \mu$ is rejected and the alternative hypothesis $H_1 : \bar{x} \neq \mu$ is valid if the calculated t-value exceeds the quantiles of the t-distribution at the chosen statistical significance level P and the degrees of freedom df $= n - 1$. In some regulations, such as pharmaceutical analysis, six replicates are required. Note that in order to calculate the standard deviation of the replicates, the measured values must be normally distributed, free of outliers, and the data in chronological order must show no trend.

Challenge 5.7.2-1

In a laboratory, the trueness of a new HPLC method must be checked by comparison of the measured mean value of a drug with the certified reference substance whose amount is $c = 97.7\%$ (w/w).

The results obtained by six replicates are:

c in % (w/w)	97.3	97.8	97.5	98.0	97.2	97.4

Check if the new HPLC method is valid for determinating the assay of the drug.

Solution to Challenge 5.7.2-1

Inspection of the measured values shows that there is no trend in the data, and thus a statistical test is not necessary.

The check for normal distribution is carried out by the David test (see Sect. 3.2.1).

The test value according to 3.2.1-1 is:

$$\hat{q}_r = \frac{x_{\max} - x_{\min}}{s} = \frac{98.0 - 97.2}{0.3077} = 2.600.$$

The test value lies between the lower limit (2.28) and the upper limit (3.012) of the David table at the significance level $P = 95\%$ and $n = 6$, which means that the data are normally distributed at the chosen significance level.

For checking an outlier with the Dixon test, (5.7.2-1) must be used for $n = 6$:

$$\hat{Q} = \left| \frac{x_1^* - x_2}{x_1^* - x_n} \right|. \qquad (5.7.2\text{-}1)$$

The calculated test values are $\hat{Q}_{x_{\min}} = 0.125$ and $\hat{Q}_{x_{\max}} = 0.250$ for the lowest and the highest measuring values, respectively. None of the test values

(continued)

exceeds the critical values at the significance level $Q(P = 95\%, n = 6) = 0.560$, and therefore no outlier is present in the data set.

The test value calculated by (5.7.2-2) with the mean value $\bar{x} = 97.535\%$ (w/w)

$$\hat{t} = \frac{|97.535 - 97.7|}{0.3077} \cdot \sqrt{6} = 1.327 \qquad (5.7.2\text{-}2)$$

does not exceed the critical value $t(P = 95\%, \mathrm{df} = 5) = 2.571$, and therefore the null hypothesis $H_0 : \bar{x} = \mu$ is valid, which means the new HPLC method may be applied for the determination of the assay of the drug.

5.7.3 Recovery Rate

If the value of the recovery rate Rr% calculated by

$$\mathrm{Rr}\% = \frac{\hat{\bar{x}}}{\mu} \cdot 100, \qquad (5.7.3\text{-}1)$$

in which $\hat{\bar{x}}$ is the observed mean value of a sample and μ is the known true value, is nearly $\mathrm{Rr}\% = 100$, then no systematic errors are present.

The estimation of the recovery rate should be carried out at two or more different concentration levels. A false result can be obtained if the concentration level used is close to a point of intersection of the measured with the hypothetical true calibration line obtained with error-free calibration solutions, as Fig. 5.7.3-1 shows. The predicted response $\hat{y}_2$ gives a correct value x_2 because the error-free calibration line (1) crosses the real but erroneous calibration line (2). Thus, concentration x_2 used for testing the trueness of the regression parameters would give a recovery rate of approximately 100%. But the response values $\hat{y}_1$ and $\hat{y}_3$ obtained with very

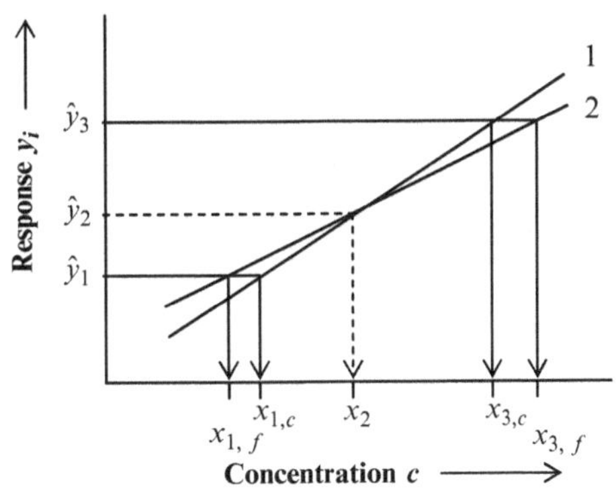

Fig. 5.7.3-1 Influence of three different calibrations on the result of the check for trueness. (1) calibration line obtained by error-free calibration solutions; (2) calibration line of real samples

different concentrations will give false analytical results $x_{1,f}$ and $x_{3,f}$, respectively. Note that $x_{1,c}$ and $x_{3,c}$, respectively, are the correct results.

The question is which value of the recovery rate is the criterion for trueness using this test method?

If no regulatory requirements are given, the accepted range for the recovery rate can be determined. Remember that an analytical result is true if the predicted value $\hat{x}$ is inside its confidence interval $\hat{x} \pm \mathrm{CI}(\hat{x})$. Therefore, the range of recovery rates which will be accepted by the test of trueness which is given by (5.7.3-2) and (5.7.3-3):

$$\mathrm{Rr}_{\min}\% = \frac{(\hat{x} - \mathrm{CI}(\hat{x}))}{\mu} \cdot 100 \quad (5.7.3\text{-}2)$$

$$\mathrm{Rr}_{\max}\% = \frac{(\hat{x} + \mathrm{CI}(\hat{x}))}{\mu} \cdot 100. \quad (5.7.3\text{-}3)$$

The confidence interval $\mathrm{CI}(\hat{x})$

$$\mathrm{CI}(\hat{x}) = s_{x.o} \cdot t(P = 95\%, \mathrm{df}) \quad (4.2\text{-}17)$$

can be calculated by the known validation parameters.

Challenge 5.7.3-1

The validated method for HPLC determination of the assay of an API is carried out in routine analysis with autosampling. The regression coefficients used for the determination of the analytical results are calculated by the software of the HPLC equipment using five calibration standard solutions at positions one to five of the autosampler. The subsequent places are occupied by vials of the samples. However in the pharmaceutical analysis the test for the trueness of the regression coefficients determined at the beginning of the analytical run must be carried out after every five samples in order to check whether the regression coefficients are still suitable. For this check, the recovery rate procedure is an appropriate test. After each fifth sample a validation sample with a known concentration μ is measured by the same analytical procedure and the predicted value $\hat{x}$ is calculated using the regression coefficients of the software. According to (5.7.3-1), the recovery rate is calculated from the predicted value $\hat{x}$ and the known value μ. If the recovery rate does not exceed the limits of the recovery rate determined by the validation parameters, the regression coefficients are valid for calculating true analytical results.

Let us assume that the following parameters were determined for HPLC determination of an API in the course of method validation for the working

(continued)

Table 5.7.3-1 Concentration of the validation solutions μ ordered by their position in the autosampler and the mean response $\bar{y}$ in counts obtained by two replicates

Validation sample VAL	μ in g L^{-1}	$\bar{y}$ in counts
1	4	7,715
2	16	33,746
3	10	21,305
4	5	10,463
5	18	39,034
6	9	18,678
7	12	25,609
8	6	13,456

range 3.750–20.635 g L^{-1}: intercept $a_0 = -725$ counts, slope $a_1 = 2,173$ counts L mg^{-1}, calibration error $s_{y.x} = 523$ counts, and number of calibration standards $n_c = 10$, mean value of the response $\bar{y} = 14804$ counts, and the sum of squares $SS_{xx} = 290\ g^2\ L^{-2}$.

In order to analyze 40 samples, eight validation samples (VAL) are used for testing the trueness of the regression coefficients. The concentrations chosen and the measured responses y are listed in Table 5.7.3-1.

Check whether the regression coefficients determined by the software are valid for the whole run at the significance level $P = 95\%$. Note that the analytical result obtained by the software is considered as true for the five samples which are positioned between the validation samples whose recovery rates lie inside the lower and upper limits.

Solution to Challenge 5.7.3-1

The check for trueness is best realized by the evaluation of the recovery rates.

The required *limits* of the recovery rates for the evaluation of trueness are calculated by (5.7.3-2) and (5.7.3-3). The required confidence interval $CI(\hat{x})$ is calculated by (4.2-17).

With the t-factor $t(P = 95\%, \text{df} = 8) = 2.306$, the confidence interval $CI(\hat{x})$ and, thus, the limits of the recovery rates can be calculated for each predicted value $\hat{x}$.

The results are summarized in Table 5.7.3-2.

The results given in Table 5.7.3-2 show that the recovery rates of the eight validation samples lie inside the limits of the recovery rates, which means that the analytical results of all 40 samples calculated from the regression coefficients of the software are true.

Note that in order to avoid false results because of the choice of an unfavorable validation concentration such as x_2 in Fig. 5.7.3-1, the concentration of the validation samples should vary over the whole working range.

Table 5.7.3-2 Predicted concentrations $\hat{x}$ in g L^{-1}, the lower and upper limits of the recovery rates $Rr_{lower}\%$ and $Rr_{upper}\%$, respectively, the calculated recovery rates of the validation samples (Rr%), and the results of the trueness check

	VAL 1	VAL 2	VAL 3	VAL 4
μ in g L^{-1}	4	16	10	5
$\hat{x}$ in g L^{-1}	3.884	15.863	10.138	5.149
Rr in %	97.1	99.2	101.4	103.0
$\hat{x} - \mathrm{CI}(\hat{x})$ in g L^{-1}	3.292	15.216	9.548	4.563
Rr_{lower} in %	82.3	95.1	95.5	91.3
$\hat{x} + \mathrm{CI}(\hat{x})$ in g L^{-1}	4.476	16.511	10.728	5.734
Rr_{upper} in %	111.9	103.2	107.3	114.7
Result	True	True	True	True
	VAL 5	VAL 6	VAL 7	VAL 8
μ in g L^{-1}	18	9	12	6
$\hat{x}$ in g L^{-1}	18.297	8.929	12.119	6.526
Rr in %	101.7	99.2	101.0	108.8
$\hat{x} - \mathrm{CI}(\hat{x})$ in g L^{-1}	17.611	8.344	11.514	5.944
Rr_{lower}in %	97.8	92.8	96.0	99.1
$\hat{x} + \mathrm{CI}(\hat{x})$ in g L^{-1}	18.970	9.503	12.712	7.097
Rr_{upper} in %	105.5	105.7	106.0	118.5
Result	True	True	True	True

5.7.4 *Recovery Rate of Stocked Samples*

If the influence of the matrix is unknown, the recovery rate can be determined by a sample which is stocked with a known amount of the analyte:

$$Rr\% = \frac{\hat{x}_{\text{total}} - \hat{x}_s}{x_{\text{add}}} \cdot 100. \qquad (5.7.4\text{-}1)$$

Two determinations must be carried out by the same procedure: the amount of the sample $\hat{x}_s$ and then, after addition of the amount x_{add}, the total amount $\hat{x}_{\text{total}}$. The recovery rate is calculated by (5.7.3-5). The limits of the recovery rate can be evaluated as described in Sect. 5.7.3.

Challenge 5.7.4-1

The validation of the determination of Cd by flame AAS in waste water from a measuring station in the range 2–9 mg L^{-1} was verified by ten calibration standard solutions. The *linearity* of the calibration line was checked and tests for *outliers* were negative but in order to apply this method in AQA the check for trueness must still be carried out, i.e. does the matrix of the waste water influence the regression coefficients determined by matrix-free solutions? The test using the recovery function described in the next section cannot be

(continued)

used because the components of the matrix are unknown, but the check using the recovery rate of stocked samples is applicable.

The regression coefficients obtained by matrix-free solutions are intercept $a_0 = -0.00039$ and slope $a_1 = 0.1090\ \mathrm{L\,mg^{-1}}$.

The preparation of the samples was carried out as follows: 90 mL waste water was added to a 100 mL volumetric flask and the flask filled up with distilled water. This sample was used for the determination of the value $\hat{x}_s$ in (5.7.4-1). The mean value of the measured absorbance of the waste water sample obtained by the same procedure as used for the determination of the regression coefficients is $\bar{y}_s = 0.5324$.

90 mL waste water was also added to two other 100 mL volumetric flasks. After the addition of

(a) 2 mL
(b) 5 mL

of a stock solution with $c_{\mathrm{stock}} = 300\ \mathrm{mg\,L^{-1}}$ Cd, the flasks were filled up with distilled water. These samples were used to determine two values $\hat{x}_{\mathrm{total}}$ with different concentrations.

The mean values of the measured absorbance of the stocked samples obtained by the same procedure as used for the determination of the regression coefficients are:

(a) Sample 1: $\bar{y}_{\mathrm{total},a} = 0.5964$ (response of the waste water sample stocked by 2 mL stock solution)
(b) Sample 2: $\bar{y}_{\mathrm{total},b} = 0.7035$ (response of the waste water sample stocked by 5 mL stock solution).

Check whether the matrix influences the regression coefficients. The limit value of the trueness should be given if the recovery rate lies in the range 95.0–105.0%.

Solution to Challenge 5.7.4-1

The predicted concentration of the non-stocked waste water solution calculated according to (4.2-15) is

$$\hat{x}_s = \frac{(0.5321 + 0.00039)}{0.1090\ \mathrm{L\ mg^{-1}}} = 4.89\,\mathrm{mg\,L^{-1}\ Cd}. \qquad (5.7.4\text{-}2)$$

The intermediate quantities and results are presented in Table 5.7.4-1.

As the results in Table 5.7.4-1 show, the recovery rate obtained by both stock solutions is inside the required range of 95.0–105.0%. This means that the predicted values calculated by the regression coefficients which were

(continued)

Table 5.7.4-1 Intermediate quantities and results of the test of trueness using the recovery rate of stocked samples

	Sample 1	Sample 2
V_{added} in mL	2	5
m_{added} in mg L^{-1}	6	15
$\bar{y}_{\mathrm{total}}$	1.1753	2.2459
$\hat{x}_{\mathrm{total}}$ in mg L^{-1}	10.786	20.608
R_r in %	98.3	104.8

obtained by matrix-free calibration solutions are also correct with waste water solutions. The method is validated for the determination of Cd in samples from the measuring station.

5.7.5 *Recovery Function*

The check of trueness by the recovery function covers not only individual points, as with the recovery rate, but also the *whole* calibration function. The application of this procedure requires knowledge of the components of the matrix which possibly have an influence on the regression coefficients. This is always given in pharmaceutical analysis because the placebo of each drug is known and available. Therefore, in the course of pharmaceutical analysis the application of the recovery function for checking trueness is popular.

Apart from the analysis of pharmaceutical products, the matrix is unknown in detail. But if the matrix can be simulated by the components which possibly influence the regression coefficients, then the recovery function can also be applied for such samples. For example, in order to check the trueness of the determination of nitrite-N in waste water containing high levels of iron, all calibration solutions can be spiked by addition of iron in such a concentration that the iron concentration corresponds to that of the waste water.

To apply the check using the recovery function, the calibration function is first determined by matrix-free solutions:

$$\hat{y} = a_{0,c} + a_{1,c} \cdot x. \qquad (5.7.5\text{-}1)$$

All calibration solutions are then spiked by the components of the matrix and are analyzed by the same procedure. Using the measured response obtained by the matrix-spiked calibration solutions y_m and the regression coefficients $a_{0,c}$ and $a_{1,c}$ determined by the matrix-free solutions, the predicted concentrations $\hat{x}_m$ are calculated by (5.7.5-2):

$$\hat{x}_m = \frac{y_m - a_{0,c}}{a_{1,c}}. \quad (5.7.5\text{-}2)$$

The relationship between the predicted concentration obtained by the matrix-spiked solutions $\hat{x}_m$ and the concentrations of the calibration solutions x_c is the so-called recovery function:

$$\hat{x}_m = a_{0,m} + a_{1,m} \cdot x_c. \quad (5.7.5\text{-}3)$$

Ideally, the recovery function should have intercept $a_{0,m} = 0$ and slope $a_{1,m} = 1$. But the matrix might not influence the *precision* of the method, which must be checked.

Because of the same measurement units of the variances, the residual standard deviation $s_{x.y,m}$ in units of concentration should correspond to the standard deviation of the analytical process $s_{x.0,c}$ which has been determined with matrix-free solutions. This last requirement must be checked.

The hypotheses

$$\mathrm{H}_0 : s^2_{y.x,m} = s^2_{x.0,c}$$

$$\mathrm{H}_1 : s^2_{y.x,m} \neq s^2_{x.o,c}$$

are checked by an F-test:

$$\hat{F} = \frac{s^2_{y.x,m}}{s^2_{x.0,c}}. \quad (5.7.5\text{-}4)$$

If the test value $\hat{F}$ does not exceed the critical value $F(P = 99\%, \mathrm{df}_1 = \mathrm{df}_2 = n_c - 2)$, then the null hypothesis H_0 is valid, which means that the matrix does not significantly influence the precision of the analytical procedure and evaluation of the regression coefficients of the recovery function is possible.

The following results are obtained:

A *constant systematic error* is confirmed at the chosen significance level P if the confidence interval of the intercept of the recovery function $\mathrm{CI}(a_{o,m})$calculated by (5.7.5-5)

$$\mathrm{CI}(a_{o,m}) = a_{0,m} \pm t(P, \mathrm{df} = n_c - 2) \cdot s_{a_{0,m}} \quad (5.7.5\text{-}5)$$

does not include zero.

The standard deviation of the intercept $s_{a_{0o,m}}$is calculated by (4.2-11).

A *proportional systematic error* is confirmed at the chosen significance level P if the confidence interval of the slope of the recovery functions $\mathrm{CI}(a_{1,m})$ calculated by (5.7.5-6)

$$\mathrm{CI}(a_{1,m}) = a_{1,m} \pm t(P, \mathrm{df} = n_c - 2) \cdot s_{a_{1,m}} \tag{5.7.5-6}$$

does not include the value 1.

The standard deviation of the slope $s_{a_{1,m}}$ is calculated by (4.2-13).

Note that the required values $s_{a_{0,m}}$ and $s_{a_{1,m}}$ are obtained by the matrix of the Excel function LINEST.

Thus, the check using the recovery function not only reveals the information that a systematic error is present but the test result distinguishes between the different kinds of error. This can be very helpful in searching for the sources of errors.

Note that if the null hypothesis H_0 has to be rejected, then the alternative hypothesis is valid, which means that the matrix has a significant influence on the precision of the analytical procedure. In that case, information as to the presence of a systematic error cannot be obtained, the reason for the worsening of the precision caused by the matrix has to be sought and, after removal of the cause, the test procedure has to be repeated.

Challenge 5.7.5-1

Let us return to the validation of the HPLC method for the determination of the assay of an API begun in Challenge 4.5-1. After checking linearity and outliers in the previous Challenges, the test of trueness must be carried out.

After addition of all placebo components of the drug into the same calibration solutions x_1–x_{10} used for the estimation of the validation parameters and given in Challenge 5.4-1, the HPLC analysis was repeated using the same procedure. The response values $y_{m,i}$ obtained by the matrix-spiked solutions are presented in Table 5.7.5-1.

(continued)

Table 5.7.5-1 Calibration data sets for the test of trueness by the recovery function

Level	c_i in g L^{-1}	Measured response y_i in counts	
		Without placebo	With placebo
1	3.750	7,370	7,655
2	5.625	11,648	12,005
3	7.500	15,980	15,985
4	9.375	19,615	19,665
5	11.250	23,935	23,922
6	13.125	27448	27429
7	15.000	31,167	31,485
8	16.875	35,160	35,056
9	18.750	40,088	39,566
10	20.625	44,575	45,155

(a) Check whether the placebo significantly influences the regression coefficients or, in other words, test the trueness.
(b) If both tests will show that the variances are homogeneous and the placebo does not influence the regression coefficients, the method can be applied in order to determine the assay of drug samples.

Let us assume that the proportion of API in the tablets is 65%. The solution obtained by dissolving ten tablets (2,025 mg) in 100 mL of the HPLC eluent was analyzed according to the procedure used in the method validation. The measured y-values obtained by two replicates were $y_1 = 27,583$ counts and $y_2 = 27,562$ counts.
Calculate the mean value of ten tablets and state the result both $\hat{\bar{x}} \pm \mathrm{CI}(\hat{\bar{x}})$ in $\mathrm{g\,L^{-1}}$. Check whether the confidence interval includes the required amount of API in ten tablets.

Solution to Challenge 5.7.5-1

(a) Using the regression coefficients obtained by Excel function LINEST, $a_{0,c} = -616.315$ counts and $a_{1,c} = 2,159.173$ counts $\mathrm{L\,g^{-1}}$, the predicted concentrations $\hat{x}_m = c_{m,i}$ calculated by (5.7.5-2) are listed in Table 5.7.5-2 and the recovery function is shown in Fig. 5.7.5-1.
The regression coefficients of the recovery function with their standard deviations and the residual standard deviation obtained by Excel function LINEST are summarized in Table 5.7.5-3.
Check for the precision according to (5.7.5-4):
The analytical standard deviation of the calibration function $s_{x.0,c}$ is calculated by (4.2-9) with the residual standard deviation $s_{y.x,c} = 449.186$ counts, and also obtained by the Excel function LINEST giving $s_{x.0} = 0.2080\ \mathrm{g\,L^{-1}}$.
The test value calculated by (5.7.5-7) is

(continued)

Table 5.7.5-2 Predicted concentrations $\hat{x}_m = c_{m,i}$ for the placebo-spiked calibration solutions

Level	c_i in g L^{-1}	$c_{m,i}$ in g L^{-1}
1	3.750	3.831
2	5.625	5.845
3	7.500	7.689
4	9.375	9.393
5	11.250	11.365
6	13.125	12.989
7	15.000	14.867
8	16.875	16.521
9	18.750	18.610
10	20.625	21.199

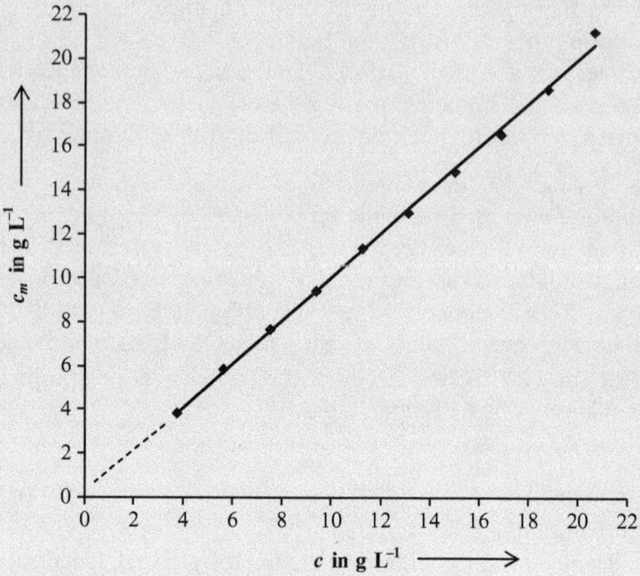

Fig. 5.7.5-1 Recovery function

Table 5.7.5-3 Regression parameters of the recovery function

Intercept $a_{0,m}$ in $g\,L^{-1}$	0.1026
Standard deviation of the intercept $s_{a_{0,m}}$ in $g\,L^{-1}$	0.21216
Slope $a_{1,m}$	0.9951
Standard deviation of the slope $s_{a1,m}$	0.01592
Residual standard deviation $s_{y.x,m}$ in $g\,L^{-1}$	0.2712

$$\hat{F} = \frac{s^2_{x.y,m}}{s^2_{x.0,c}} = \frac{0.2712^2}{0.2080^2} = 1.700 \qquad (5.7.5\text{-}7)$$

The table value $F(P = 99\%, \mathrm{df}_1 = \mathrm{df}_2 = 8) = 6.029$ is greater than the test value $\hat{F}$ which means that the placebo of the drug does not significantly influence the precision of the analytical procedure. Therefore, a check for systematic errors is possible.

Test for a *constant systematic* error:

The range of the confidence interval of the intercept of the recovery function calculated by (5.7.5-5)

$$\mathrm{CI}(a_{0,m}) = 0.1026 \pm 0.21216 \cdot 2.306 \qquad (5.7.5\text{-}8)$$

(continued)

is from −0.3866 to 0.5919. The value zero is included; therefore, the placebo does not cause a constant systematic error.
Test for a *proportional systematic* error:
The confidence interval of the slope of the recovery function calculated by (5.7.5-6)

$$\mathrm{CI}(a_{1,m}) = 0.9951 \pm 0.01592 \cdot 2.306 \tag{5.7.5-9}$$

includes the value 1; therefore, a proportional systematic error cannot be detected.
Thus the placebo does not cause a systematic error at the chosen significance level $P = 95\%$. The analytical results obtained with this method are correct.

(b) The amount of API in 0.1 L eluent is 2.025 g · 0.65 = 1.316 g.
With the regression coefficients $a_{0,c}$ and $a_{1,c}$ given above and the mean value of the measured response $\bar{y} = 27,572.5$ counts, the predicted concentration is $\hat{x} = 13.055\ \mathrm{g\,L^{-1}}$.
The confidence interval calculated by (4.2-17) with the regression coefficients and the analytical error given above as well as $n_a = 2$, $n_c = 10$, $\mathrm{df} = 8$, $\mathrm{SS}_{xx} = 290.04$, and $t(P = 95\%, \mathrm{df} = 8) = 2.306$ is

$$\mathrm{CI}(\hat{x}) = 13.055 \pm 0.3724\ \mathrm{g\,L^{-1}}. \tag{5.7.5-10}$$

Thus, the required amount of API of the ten tablets (1.316 g) is included in the range 1.268–1.343 g.

5.7.6 Standard Addition Procedure

If the matrix is unknown and the check for trueness cannot be made using the recovery function, the test for a proportional systematic error can be verified by the standard addition procedure.

A representative sample is stocked up with the analyte at six or more levels up to twofold concentration of the analyte. The non-stocked and the stocked samples are analyzed by the same procedure and the slope $a_{1,\mathrm{add}}$ of the calibration function is calculated. The proportional systematic error is checked by comparison of the slope $a_{1,c}$ of the calibration function obtained from matrix-free standard solutions and the slope $a_{1,\mathrm{add}}$ of the calibration function obtained by stocked standard solutions. The hypotheses

$$\mathrm{H}_0: a_{1,c} = a_{1,\mathrm{add}}$$
$$\mathrm{H}_1: a_{1,c} \neq a_{1,\mathrm{add}}$$

are checked by the *t*-test.

The test value is calculated by (5.7.6-1)

$$\hat{t} = \frac{|a_{1,c} - a_{1,\mathrm{add}}|}{s_p} \cdot \sqrt{\frac{n_c \cdot n_{\mathrm{add}}}{n_c + n_{\mathrm{add}}}}. \tag{5.7.6-1}$$

The pooled standard deviation s_p is given by (5.7.6-2)

$$s_\mathrm{p} = \sqrt{\frac{(n_c - 2) \cdot s_{a_{1,c}}^2 + (n_{\mathrm{add}} - 2) \cdot s_{a_{1,\mathrm{add}}}^2}{n_c + n_{\mathrm{add}} - 4}}, \tag{5.7.6-2}$$

in which $s_{a_{1,c}}$ and $s_{a_{1,\mathrm{add}}}$ are the standard deviations of the slope $a_{1,c}$ of the calibration function from the matrix-free solutions and the slope $a_{1,\mathrm{add}}$ of the calibration function from the stocked solutions, respectively, obtained by the respective number of calibration standards n_c and n_{add}.

The null hypothesis $\mathrm{H}_0 : a_{1,c} = a_{1,\mathrm{add}}$ is rejected and a systematic error is confirmed if the test value $\hat{t}$ exceeds the two-sided critical t-value at the chosen significance level P and degrees of freedom $\mathrm{df} = n_c + n_{\mathrm{add}} - 4$.

The requirements for this test are:

1. No significant change in precision
2. Linearity of the calibration function

The check for requirement (1) is performed by comparing the calibration errors obtained by the normal calibration $s_{y,x,c}$ and the stocked procedure $s_{y.x,\mathrm{add}}$ with an F-test:

$$\hat{F} = \frac{s_{y.x,\mathrm{add}}^2}{s_{y.x,c}^2}. \tag{5.7.6-3}$$

The matrix does not significantly influence the precision if the $\hat{F}$-value does not exceed the critical value $F(P = 99\%, \mathrm{df}_{\mathrm{add}} = n_{\mathrm{add}} - 2, \mathrm{df}_c = n_c - 2)$.

The linearity of the calibration line (2) is checked by tests described in Sect. 5.3.

Note that a constant systematic error cannot be proved by the standard addition method.

Challenge 5.7.6-1

Let us continue the validation of the flame AAS method for the determination of Cd in waste water which was begun in Challenge 5.5-1. Because the matrix of the waste water samples is unknown, the recovery function by matrix-simulated calibration standards cannot be applied, but the standard addition method can be used in order to check the method for trueness.

The data set obtained for the calibration in distilled water is presented in Table 5.7.6-1.

(continued)

Table 5.7.6-1 Data set for the calibration of the determination of Cd by flame AAS in matrix-free solutions

c in mg L^{-1}	2	3	4	5	6	7
A	0.2168	0.3241	0.4468	0.5422	0.6159	0.7121

Table 5.7.6-2 Preparation of the stocked solution and the mean values of the absorbance $\bar{y}_i$ obtained by two replicates

Added volume (V_{add}) of the stock solution (c_{st} = 25 mg L^{-1}) in mL						
0	0.5	1.0	1.5	2.0	2.5	3.0
Mean value of the absorbance $\bar{y}_i$ obtained by two replicates						
0.3275	0.3658	0.4271	0.4758	0.5249	0.5784	0.6298

(a) For a representative waste water sample the mean measured absorbance $\bar{y} = 0.33585$ was obtained by two replicates.
Calculate the predicted value $\hat{x}$ using the regression coefficients obtained by the calibration parameters.

(b) The stocked solutions which are needed for the application of the standard addition method were prepared as follows:
Seven 25 mL volumetric flasks were each filled with 20 mL waste water. Then, the volumes of a stock solution (c_{st} = 25 mg L^{-1} Cd) given in Table 5.7.6-2 were added, and the flasks were filled with distilled water. The solutions were analyzed by the same procedure as was used for the calibration method.
The mean values of the absorbance $\bar{y}_i$ obtained by two replicates are given in Table 5.7.6-2.
1. Test the linearity of the calibration line.
2. Check whether the matrix influences the precision.
3. Check the trueness of the method.

Solution to Challenge 5.7.6-1

(a) The regression parameters obtained by the Excel function are
$a_{0,c} = 0.0331$, $a_{1,c} = 0.0985\ \mathrm{L\,mg^{-1}}$, $s_{y.x,c} = 0.0161$, df = 4.
The concentration of Cd calculated by (4.2-15) is $\hat{x} = 3.07\ \mathrm{mgL^{-1}}$. Note that this result may be not correct because the check for trueness was not yet carried out, but this value is useful in choosing the required stocked concentrations which should be stocked up to the twofold concentration. Thus, the highest stocked concentration should be $c_n = 3\ \mathrm{mg\,L^{-1}}$.

(continued)

The check for trueness is verified by the standard addition procedure.

(b) The concentrations of the stocked solutions calculated by

$$c_{\text{add}} = \frac{c_{\text{st}} \cdot V_{\text{add}}}{V_{\text{flask}}} = \frac{25\,\text{mg}\,\text{L}^{-1} \cdot V_{\text{add}}\,\text{mL}}{25\ \text{mL}} \qquad (5.7.6\text{-}4)$$

with the added volume V_{add} given in Table 5.7.6-2 are listed in Table 5.7.6-3 together with the mean values of the measured absorbance $\bar{y}_i$ of the i stocked solutions.

The calibration function obtained by the stocked concentrations is shown in Fig. 5.7.6-1. The regression parameters required for the tests are obtained by Excel LINEST functions for the linear and the quadratic regression model, respectively.

The calibration parameters obtained for the linear regression model are $a_{1,\text{add}} = 0.10214\ \text{mg}\,\text{L}^{-1}$, $s_{a_{1,\text{add}}} = 0.0016560\ \text{mg}\,\text{L}^{-1}$, $\text{df}_{\text{add}} = 5$.

(continued)

Table 5.7.6-3 Concentration of the stocked solutions c_{add} in $\text{mg}\,\text{L}^{-1}$ and the mean values of the measured absorbance $\bar{y}_i$

Level	c_{add} in $\text{mg}\,\text{L}^{-1}$	$\bar{y}_i(\bar{A}_i)$
1	0	0.3275
2	0.5	0.3658
3	1.0	0.4271
4	1.5	0.4758
5	2.0	0.5249
6	2.5	0.5784
7	3.0	0.6298

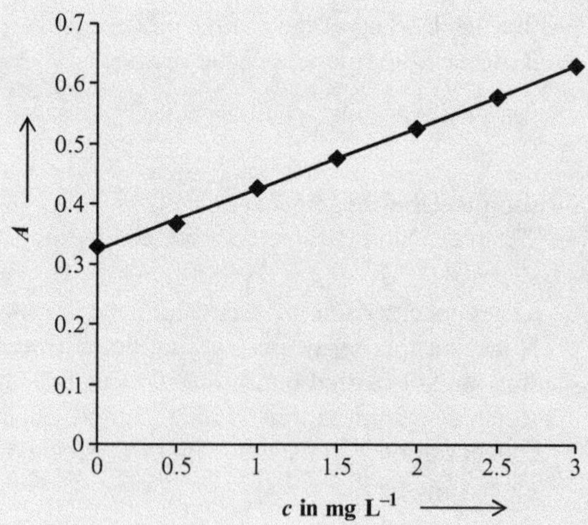

Fig. 5.7.6-1 Calibration function obtained from the stocked solutions

The regression coefficient $a_{2,\mathrm{add}}$ and its confidence interval for the quadratic regression model are

$$a_{2,\mathrm{add}} = 0.00130\ (\mathrm{mg^2\,L^{-2}},$$

$$s_{a_{2,\mathrm{add}}} = 0.00204\ \mathrm{mg^2\,L^{-2}},$$

$$s_{y.x,2,\mathrm{add}} = 0.00467.$$

1. Linearity tests (see Sect. 4.3)
   The test hypotheses
   $\mathrm{H_0}:\ a_{2,\mathrm{add}} = 0$ (the linear regression model is valid)
   $\mathrm{H_1}:\ a_{2,\mathrm{add}} \neq 0$ (the quadratic regression is the better model) can be checked by means of the confidence interval for $a_{2,\mathrm{add}}$ or by means of a t-test, or the linearity can be checked by the Mandel test. All tests provide the same result: the null hypothesis is valid, which means that the linear regression model can be applied.
   The results of the various linearity tests are given below.
   The range of the confidence interval of $a_{2,\mathrm{add}}$ calculated according to (5.3.6-1)

$$\mathrm{CI}(a_{2,\mathrm{add}}) = 0.00130 \pm 0.00204 \cdot 2.776$$

includes the value zero.
   The test value $\hat{t}$ calculated by (5.3.6-2)

$$\hat{t} = \frac{a_{2,\mathrm{add}}}{s_{a_{2,\mathrm{add}}}} = \frac{0.00130}{0.00204} = 0.638$$

does not exceed the critical value $t(P = 95\%, \mathrm{df} = 4) = 2.776$.
   The test value $\hat{F}$ of the Mandel test calculated by (5.3.4-1)

$$\hat{F} = \frac{5 \cdot 0.004381^2 + 4 \cdot 0.004667^2}{0.004667^2} = 0.406$$

does not exceed the critical value $F(P = 95\%, \mathrm{df} = 1, \mathrm{df} = 4) = 7.709$, and therefore the quadratic regression model must be rejected.
2. The hypotheses for the check on the significant influence of the matrix are:

   $\mathrm{H_0}:\ s_{y.x,\mathrm{add}} = s_{y.x,c}$
   $\mathrm{H_1}:\ s_{y.x,\mathrm{add}} > s_{y.x,c}.$

   The test value $\hat{F}$ is calculated by (5.7.6-3).

(continued)

The null hypothesis H_0 is rejected if the test value $\hat{F}$ does not exceed the one-sided critical F-value $F(P = 99\%, \mathrm{df}_{\mathrm{add}} = n_{\mathrm{add}} - 2, \mathrm{df}_{\mathrm{c}} = n_{\mathrm{c}} - 2)$. The values of the residual standard deviation are $s_{y.x,c} = 0.01611$ and $s_{y.c,\mathrm{add}} = 0.00438$. Because in the example given $s_{y.x,\mathrm{add}}$ is even smaller than $s_{y.x,c}$ the null hypothesis is valid without the F-test.

3. Trueness test
   The test for trueness is made by comparing the slopes $a_{1,c}$ and $a_{1,\mathrm{add}}$ according to (5.7.6-1) and (5.7.6-2):

$$s_{\mathrm{p}} = \sqrt{\frac{(4 \cdot 0.003851)^2 + (5 \cdot 0.001656)^2}{6 + 7 - 4}} = 0.002849$$

$$\hat{t} = \frac{|0.09849 - 0.10214|}{0.002849} \cdot \sqrt{\frac{6 \cdot 7}{6 + 7}} = 2.298.$$

The test value $\hat{t}$ does not exceed the critical value $t(P = 99\%, \mathrm{df} = 9) = 3.25$.

Thus, a proportional systematic error could not be confirmed at the chosen significance level $P = 95\%$.

Note that information on a constant systematic error cannot be obtained by this method.

5.7.7 Test by Method Comparison

Let us assume that a validated method has to be replaced by another method, for example, the photometric determination of Cd by flame AAS, or that the validated method must be changed, for example, for another matrix or for a larger working range; in all these cases a new validation or a revalidation must be carried out.

In order to avoid this extensive validation procedure, comparison of the results obtained by the new method and by the validated method is very useful. If no differences between results obtained from the same real samples are detected, both methods are equivalent and the old method can be substituted by the new method for routine analysis. However, the study of relationships between two variables that are measured quantities obtained by two methods requires regression methods that take the error in both variables into account.

Up to now it has been assumed that only the response variable y is subject to error and that the variable $x = c$ is known without error, but on comparison of paired results obtained by different methods the assumption that x is error-free is not justified. If both variables are subject to error, *model II regression techniques* must be applied. The regression analysis according to the *model I regression technique* assumes that the independent variable is not subject to error. A special application is the *calibration* which is described in Sect. 4.2.

Remember, in the ordinary least squares (OLS) regression method justified by the requirement that the error in the x-values is neglected, the regression coefficients are estimated by minimizing $\sum e_i^2$, which is the sum of squares of the distances between the data points and the regression line *parallel* to the y-axis (see e_i in Fig. 5.7.7-1).

However, if both variables are affected by random errors an unbiased estimation of the regression coefficients is obtained by minimizing $\sum d_i^2$, which is the sum of squares of the *perpendicular* distances from the data points to the regression line (see d_i in Fig. 5.7.7-1). This method is called *orthogonal distance regression* (ODR).

Although one can find various procedures using ODR, in all estimations the covariances have to be considered. However, in the course of AQA the procedures given in DIN 38402-71 [18] must be applied. There are some general requirements in the application of this directive:

- There must be no significant difference between the precision of both analytical procedures in matrix-free solutions, confirmed by an F-test.
- The data sets have to be free of outliers, confirmed by the Grubbs test, independent of the size of the data set.
- One or more replicates are allowed, but the procedure has to be equivalent in both methods.
- Agreement of the working ranges of both methods is not required because the tests are carried out only by analytical results including the preparation of the samples.

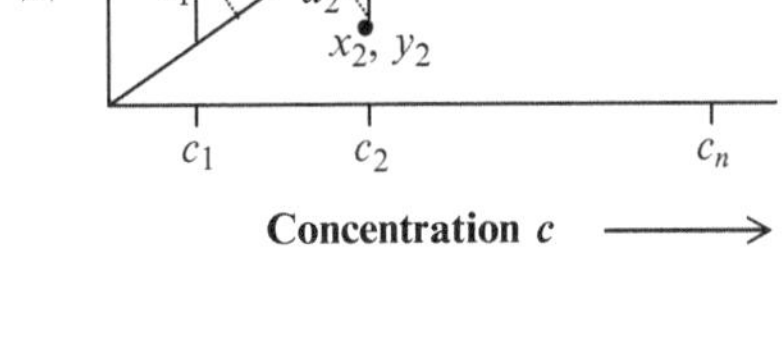

Fig. 5.7.7-1 The residuals e_i and d_i at the point x_i, y_i in the least squares (LS) regression method and in the orthogonal distance regression (ODR) method

DIN [18] emphasizes the evidence of the equivalence between analytical results obtained by samples with the *same* matrix and samples with *different* matrices. The individual steps of the procedure will be given in both cases. In the following, the index "*R*" is used for the data which were obtained by the reference procedure and "*C*" refers to data of the comparative or alternative method.

(a) *Check for equivalence of results obtained by one matrix*
The requirement of this test method is the statistical equivalence of the precision of both methods in matrix-free samples, confirmed by an F-test. The proof of equivalence of both methods is verified by a t-test of the analytical results obtained by real samples with at least six replicates under repeatability conditions.
The test parameters are calculated by (5.7.7-1)–(5.7.7-3).
1. Comparison of the precision in matrix-free solutions by an F-test
If the relative standard deviation of the comparative method $s_{\mathrm{r,C}}$ is greater than that of the reference method $s_{\mathrm{r,R}}$, then the significance is checked by the F-test. If the test value calculated by

$$\hat{F} = \frac{s_{\mathrm{r,C}}^2}{s_{\mathrm{r,R}}^2} \qquad (5.7.7\text{-}1)$$

does not exceed the critical value $F(P = 99\%, \mathrm{df_C} = \mathrm{df_R} = n - 2)$, then the precision of both analytical procedures is equal at the significance level $P = 99\%$.
2. Check on the equivalence by means of real samples
An aliquot of a representative sample is analyzed by the comparative and the reference method using at least six replicates. Each data set is checked for outliers using the Grubbs test at the significance level $P = 95\%$ (see Sect. 3.2.3), whereby each data set may not have more than one outlier which is to be rejected. The equivalence of the precision is checked by an F-test according to (5.7.7-1) at the significance level P = 99%. Note that the variance of the comparative method is permitted to be smaller than that of the reference method. After testing homogeneity of variances, the mean values $\bar{x}_{\mathrm{C}}$ and $\bar{x}_{\mathrm{R}}$ obtained by the comparative and the reference method, respectively, are checked by the mean value t-test according to (5.7.7-2):

$$\hat{t} = \frac{|\bar{x}_{\mathrm{R}} - \bar{x}_{\mathrm{C}}|}{s_{\mathrm{p}}} \sqrt{\frac{n_{\mathrm{R}} \cdot n_{\mathrm{C}}}{n_{\mathrm{R}} + n_{\mathrm{C}}}}. \qquad (5.7.7\text{-}2)$$

The pooled standard deviation is calculated by (5.7.7-3):

$$s_{\mathrm{p}} = \sqrt{\frac{(n_{\mathrm{R}} - 1) \cdot s_{\mathrm{R}}^2 + (n_{\mathrm{C}} - 1) \cdot s_{\mathrm{C}}^2}{n_{\mathrm{R}} + n_{\mathrm{C}} - 2}}. \qquad (5.7.7\text{-}3)$$

Note that if the variance of the comparative method is smaller than that of the reference method, then, instead of s_{p} in (5.7.7-3), the larger standard deviation of the reference method is used.

If the test value $\hat{t}$ is smaller than the critical value $t(P = 99\%, \mathrm{df} = n_{\mathrm{R}} + n_{\mathrm{C}} - 2)$, the analytical results are assumed to be equal at the significance level $P = 99\%$.

For a wide working range, the mean value t-test must be carried out at both ends and in the middle of the working range, at the very least.

(b) *Check for equivalence of results obtained by various matrices*

After testing the equivalence of the relative standard deviations of both analytical methods with matrix-free solutions as described above, the check for equivalence of analytical results obtained by both methods must be carried out, either by orthogonal regression or by the difference method.

1. Using orthogonal regression

   The requirements are as follows:

   - At least $n = 30$ real samples of various matrices and various concentrations are analyzed by the reference and the comparative or alternative method.
   - For the highest analytical result obtained by the reference method, is (5.7.7-4) valid if

$$5 \cdot \bar{x}_{\min} < \bar{x}_{\max} < 100 \cdot \bar{x}_{\min}. \qquad (5.7.7\text{-}4)$$

   If is $\bar{x}_{\max} < 5 \cdot \bar{x}_{\min}$, then the difference method must be applied, and if $\bar{x}_{\max} > 100 \cdot \bar{x}_{\min}$ the working range must be split.

   - One or more replications may be done but the number of replicates must be same in both methods.
   - The quotient Q_{mc} calculated by each pair of values

$$Q_{\mathrm{mc}} = \frac{x_{\mathrm{C},i}}{x_{\mathrm{R},i}} \qquad (5.7.7\text{-}5)$$

   is tested by outliers using the Grubbs test (see Sect. 3.2.3)

$$\hat{r}_m = \frac{\left|Q_{\mathrm{mc}}^* - \bar{Q}_{\mathrm{mc}}\right|}{s_{Q_{\mathrm{mc}}}} \qquad (5.7.7\text{-}6)$$

with

$$\bar{Q}_{\mathrm{mc}} = \frac{\sum Q_{\mathrm{mc},i}}{n} \tag{5.7.7-7}$$

and

$$s_{Q_{\mathrm{mc}}} = \frac{\sqrt{\sum (Q_{\mathrm{mc},i} - \bar{Q}_{\mathrm{mc}})^2}}{n-1}. \tag{5.7.7 8}$$

The test value $\hat{r}_m$ is compared with the critical value $r_m(P = 95\%, n)$. If this value is smaller than the test value $\hat{r}_m$ the suspect pair of values has to be rejected, but more than one outlier in each data set is not allowed. The regression coefficients are calculated by the outlier-free data set of the pairs of values using the orthogonal regression method.
Slope:

$$a_1 = \frac{s_{\mathrm{C}}}{s_{\mathrm{R}}} \tag{5.7.7-9}$$

with the standard deviation of the data obtained by the comparative method

$$s_{\mathrm{C}} = \sqrt{\frac{\sum (x_{\mathrm{C},i} - \bar{x}_{\mathrm{C}})^2}{n-1}} \tag{5.7.7-10}$$

and the standard deviation of the data obtained by the reference method

$$s_{\mathrm{R}} = \sqrt{\frac{\sum (x_{\mathrm{R},i} - \bar{x}_{\mathrm{R}})^2}{n-1}}. \tag{5.7.7-11}$$

Intercept:

$$a_0 = \bar{x}_{\mathrm{C}} - a_1 \cdot \bar{x}_{\mathrm{R}}. \tag{5.7.7-12}$$

Check for *proportional systematic* error:
The check is carried out by a χ^2-test.
The test value is calculated by

$$\hat{\chi}^2 = n \cdot \ln\left(\frac{s^4 - s_{RC}^4}{s_R^2 \cdot s_C^2 - s_{RC}^4}\right) \tag{5.7.7-13}$$

with the geometric mean of the variances of the data obtained by the reference and the comparative methods

$$s = \sqrt{\frac{1}{2} \cdot \left(s_R^2 + s_C^2\right)} \tag{5.7.7-14}$$

and the covariances

$$s_{RC} = \sqrt{\frac{\sum (x_{R,i} - \bar{x}_R) \cdot (x_{C,i} - \bar{x}_C)}{n-1}}, \tag{5.7.7-15}$$

which can also be obtained by the Excel function =COVAR(Matrix 1, Matrix 2). Note that there is a small difference between the values of the covariance obtained by the Excel function and (5.7.7-15) because the degrees of freedom (denomination) is n instead of $n - 1$.
The data follow a chi-square distribution with $n > 20$. Because the critical value is $\chi^2(P = 95\%, \mathrm{df} = 1) = 3.8$, a proportional systematic error is confirmed at the significance level $P = 95\%$ if the calculated value $\hat{\chi}^2$ is greater than 3.8.
Check for *constant systematic* error
A constant systematical error shifts the regression line parallel to the angle bisector. This shift $\bar{D}$ corresponds to the difference of the means. The bias is:

$$\bar{D} = \bar{x}_C - \bar{x}_R. \tag{5.7.7-16}$$

The presence of a constant systematical error is checked by a t-test

$$\hat{t} = \frac{|\bar{x}_C - \bar{x}_R|}{s_D} \cdot \sqrt{n} \tag{5.7.7-17}$$

with the standard deviation

$$s_D = \sqrt{\frac{\sum (D_i - \bar{D})^2}{n-1}}. \tag{5.7.7-18}$$

If the test value $\hat{t}$ is greater than the critical value $t(P = 99\%, \mathrm{df} = n - 1)$ a constant significant error is confirmed at the significance level $P = 99\%$.

2. Using the difference method
   The following conditions must be fulfilled:

- There must be $n \geq 30$ pairs of values.
- One or more replicates are necessary but the number of replicates must be the same for both methods. For each pair of values the difference D_i is calculated:

$$D_i = \bar{x}_{\mathrm{R},i} - \bar{x}_{\mathrm{C},i}. \qquad (5.7.7\text{-}19)$$

- For the highest analytical result $\bar{x}_{\max}$ is valid:

$$\bar{x}_{\max} < 5 \cdot \bar{x}_{\min} \qquad (5.7.7\text{-}20)$$

- Only one pair of values may be rejected as an outlier confirmed by the Grubbs test:
  the test value $\hat{r}_m$ is calculated by

$$\hat{r}_m = \frac{|D^* - \bar{D}|}{s_D} \qquad (5.7.7\text{-}21)$$

  with

$$\bar{D} = \frac{\sum D_i}{n} \qquad (5.7.7\text{-}22)$$

  and

$$s_D = \sqrt{\frac{\sum (D_i - \bar{D})^2}{n-1}}. \qquad (5.7.7\text{-}23)$$

  The test value $\hat{r}_m$ is compared with the critical value $r_m(P = 95\%, n)$. The suspect difference D^*, which is either the highest or the lowest value, is identified as an outlier if the test value is greater than the critical value. This pair of values must be rejected from the data set. Note that only one pair of values can be rejected as an outlier.
  The check for the equivalence of the analytical results is performed by a t-test. The calculated test value $\hat{t}$

$$\hat{t} = \frac{|\bar{D}|}{s_D} \cdot \sqrt{n} \qquad (5.7.7\text{-}24)$$

is compared with the critical value $t(P = 99\%, \mathrm{df} = n - 1)$. If the test value $\hat{t}$ exceeds the critical value, then the results are not equal and a systematic error is present.

Challenge 5.7.7-1

The validated photometric determination of nitrite-N in surface water according to EU-Norm [4] should be replaced by ion chromatography (IC) [19]. Check whether both methods are equivalent by means of the analytical results obtained by the reference method "photometry" and the comparative method "IC" given in Tables 5.7.7-1 and 5.7.7-2.

The analytical error of the reference method determined in the course of validation in the working range 0.025–0.25 mg L^{-1} is $s_{x.0} = 0.001834\ \mathrm{mg\,L^{-1}N}$ with mean value $\bar{x} = 0.15\ \mathrm{mg\,L^{-1}}$ and df = 8 degrees of freedom.

(continued)

Table 5.7.7-1 Analytical results of 32 samples of surface water obtained by the reference method "photometry" and the measured peak areas A (y_i) in counts using the alternative method "IC"

Sample no.	Photometry c_i in mg L^{-1} N	IC A in counts
1	0.206	69,977
2	0.053	16,954
3	0.119	37,478
4	0.151	55,832
5	0.221	80,814
6	0.076	23,202
7	0.176	62,969
8	0.232	82,343
9	0.214	79,284
10	0.046	15,555
11	0.106	32,889
12	0.185	69,597
13	0.123	41,047
14	0.134	45,125
15	0.241	87,951
16	0.140	52,263
17	0.063	21,673
18	0.216	78,264
19	0.157	71,637
20	0.081	23,202
21	0.059	16,065
22	0.193	72,656
23	0.203	71,127
24	0.102	36,458
25	0.070	21,673
26	0.157	59,401
27	0.095	34,929
28	0.090	33,909
29	0.238	83,363
30	0.140	51,243
31	0.064	19,124
32	0.206	77,755

Table 5.7.7-2 Calibration data of the ion chromatographic determination of nitrite-N

Level	c_i in mg L^{-1}	y_i in counts
1	0.04	14,506
2	0.08	27,969
3	0.12	45,657
4	0.16	58,938
5	0.20	74,886
6	0.24	91,034
7	0.28	105,473

Calculate the concentration of nitrite-N using the regression parameters of the alternative method calculated from the calibration data given in Table 5.7.7-2.

Solution to Challenge 5.7.7-1

The regression parameters of the alternative method obtained from the data given in Table 5.7.7-2 using Excel function LINEST are:

intercept $a_{0,\mathrm{C}} = -1,399.57$ counts,

slope $a_{1,\mathrm{C}} = 382,375$ counts $\mathrm{L\,mg^{-1}}$,

residual standard deviation $s_{y.x,\mathrm{C}} = 944.48$ counts.

The analytical error calculated by (4.5-9) is $s_{x.0,\mathrm{C}} = 0.00247\ \mathrm{mg\,L^{-1}}$. The relative standard deviation of the alternative method is $s_{\mathrm{r,C}}\% = 1.54$ obtained with $\bar{x}_\mathrm{C} = 0.16\ \mathrm{mg\,L^{-1}}$, and that of the reference method is $s_{\mathrm{r,R}}\% = 1.22$ calculated from the data given above.

1. Comparison of the precision of both methods in matrix-free solutions

The test value $\hat{F}$ calculated by (3.3-1) is

$$\hat{F} = \frac{0.0154^2}{0.0122^2} = 1.594. \qquad (5.7.7\text{-}25)$$

The critical value $F(P = 99\%, \mathrm{df}_1 = 5, \mathrm{df}_2 = 8) = 6.632$ is greater than the test value $\hat{F}$, and therefore the precision in matrix-free solutions is comparable.

2. Calculation of the analytical results obtained by the alternative method

The predicted values $\hat{x}_{i,\mathrm{C}}$ of the 32 samples determined by the alternative method are calculated from the regression coefficients given above. The results are presented in Table 5.7.7-3.

(continued)

Table 5.7.7-3 Analytical results of surface water samples in mg L^{-1} obtained by the reference and the alternative method and their quotients $Q_{\mathrm{mc},i}$ calculated by (5.7.7-5). The values of sample number 19 (in italics) confirmed as an outlier are rejected from the data for further calculations

Sample no.	Photometry	IC	$Q_{\mathrm{mc},i}$
1	0.206	0.187	0.9070
2	0.053	0.048	0.9022
3	0.119	0.102	0.8544
4	0.151	0.150	0.9899
5	0.221	0.215	0.9720
6	0.076	0.064	0.8510
7	0.176	0.168	0.9543
8	0.232	0.219	0.9424
9	0.214	0.211	0.9851
10	0.046	0.044	0.9597
11	0.106	0.090	0.8428
12	0.185	0.186	1.0047
13	0.123	0.111	0.9010
14	0.134	0.122	0.9053
15	0.241	0.234	0.9704
16	0.140	0.140	1.0024
17	0.063	0.060	0.9578
18	0.216	0.208	0.9663
19	0.157	0.191	*1.2182*
20	0.081	0.064	0.7924
21	0.059	0.046	0.7768
22	0.193	0.194	1.0024
23	0.203	0.190	0.9344
24	0.102	0.099	0.9688
25	0.070	0.060	0.8620
26	0.157	0.159	1.0141
27	0.095	0.095	0.9980
28	0.090	0.092	1.0306
29	0.238	0.222	0.9314
30	0.140	0.138	0.9834
31	0.064	0.054	0.8334
32	0.206	0.207	1.0059

3. Check for outliers in the data set of pairs of values

The quotients $Q_{\mathrm{mc},i}$ calculated by (5.7.7-5) are listed in Table 5.7.7-3. The smallest quotient $Q_{\mathrm{mc,min}} = 0.7768$ and the greatest $Q_{\mathrm{mc,max}} = 1.2182$ are checked as outliers by the Grubbs test. The test values calculated by (5.7.7-6) with $\bar{Q}_{\mathrm{mc}} = 0.9444$ and $s_{Q_{\mathrm{mc}}} = 0.08364$ are $\hat{r}_{m,Q_{\mathrm{mc,min}}} = 2.004$ and $\hat{r}_{m,Q_{\mathrm{mc,max}}} = 3.273$, respectively. Thus, the greatest quotient exceeds the critical value $r_m(P = 95\%, n = 32) = 2.773$. The pair of values number 19 is confirmed as an outlier and it must be rejected from the data set. Further outliers are not found, and therefore the test for equivalence can be carried out. The number of samples is thus $n = 31$.

(continued)

Because $x_{\max,R} = 0.241 > 5 \cdot x_{\min,R} = 5 \cdot 0.046 = 0.23$, orthogonal regression must be used for testing the results regarding equivalence.

4. Calculation of the regression coefficients

The orthogonal regression coefficients are calculated by (5.7.7-9)–(5.7.7-12) using the outlier-free data sets of Table 5.7.7-3. The parameters are:
slope $a_1 = 1.00281$, intercept $a_0 = 0.000378\,\text{mg L}^{-1}$ calculated with the standard deviations $s_R = 0.0633\ \text{mg L}^{-1}$, $s_C = 0.0635\ \text{mg L}^{-1}$, $\bar{x}_C = 0.135\ \text{mg L}^{-1}$, and $\bar{x}_R = 0.142\ \text{mg L}^{-1}$.
The function $c_{\text{comparative}} = f(c_{\text{reference}})$ is shown in Fig. 5.7.7-2.

5. Check for *proportional* systematic error

The intermediate quantities for the calculation of the test value $\hat{\chi}^2$ according to (5.7.7-13) are summarized in Table 5.7.7-4. A proportional systematic error is not detected because the test value $\hat{\chi}^2$ is much smaller than the critical value 3.8. Note that this result should be expected because the slope $a_1 = 1.00281$ is close to 1.0.

6. Check for *constant* systematic error

The intermediate quantities for the calculation of the test value $\hat{t}$ according to (5.7.7-17) are also given in Table 5.7.7-4.

(continued)

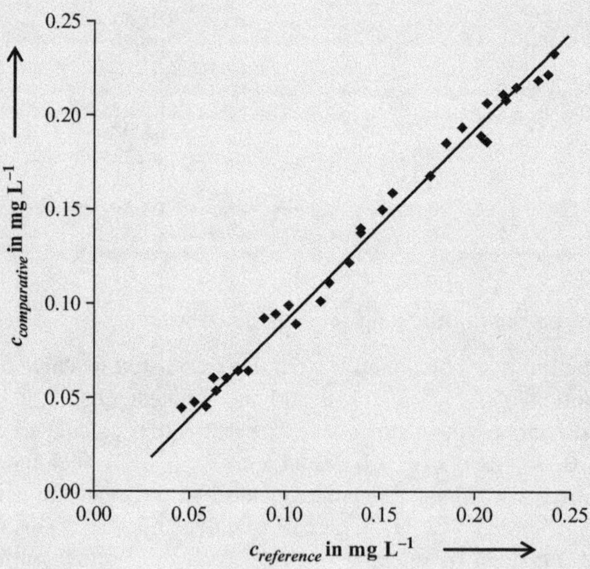

Fig. 5.7.7-2 The function $c_{\text{comparative}} = f(c_{\text{reference}})$ after rejecting the outlier pair of sample number 19

Table 5.7.7-4 Intermediate quantities for the calculation of the test values $\hat{\chi}^2$ and $\hat{t}$, respectively

Parameter	Equation	Excel function	Value
s_R^2	(5.7.7-11)	=VAR(*R*)	0.004003
s_C^2	(5.7.7-10)	=VAR(*C*)	0.004027
s	(5.7.7-14)		0.063368
s_{RC}^2		=COVAR(*R*,*C*)	0.003864
$\bar{x}_R$		=AVERAGE(*R*)	0.142
$\bar{x}_C$		=AVERAGE(*C*)	0.135
$\sum (x_{R,i} - \bar{x}_R)(x_{C,i} - \bar{x}_C)$			0.119793
s_{RC}^2	(5.7.7-15)		0.003993
$\hat{\chi}^2$	(5.7.7-13)		0.003293
$\bar{D}$	(5.7.7-16)		−0.007217
$\sum (D_i - \bar{D})^2$			0.007804
s_D	(5.7.7-18)		0.00695
$\hat{t}$	(5.7.7-17)		6.002
$t(P = 99\%, \text{df} = 30)$		=TINV(1%,30)	2.750

The test value $\hat{t}$ exceeds the critical value at the significance level $P = 99\%$, which means that a constant systematic error is present. Thus, the comparative method is not equivalent to the reference method, and the photometric reference method cannot be replaced by the comparative IC method without further tests.

Challenge 5.7.7-2

In a laboratory the determination of atrazine in seepage water is carried out by a validated GC method. Information as to whether an analysis is generally necessary because of elevated concentrations should be obtained by a quick test, for which ELISA was chosen.

With ELISA, about hundred samples can be analyzed in less than an hour, and it can therefore be used as a screening method in order to identify the samples whose content must be determined again by the validated GC method. In order to use ELISA, the test of trueness of this method must be checked, which can be accomplished by checking the equivalence of the ELISA method (the comparative method C) with the validated GC method (the reference method R) on one matrix.

The analytical results of a split seepage water sample analyzed by both methods are given in Table 5.7.7-5.

Using matrix-free samples the following results were obtained for testing the precision of the methods.

(continued)

Table 5.7.7-5 Predicted $\hat{x}$ values obtained by a seepage water sample analyzed by the reference method GC and the comparative method ELISA with n_i replicates

n_i	Reference method R $\hat{x}_i$ in ppb (w/w)	Comparative method C $\hat{x}_i$ in ppb (w/w)
1	5.25	3.29
2	3.03	4.64
3	3.82	3.02
4	4.58	3.25
5	3.11	4.21
6	4.52	3.32
7	3.13	

Reference method:

$$s_{r,R}\% = 22.5,$$

$$\mathrm{df}_R = 8$$

Comparative method:

$$s_{r,C}\% = 32.4,$$

$$\mathrm{df}_C = 6$$

Estimate whether the large values of the precision obtained in matrix-free solutions can be accepted.

Check whether the comparative method ELISA is equivalent to the reference method GC.

Solution to Challenge 5.7.7-2

According to (5.4-2), the values of the relative standard deviations under repeatability conditions lie at approximately 30% of the concentration in the ppb-range (see Table 5.2-6). Therefore, the large values of $s_r\%$ can be accepted.

Because the precision of the comparative method is greater than that of the reference method in matrix-free solutions, the homogeneity of precision of the methods must be checked by an F-test according to (5.7.7-1). The test value

$$\hat{F} = \frac{s_{r,C}^2}{s_{r,R}^2} = \left(\frac{32.4}{22.5}\right)^2 = 2.074 \tag{5.7.7-26}$$

(continued)

Table 5.7.7-6 Intermediate quantities and results for the Grubbs outlier test calculated by (3.2.3-2)

Reference method R			
$\bar{x}_R$	3.920	s_R	0.880
$x_{min,R}$	3.03	$\hat{r}_{m,R,min}$	1.011
$x_{max,R}$	5.25	$\hat{r}_{m,R,max}$	1.511
$r_m(P = 95\%, n_R = 7)$			1.938
Comparative method C			
$\bar{x}_C$	3.622	s_C	0.646
$x_{min,C}$	3.02	$\hat{r}_{m,C,min}$	0.932
$x_{max,C}$	4.64	$\hat{r}_{m,C,max}$	1.577
$r_m(P = 95\%, n_C = 6)$			1.822

is smaller than the critical value $F(P = 99\%, \mathrm{df}_C = 6, \mathrm{df}_R = 8) = 6.371$. Thus, the precision of the methods checked with matrix-free samples is not statistically different.

Next, the data sets must be checked for outliers using the Grubbs test. The results are summarized in Table 5.7.7-6.

As Table 5.7.7-6 shows, no outlier is detected.

The equivalence test for one matrix is checked by a t-test. After checking the homogeneity of the precision by an F-test, the test value $\hat{t}$ is calculated by (5.7.7-2) and (5.7.7-3).

The homogeneity of the variances is confirmed because the test value calculated by (5.7.7-1) $\hat{F} = 0.538$ is much smaller than the critical value $F(P = 95\%, \mathrm{df}_C = 5, \mathrm{df}_R = 6) = 8.746$. Thus, the t-test can be carried out. The test value calculated by (5.7.7-2) with $s_p = 0.78225$ is $\hat{t} = 0.686$. The critical value $t(P = 99\%, \mathrm{df}_R + \mathrm{df}_C) = 3.106$ is greater than the test value $\hat{t}$. Thus, the comparative method ELISA is equivalent to the reference method GC and can be used as a screening method for the determination of atrazine in seepage water.

Challenge 5.7.7-3

The validated reference method for the photometric determination of Cd in waste water, must be replaced by flame AAS because of the use of harmful carbon tetrachloride. The equivalence of the comparative method should be checked by orthogonal regression.

The check for homogeneity of precision of both methods in matrix-free solutions is accomplished by the results obtained by the two calibration data sets shown in Table 5.7.7-7.

(continued)

Table 5.7.7-7 Data sets of absorbances measured with two replicates in matrix-free solutions for the reference and comparative methods, respectively

Reference method R – Photometry			
Standard	c in mg L^{-1}	$A_{1,i}$	$A_{2,i}$
1	0.3	0.1502	0.1501
2	0.6	0.2427	0.2428
3	0.9	0.3579	0.3579
4	1.2	0.4778	0.4776
5	1.5	0.5811	0.5809
6	1.8	0.6907	0.6909
Comparative method C – Flame AAS			
Standard	c in mg L^{-1}	$A_{1,i}$	$A_{2,i}$
1	0.2	0.1336	0.1334
2	0.4	0.2784	0.2781
3	0.6	0.4001	0.4008
4	0.8	0.5417	0.5421
5	1.0	0.6782	0.6786
6	1.2	0.8141	0.8139
7	1.4	0.9539	0.9539
8	1.6	1.0667	1.0661
9	1.8	1.2115	1.2111
10	2.0	1.3599	1.3595

Table 5.7.7-8 Mean values of absorbances measured with two replicates of 30 real waste water samples obtained by the reference and comparative methods, respectively

Sample no.	Photometry	Flame AAS
	$\bar{A}_{R,i}$	$\bar{A}_{C,i}$
1	0.5702	0.9665
2	0.1718	0.2444
3	0.3436	0.5616
4	0.4277	0.7100
5	0.6104	1.0407
6	0.2303	0.3659
7	0.4935	0.8045
8	0.6397	1.0947
9	0.5922	1.0204
10	0.1836	0.2377
11	0.3107	0.5009
12	0.5154	0.8922
13	0.3546	0.5818
14	0.3838	0.6358
15	0.6416	1.1487
16	0.3984	0.6628
17	0.1974	0.2984
18	0.5958	1.0137
19	0.4423	0.7843
20	0.2449	0.3861
21	0.1864	0.2714

(*continued*)

Table 5.7.7-8 (continued)

Sample no.	Photometry	Flame AAS
	$\bar{A}_{R,i}$	$\bar{A}_{C,i}$
22	0.5373	0.9057
23	0.5629	0.9597
24	0.2998	0.5009
25	0.2157	0.3456
26	0.4423	0.6223
27	0.2815	0.4536
28	0.2669	0.4199
29	0.6543	1.1352
30	0.3984	0.6493

The values of the measured absorbance A_i of 30 waste water samples are given in Table 5.7.7-8. The analytical results of the waste water samples are calculated with the regression parameters obtained by the matrix-free calibration solutions given in Table 5.7.7-7. Do not forget to test the linearity of the regression functions by an appropriate method.

Check the equivalence of the comparative method with the reference method and decide whether flame AAS can be applied for the determination of Cd in waste water.

Solution to Challenge 5.7.7-3

First, let us determine the parameters of the linear regression function and test the linearity.

The regression parameters must be calculated by the mean values of the absorbances measured with two replicates listed in Table 5.7.7-7. Because no twofold determinations were carried out except twofold measurement of the absorbance, the mean value must be calculated for each standard given in Table 5.7.7-9. Thus, the degrees of freedom are $df_{c,R} = 4$ and $df_{c,C} = 8$ for the reference and comparative methods, respectively.

The parameters of the linear regression function calculated by Excel functions are given in Table 5.7.7-10.

Note that it is not necessary to test the homogeneity of the precision because the precision of the comparative method expressed as the relative standard deviation $s_{r,C}\% = 1.04$ is smaller than that of the reference method $s_{r,R}\% = 1.85$.

The Mandel test for checking linearity could be applied but this test requires at least seven calibration standards, which is not the case in the reference method. Therefore, the test for linearity of the reference method is carried out by checking the significance of the quadratic regression

(continued)

Table 5.7.7-9 Data sets for the calculation of the regression parameters of the reference and comparative methods, respectively

Reference method R – Photometry		
Standard	c in mg L^{-1}	$\bar{A}_i$
1	0.3	0.1502
2	0.6	0.2428
3	0.9	0.3579
4	1.2	0.4777
5	1.5	0.5810
6	1.8	0.6908
Comparative method C – Flame AAS		
1	0.2	0.1335
2	0.4	0.2783
3	0.6	0.4005
4	0.8	0.5419
5	1.0	0.6784
6	1.2	0.8140
7	1.4	0.9539
8	1.6	1.0664
9	1.8	1.2113
10	2.0	1.3597

Table 5.7.7-10 Parameters of the linear regression function of the reference and comparative methods

Parameter	Reference method R	Comparative method C
Intercept a_0	0.03294	0.00150
Slope a_1 in L mg^{-1}	0.36550	0.67480
Residual error $s_{y.x}$	0.00709	0.00773
Mean value of c $\bar{x}$ in mg L^{-1}	1.05	1.10
Analytical error $s_{x.0}$ in mg L^{-1}	0.01940	0.01145
s_r in %	1.85	1.04
df	4	8

coefficient a_2 (see Sect. 5.3.6). The required parameters a_2 and s_{a_2} are obtained by Excel function LINEST. The test value calculated by (5.3.6-2)

$$\hat{t} = \left|\frac{a_2}{s_{a_2}}\right| = \frac{0.00766}{0.01422} = 0.539 \qquad (5.7.7\text{-}28)$$

does not exceed the critical value $t(P = 95\%, \mathrm{df_R} = 3) = 3.182$, and thus a_2 cannot be distinguished from zero. The linearity of the regression function of the reference method is confirmed.

With $n_{c,\mathrm{C}} = 10$, the linearity of the regression function of the comparative method can by checked by the Mandel test (see Sect. 5.3.4). The residual standard deviation of the quadratic regression function calculated by Excel

(continued)

Table 5.7.7-11 Analytical results given in mg L^{-1} of the waste water samples and the difference of each pair of values $D_i = D_{R,i} - D_{C,i}$

Sample	Method R	Method C	D_i
1	1.47	1.43	0.04
2	0.38	0.36	0.02
3	0.85	0.83	0.02
4	1.08	1.05	0.03
5	1.58	1.54	0.04
6	0.54	0.54	0.00
7	1.26	1.19	0.07
8	1.66	1.62	0.04
9	1.53	1.51	0.02
10	0.41	0.35	0.06
11	0.76	0.74	0.02
12	1.32	1.32	0.00
13	0.88	0.86	0.02
14	0.96	0.94	0.02
15	1.67	1.70	−0.03
16	1.00	0.98	0.02
17	0.45	0.44	0.01
18	1.54	1.50	0.04
19	1.12	1.16	−0.04
20	0.58	0.57	0.01
21	0.42	0.40	0.02
22	1.38	1.34	0.04
23	1.45	1.42	0.03
24	0.73	0.74	−0.01
25	0.50	0.51	−0.01
26	1.12	0.92	0.20
27	0.68	0.67	0.01
28	0.64	0.62	0.02
29	1.70	1.68	0.02
30	1.00	0.96	0.04

function LINEST is $s_{y.x,2} = 0.008247$. The test value $\hat{F} = 0.0233$ calculated according to (5.3.4-1) is much smaller than the critical value $F(P = 99\%, \mathrm{df}_1 = 1, \mathrm{df}_2 = 7) = 12.246$, which means that the assumed linearity is confirmed. Thus, the regression parameters can be used to calculate the analytical results of the waste water samples which are presented in Table 5.7.7-11.

Because of the condition $x_{max} < 5 \cdot x_{min}$ for the reference and comparative methods the *difference* method must be applied for testing the equivalence of both methods. The differences $D_i = \bar{x}_{R,i} - \bar{x}_{C,i}$ calculated according to (5.7.7-19) are given in Table 5.7.7-11.

Next, the differences D_i have to be checked for an outlier by the Grubbs test. According to the requirements of EURO-Norm, the data sets may not include more than one outlier. The test values are calculated for the lowest and highest D_i-values $|D_{min}| = 0.04$ and $|D_{max}| = 0.20$, respectively, using

(continued)

(3.2.3-2) with $\bar{D} = 0.0256$ and the standard deviation obtained by Excel function LINEST $s_D = 0.040612$ giving $\hat{r}_{m,D_{\min}} = 0.355$ and $\hat{r}_{m,D_{\max}} = 4.295$. The critical value is $r_m(P = 95\%, n = 30) = 2.745$, which means that the highest difference $D_{\max} = 0.20$ obtained by sample number 26 is detected as an outlier at the significance level $P = 95\%$. After the rejection of this pair of values, no other outliers can be detected. Therefore, the data set can be used for the *t*-test. The test value calculated according to (5.7.7-24) with $s_D = 0.02418$ obtained with the outlier-free data set is $\hat{t} = 4.356$. The critical value $t(P = 99\%, \mathrm{df} = 28) = 2.763$ is smaller than the test value $\hat{t}$, and thus the comparative method is not equivalent to the reference method at the significance level $P = 99\%$. The photometric determination of Cd in waste water cannot be substituted by flame AAS.

5.7.8 Standard Addition Method

Direct calibration cannot be used to determine an analyte in a sample if it is confirmed that the sample matrix interferes with the determination. The question is: what can be done if a significant error has been detected?

There are some possibilities for elimating the influence of the matrix, for example:

- Separation of matrix components or the analyte by means of solid-phase extraction (SPE) [20]
- Using solid-phase micro-extraction (SPME) techniques [21]
- Application of headspace GC (HS-GC) [22] for the determination of volatile organic compounds

However, a potential solution to this problem is to apply the *standard addition* method [23].

In the standard addition method small known concentrations of the analyte to be determined are added to aliquots of the unknown samples. These spiked samples, as well as the unspiked, are measured by the same procedure.

In AQA there are some requirements for the application of the standard addition method [13]:

- In order to minimize expense, at least four stocked samples should be prepared by aliquot concentration steps up to the final stocked sample whose concentration is about double the content of the analyte. Therefore, the approximate concentration of the analyte must be known or it must be determined by the direct calibration method.
- In order to change the matrix effects, the stocked volume should be small in comparison to the sample solution and the volume of the sample must be the same for all stock solutions.

- Linearity and homogeneity of variances must be present over the working range.
- The unspiked sample is "stocked" with water up to the same volume as the stocked solutions.
- For the unstocked and stocked samples, the measurement values y_0, y_1 to y_4 are determined and the blank y_{bl} is also measured.

A typical plot of the stocked concentration as a function of the measured response is shown in Fig. 5.7.8-1 with the stock calibration function

$$\hat{y} = \hat{y}_0 + a_{1,\text{add}} \cdot x. \tag{5.7.8-1}$$

The least squares regression line is obtained in the usual way and the content of the analyte $\hat{x}$ in the sample is obtained by extrapolating the line to the abscissa ($y = 0$). The negative intercept on the concentration axis corresponds to $-\hat{x}$ adjusted by the blank y_{bl}.

The predicted value $\hat{x}$ is calculated by (5.7.8-2):

$$\hat{x} = \frac{\hat{y}_0 - y_{\text{bl}}}{a_{1,\text{add}}}. \tag{5.7.8-2}$$

Because the concentration of the sample was changed by filling up steps with the volumeV_{add}, the change in the concentration must be considered using the volume factor

$$\hat{x} = \frac{\hat{y}_0 - y_{\text{bl}}}{a_{1,\text{add}}} \cdot \frac{V_{\text{flask}}}{V_{\text{sample}}}. \tag{5.7.8-3}$$

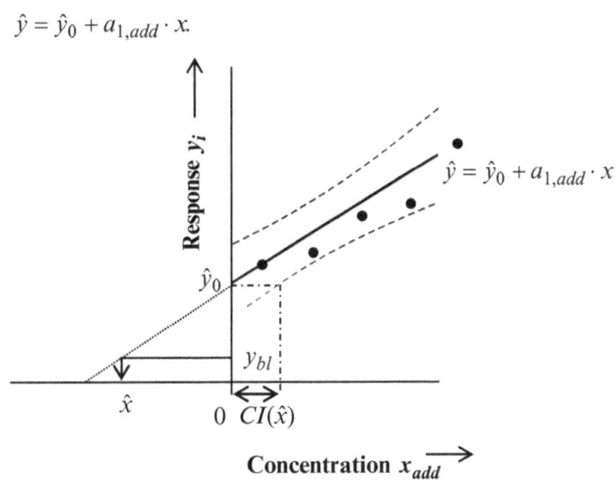

Fig. 5.7.8-1 Graphical illustration of the estimation of the predicted value $\hat{x}$ with its confidence interval CI($\hat{x}$) by the standard addition method

The confidence interval calculated at $\hat{y}_0$ is calculated by (5.7.8-4):

$$\mathrm{CI}(\hat{x}) = \frac{s_{y.x,\mathrm{add}}}{a_{1,\mathrm{add}}} \cdot t(P, \mathrm{df} = n-2) \cdot \sqrt{1 + \frac{1}{n} + \frac{(\hat{y}_0 - \bar{y})^2}{a_{1,\mathrm{add}}^2 \cdot \mathrm{SS}_{xx}}}. \quad (5.7.8\text{-}4)$$

Because the matrix can also have an effect on the precision of the analytical result, one should check whether the predicted value $\hat{x}$ differs from the concentration $\hat{x} = 0$. This is checked by calculation of the test value x_p [13]:

$$x_\mathrm{p} = 2 \cdot \frac{s_{y.x,\mathrm{add}}}{a_{1,\mathrm{add}}} \cdot t(\bar{P}_{\mathrm{one-sided}}, \mathrm{df} = n-2) \cdot \sqrt{1 + \frac{1}{n} + \frac{(y_p - \bar{y})^2}{a_{1,\mathrm{add}}^2 \cdot \mathrm{SS}_{xx}}} \quad (5.7.8\text{-}5)$$

with

$$y_\mathrm{p} = \hat{y}_0 + s_{y.x,\mathrm{add}} \cdot t(\bar{P}_{\mathrm{one-sided}}, \mathrm{df}) \cdot \sqrt{1 + \frac{1}{n} + \frac{\bar{x}^2}{\mathrm{SS}_{xx}}}. \quad (5.7.8\text{-}6)$$

Note that the symbols are the same as those explained in Sect. 4.2. The subscript index $_\mathrm{add}$ refers to the standard addition procedure.

If $x_\mathrm{p} < \hat{x}$, then the obtained value $\hat{x}$ is different from the concentration $x = 0$ at the chosen significance level P. However, for $x_\mathrm{p} > \hat{x}$ the calculated value $\hat{x}$ cannot be significantly distinguished from $x = 0$.

Challenge 5.7.8-1

Let us return to the determination of Cd in waste water by flame AAS. After confirming a systematic error, the standard addition method was chosen as an alternative method.

The measured value obtained by a representative sample was $y = 0.4401$. The regression coefficients determined by the calibration methods are $a_0 = 0.00015$, and $a_1 = 0.6748\,\mathrm{L\,mg^{-1}}$. Clearly, these values are not correct because the matrix changed the regression coefficients, but they can be used in order to choose the range of the stock solutions.

In order to apply the standard addition method, five stock samples were prepared as follows:

80 mL sample solution was added to five 100 mL volumetric flasks. The volumes V_add of a stock solution given in Table 5.7.8-1 were then added, the flasks were filled with water and the absorbance A was measured. The stock solution was prepared by dissolving 6.52314 mg $CdCl_2$ in 1 L water.

The constants are $M_{\mathrm{CdCl_2}} = 183.3\ \mathrm{g\,mol^{-1}}$, $M_\mathrm{Cd} = 112.4$ u

The value of the blank was determined as $y_\mathrm{bl} = 0.0042$.

(continued)

Table 5.7.8-1 Added volumes V_{add} of the stock solution for the preparation of the five stocked calibration solutions

Calibration solutions	V_{add} in mL	$y_i = A_i$
ADD_1	0	0.3529
ADD_2	4	0.4953
ADD_3	8	0.6487
ADD_4	12	0.7854
ADD_5	16	0.9308

Calculate the analytical result with the confidence interval of the waste water sample, construct the calibration function with the confidence intervals, and check whether the analytical result differs from zero.

Solution to Challenge 5.7.8-1

The concentration of the stock solution c_{stock} calculated according to (4.5-1a) is $c_{stock} = 4 \text{ mg L}^{-1}$ Cd. The added masses and the added concentrations are given in Table 5.7.8-2.

Calculation of the regression parameters:

The regression coefficients and the intermediate quantities used for the calculation of the predicted value $\hat{x}$ with the confidence interval $\text{CI}(\hat{x})$ calculated by appropriate Excel functions are summarized in Table 5.7.8-3.

The predicted value $\hat{x}$ calculated by (5.7.8-2) is $\hat{x} = 0.3862 \text{ mg L}^{-1}$. Consideration of the volume factor f_V according to (5.7.8-10) yields the concentration of the analyte in the sample, which is 0.4828 mg L^{-1}.

(continued)

Table 5.7.8-2 Calibration data for the standard addition analysis

Calibration solutions	m_{add} in mg	c_{add} in mg L^{-1}	$y_i = A_i$
ADD_1	0	0	0.3529
ADD_2	0.016	0.16	0.4953
ADD_3	0.032	0.32	0.6487
ADD_4	0.048	0.48	0.7854
ADD_5	0.064	0.64	0.9308

Table 5.7.8-3 Parameters of the linear regression function and the Excel functions used

$a_{0,add} = \hat{y}_0 = 0.35344 = \text{INTERCEPT}(y_i, x_i)$	$a_{1,add} = 0.90369 \text{ L mg}^{-1} = \text{SLOPE}(y_i, x_i)$
$s_{y.x,add} = 0.00404 = \text{STEYX}(y_i, x_i)$	$s_{x.0,add} = 0.00472 \text{ mg L}^{-1}$ (5.2.9)
$SS_{xx} = 0.2560 \text{ mg}^2 \text{ L}^{-2} = \text{DEVSQ}(x_i)$	$s_r\% = 1.40$ (5.2.10)
$\bar{x} = 0.320 \text{ mg L}^{-1} = \text{AVERAGE}(x_i)$	$\bar{y} = 0.6426 = \text{AVERAGE}(y_i)$
$n = 5$	$\text{df} = 3$
$t(P = 95\%, \text{df}) = 3.182 = \text{TINV}(5\%, 3)$	$t_{one-sided}(\bar{P} = 95\%, \text{df}) = 2.353 = \text{TINV}(10\%, 3)$

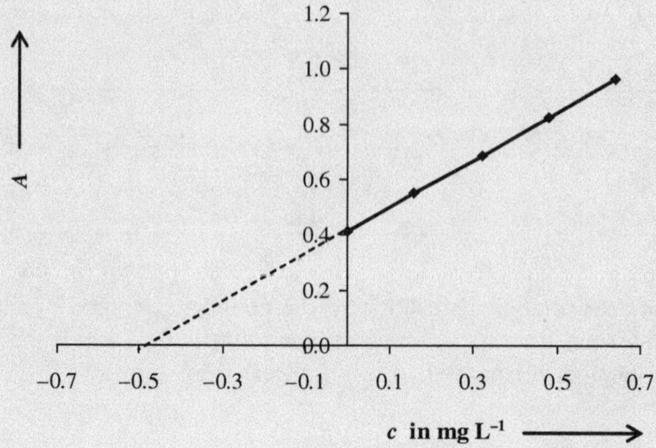

Fig. 5.7.8-2 Calibration line of the standard addition method for the determination of Cd in waste water

$$f_V = \frac{V_{\text{flask}}}{V_{\text{sample}}} = \frac{100 \text{ mL}}{80 \text{ mL}} = 1.25. \tag{5.7.8-10}$$

The confidence interval calculated by (5.7.8-4) and adjusted by the volume factor is $\text{CI}(\hat{x}) = 0.020719 \text{ mg L}^{-1}$.

The test value x_p calculated according to (5.7.8-5) and (5.7.8-6) with the intermediate $y_p = 0.3640$ and considering the volume factor is $x_p = 0.02635 \text{ mg L}^{-1}$. The test value is smaller than the predicted value $\hat{x}$ which means that $\hat{x}$ is significant different from zero. Therefore, the analytical result of the waste water sample $c = 0.483 \pm 0.021 \text{ mg L}^{-1}$ is valid.

The calibration line is shown in Fig. 5.7.8-2.

5.8 Limit of Detection and Limit of Quantification

These validation parameters are required by all regulatory agencies and guidelines but they are important only in the analysis of samples with low concentration of analytes, i.e. for trace analysis methods or for the determination of byproducts in substances. Three meaningful analytical limits can be specified [24–27]:

1. The *critical measurement value* y_c, is the lowest signal that can be detected with reasonable certainty in a given analytical procedure.
2. The *limit of detection* (detection limit) x_{LD} is the lowest concentration of the analyte that can reliably be detected with a specified level of significance.
3. The *limit of quantification* (limit of determination; quantitation limit) x_{LQ} is defined as the lowest concentration at which the measurement precision will be satisfactory for quantitative determination.

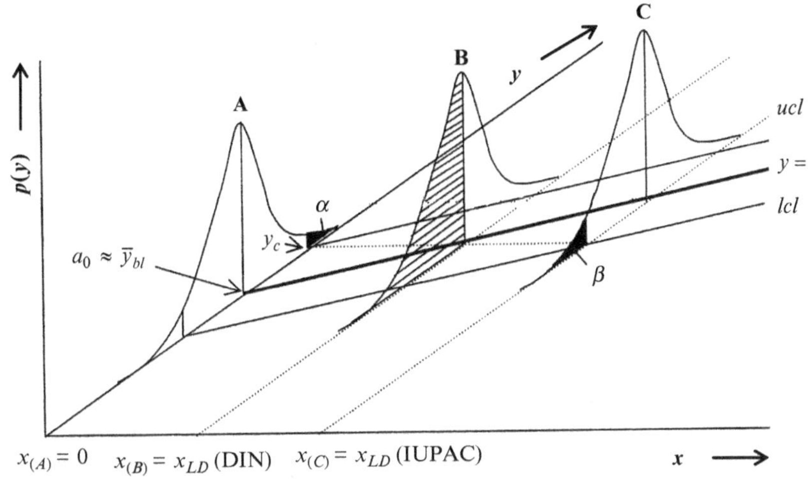

Fig. 5.8-1 Three-dimensional representation of the calibration function $y = a_0 + a_1 x$ with the limits of its two-sided upper confidence limit (*ucl*) and lower confidence limit (*lcl*) and three probability density functions $p(y)$of measured values y belonging to the analytical values x_A(**A**),x_B(**B**), and x_C(**C**). The *symbols* are explained in the text

The statistical fundamentals of y_c and x_{LD} are illustrated by Fig. 5.8-1, which shows a three-dimensional representation of the relationship between measured values and analytical values, characterized by a straight calibration line $y = a_0 + a_1 x$, its two-sided confidence interval, and the probability density function of the measured values $p(y)$ in the z-direction [25, 26]. As Fig. 5.8-1 shows, there are different definitions of the limit of detection x_{LD}, which must be explained.

Firstly, let us consider the blank measurement. The blank is the measured response of a sample which does not contain the analyte. The observations obtained by a sufficiently large number of replicates will be normally distributed, which is shown as the distribution **A** in Fig. 5.8-1. The *critical value* y_c represents the smallest measurement value that can be distinguished from that of the blank y_{bl} at a given significance level P

$$y_c = \bar{y}_{\text{bl}} + k\, s_{\text{bl}}, \tag{5.8-1}$$

where $\bar{y}_{\text{bl}}$ is the mean of repeated blank measurements, s_{bl} the standard deviation of the blank measurements, and k a constant specified by the user. For example, a value of $k = 3$ corresponds to a probability of $\alpha = 0.0013$ that a signal larger than y_c is due to a blank; thus an analyte is detected with the high probability $1 - \alpha = 0.9987$. Consequently, the probability of deciding that the analyte is present when in fact it is absent, i.e. the false positive error α, is small. The α error is the area marked in black at the tail of the probability density function in Fig. 5.8-1 (**A**). In other words, the probability of measuring a blank signal which is higher than y_c is equal to α.

However, if a signal is measured which is lower than y_c it cannot be concluded with the same significance that the analyte is not present. This is better illustrated in Fig. 5.8-2 (**B**) which represents the distribution of an infinite number of repeated measurements of a sample with a true concentration corresponding to an average response to y_c. As Fig. 5.8-2 shows, 50% of the signal observed for the analyte are smaller than the limit concentration y_c.

Thus, the statement that the analyte is absent if the measured response is smaller than y_c can be made only at the probability 50%. This is the so-called β error, which is the probability of false negative decisions. Consequently, the risk of deciding that the analyte is present when in fact it is absent, i.e. the α error, is small, whereas the probability of deciding that the analyte is absent when in fact it is present, i.e. the β error, is large. This is the situation if x_B is defined as the limit of detection, defined as "Nachweisgrenze" in the German DIN [26], using the response y_c for the calculation of the limit of detection.

In order to reduce the β error, distribution **B** in Fig. 5.8-2 must be shifted to a larger response. If the distance between the average of the blank $\bar{x}_{bl}$ and that of the shifted distribution is $2 \cdot k \cdot s_{bl}$, then the α and β errors are equal, which is the situation in distribution **C** shown in Fig. 5.8-1 or better in Fig. 5.8-3.

The value x_C is defined as the limit of detection according to ISO [27] and IUPAC [28]. Consequently, the risk for both false positive (α) and false negative (β) results is very small.

For $k = 3$ and $\alpha = \beta = 0.0013$, (5.8-1) yields for the response

$$y_c = \bar{y}_{bl} + 2 \cdot (3\, s_{bl}) = \bar{y}_{bl} + 6\, s_{bl}, \tag{5.8-2}$$

which corresponds to the limit of detection.

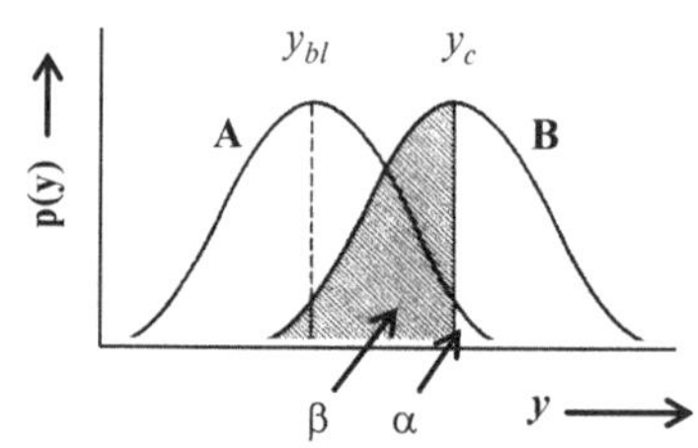

Fig. 5.8-2 Illustration of the limit of detection according to DIN [26]

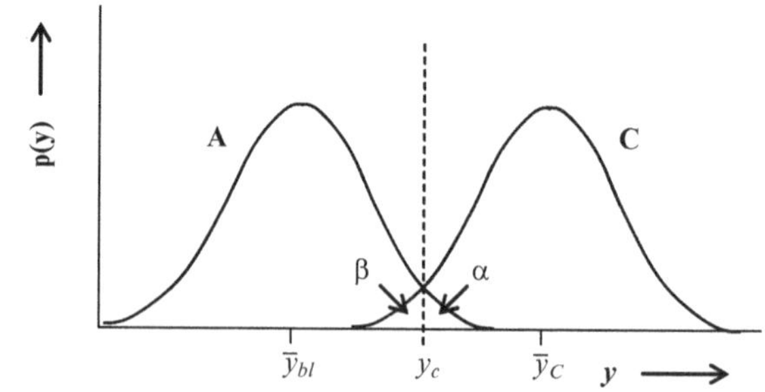

Fig. 5.8-3 Illustration of the limit of detection according to ISO [27] and IUPAC [28]

Note that the limit of detection according to ISO corresponds to the parameter "Erfassungsgrenze" in the German DIN [26], which does not have an equivalent in the ISO [27] and IUPAC[28] definitions.

It is essential to know which of the different definitions is used when documenting the limit of detection.

Two methods can be used in order to determine the critical value experimentally:

1. From replicates of the blank

$$y_c = \bar{y}_{\text{bl}} + s_{\text{bl}} \cdot t(\bar{P}, \text{df}) \cdot \sqrt{\frac{1}{n} + \frac{1}{n_r}}, \tag{5.8-3}$$

where n is the number of the blanks, n_r is the number of repetition measurements, $t(\bar{P}, \text{df})$ is the critical t-value at the chosen one-sided significance level $\bar{P}$, and the degrees of freedom $\text{df} = n - 1$.

2. From the calibration

$$y_c = a_0 + \Delta a_0 = a_0 + s_{y.x} \cdot t(\bar{P}, \text{df}) \cdot \sqrt{\frac{1}{n} + \frac{1}{n_r} + \frac{\bar{x}^2}{\text{SS}_{xx}}}, \tag{5.8-4}$$

where Δa_0 is the confidence interval of the intercept a_0, $s_{y.x}$ is the calibration error, $\bar{x}$ is the mean value of the calibration standards, and $\text{SS}_{xx}$is the sum of squares of the x-values.

To estimate the *limit of detection* x_{LD} the critical value has to be converted into a concentration value

$$x_{\text{LD}} = \frac{\Delta}{\text{Sens}}, \tag{5.8-5}$$

where Δ is the uncertainty, which is $y_c - a_0$ in the calibration method and $y_c - y_{\text{bl}}$ in the blank measurement method, and Sens is the sensitivity, which is equivalent to the slope of the calibration line a_1. Equations (5.8-1) and (5.8-5) give (5.8-6) with $k = 3$ using the limit of detection in the blank measurement method:

$$x_{\text{LD}} = \frac{k \cdot s_{\text{bl}}}{\text{Sens}}. \tag{5.8-6}$$

If the calibration method is used, the limit of detection can be calculated by (5.8-7)

$$x_{\text{LD}} = s_{0.x} \cdot t(\bar{P}, \text{df}) \cdot \sqrt{1 + \frac{1}{n} + \frac{\bar{x}^2}{\text{SS}_{xx}}}, \tag{5.8-7}$$

in which $s_{o.x}$ is the analytical error and n is the number of calibration standards, and all other symbols are explained above.

It should be noted that official regulations [26] and commercial software packages, for example SQS [29], often use (5.8-2) which corresponds to x_B in Fig. 5.8-1 for estimating the detection limit. We will adhere to this standard by using the same equation in this book too. According to (5.8-2), the value of x_{LD} determined by DIN can be easily converted into the ISO definition by multiplying by 2.

Additionally, the detection limit can be calculated from repeated measurements of a peak on a noisy baseline. This so-called signal-to-noise (*S/N*) method can only be applied to analytical procedures which exhibit baseline noise, e.g. chromatography. Determination of the *S/N* ratio is performed by comparing measured signals from samples with low concentrations of the analyte with those of blank samples and establishing the minimum concentration at which the analyte can be reliably detected; an *S/N* ratio between 3:1 and 2:1 is generally considered acceptable for estimating the detection limit [27, 28]. The signal-to-noise method is illustrated in Fig. 5.8-4.

It should be emphasized that the values of the detection limit determined by various methods are not comparable and differences up to a factor of 10 are possible [30]. Therefore, it must be specified which method has been used for the determination of the detection limit.

When using the calibration method, calibration standards with low concentrations must be used to determine the limit of detection. However, there are some limits on the concentration range of the calibration standards. The highest concentration x_n must be not greater than ten times the limit of detection [26]; this is necessary to achieve homogeneity of variances. The method commonly used to determine the homogeneity of the variances – comparing the variances of the data sets for ten replicates of the lowest concentration standard and ten of the highest concentration standard using an F-test (see Sect. 5.5) – is not applicable in this case

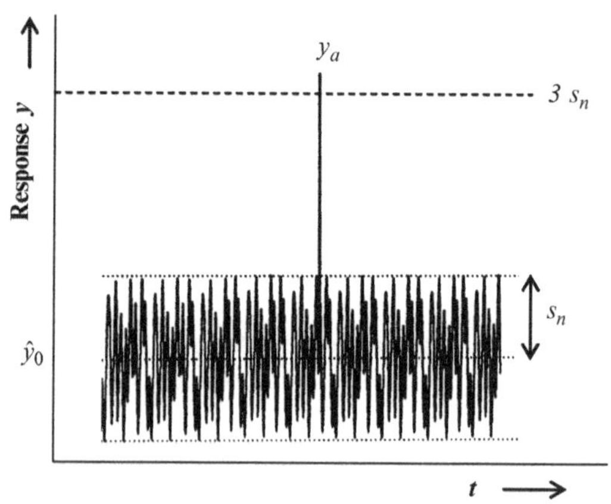

Fig. 5.8-4 The signal-to-noise method for the determination of the detection limit of the analyte *a*

since the variance measured near concentration $x \approx 0$ is very high. Because the homogeneity of the variances cannot be measured experimentally, the calibration standards must lie within a relatively narrow range of concentrations near the limit of detection. If x_n exceeds the value $10 \cdot x_{\mathrm{LD}}$, the highest calibration standard x_n must be rejected, but if more than two standards are rejected a new calibration set with smaller concentrations is necessary.

In addition, the regression line must be tested for linearity and the absence of outliers, which can be done by inspection of the residuals or by methods explained above.

The *limit of quantification* y_{LQ} ("Bestimmungsgrenze" in German) is the parameter with which the analyte can be determined quantitatively with a particular user-specified precision. Therefore, in contrast to the detection limit, the limit of quantification is a conventionally defined measure and depends on how precisely the analyte has to be determined. The precision of the result at the quantification limit is usually expressed in multiples k of the uncertainty $y_{\mathrm{LQ}}/\Delta y_{\mathrm{LQ}} = k$ and is specified by the user in advance.

For a given k, the limit of quantification x_{LQ} is calculated according to (5.8-8)

$$x_{\mathrm{LQ}} = k\, s_{0.x} \cdot t(P, \mathrm{df}) \cdot \sqrt{1 + \frac{1}{n_c} + \frac{(k\, x_{\mathrm{LD}} - \bar{x})^2}{\mathrm{SS}_{xx}}}, \tag{5.8-8}$$

in which P is the two-sided significance level. In general, the factor $k = 3.03$ is used which corresponds to the uncertainty 33%.

Regulatory guidelines such as ICH [31] also recommend the determination of the limit of quantification based on the standard deviation of the blank, which is calculated according to (5.8-5) by (5.8-9):

$$x_{\mathrm{LQ}} = \frac{10 \cdot s_{\mathrm{bl}}}{\mathrm{Sens}}. \tag{5.8-9}$$

The factor $k = 10$ in (5.8-1) is also recommended by IUPAC [28] for the calculation of the limit of quantification according to the blank method.

A quick method for the evaluation of x_{LD} is given in [26] using (5.8-10) for the estimation of the limit of detection on the basis of the standard deviation of the analytical method $s_{x,0}$. The factors Φ_n for various replicates n are listed in Table 5.8-1.

$$x_{\mathrm{LD}} = 1.2 \cdot \Phi_n(P) \cdot s_{x.0}. \tag{5.8-10}$$

Multiplication of the estimated value of x_{LD} by ten provides the highest calibration standard for the direct determination of the detection limit as a starting point for an appropriate choice of the calibration range.

Table 5.8-1 Factors Φ_n for the calculation of the limit of detection at the significance levels $P = 95\%$ and $P = 99\%$ according to the quick method for n repeated measurements [26]

n	$\Phi_n(P = 95\%)$	$\Phi_n(P = 99\%)$
4	2.8	5.1
5	2.3	4.1
6	2.2	3.6
7	2.1	3.4
8	2.0	3.2
9	2.0	3.1
10	1.9	3.0
11	1.9	2.9
12	1.9	2.9

Challenge 5.8-1

(a) The limit of detection for the photometric determination of nitrite-N must be determined based on the standard deviation of the blank. The measurement of the magnitude of analytical background response was performed by 18 replicates of solutions prepared with all solvents and reagents which are used for the analytical method.

The values of the measured absorbance A are:

0.00035	0.00031	0.00024	0.00046	0.00037	0.00051
0.00034	0.00028	0.00042	0.00212	0.00033	0.00029
0.00041	0.00038	0.00029	0.00036	0.00021	0.00028

The sensitivity of the method was determined by calibration with the results given in Table 5.8-2.

What is the *limit of detection* x_{LD} in the photometric determination of nitrite-N?

(b) The analytical method is to be validated for the determination of nitrite-N in waste water. The lowest concentration c_l which has to be determined is assumed to be $c_l = 0.4\ \mu g\, L^{-1}$.

Check whether the limit of quantification estimated by this procedure permits the use of photometric determination.

Table 5.8-2 Calibration data of the photometric determination of nitrite-N

Level i	c_i in mg L^{-1}	A_i
1	0.017	0.05256
2	0.034	0.10952
3	0.049	0.16085
4	0.065	0.21024
5	0.081	0.26342
6	0.097	0.31862

Solution to Challenge 5.8-1

(a) In order to calculate the standard deviation, the data must be checked for normal distribution by the David test (Sect. 3.2.1) and for outliers by the Dixon test (Sect. 3.2.3). The results are:

Test for *normal distribution*

Test value calculated by (3.2.1-1):	$\hat{q}_r = 4.49$
Lower critical value for $P = 95\%, n = 18$:	3.10
Upper critical value for $P = 95\%, n = 18$:	4.37

The test value is not included by the limit values. Strictly speaking, the data cannot be assumed as normally distributed. But the difference between the test and the upper limit values is only small.

Test for *outliers*

The following equation must be applied for $n = 18$ in order to test if the suspected value $A = 0.00212$ is an outlier:

$$\hat{Q} = \left|\frac{x_1^* - x_3}{x_1^* - x_{k-2}}\right| = \frac{0.00212 - 0.00046}{0.00212 - 0.0028} = 0.902. \qquad (5.8\text{-}11)$$

The critical value $Q(P = 95\%, n = 18) = 0.475$ is smaller than the test value $\hat{Q}$, and therefore the measured value $A = 0.00212$ must be rejected from the data set. Further outliers cannot be identified.

Estimation of the *limit of detection* x_{LD}

The mean value of the blank obtained by the outlier-free data set is $\bar{x}_{\text{bl}} = 0.000343$ and its standard deviation is $s_{\text{bl}} = 0.0000781$. The critical value y_c calculated by (5.8-1) with $k = 3$ is $y_c = 0.000577$. The conversion into concentration units is done using (5.8-6). The required sensitivity is the slope a_1 of the recovery function which is calculated by the data set given in Table 5.8-2. The parameter a_1 is obtained by Excel function =SLOPE(y_i, x_i) to give $a_1 = 3.30596\,\text{L}\,\text{mg}^{-1}$. Thus, the detection limit is $x_{\text{LD}} = 0.0000709\ \text{mg}\,\text{L}^{-1} = 0.0709\ \mu\text{g}\,\text{L}^{-1}$.

(b) The limit of quantification x_{LQ} must be smaller than the threshold concentration $c = 0.4\ \mu\text{g}\,\text{L}^{-1}$. The limit of quantification obtained by (5.8-6) with the factor $k = 10$ and the other values given above is $x_{\text{LQ}} = 0.236\ \mu\text{g}\,\text{L}^{-1}$. The limit of quantification is smaller than the threshold concentration, and therefore the method may be used in routine analysis according to the validation parameters "limit of quantification".

Challenge 5.8-2

A laboratory must introduce the determination of phosphate in waste water, for which photometric determination should be used according to DIN EN

(continued)

Table 5.8-3 Calibration set 1 for the determination of the limit of detection for the photometric determination of phosphorus

Standard	$V_{SSL,1}$ in mL	A
1	1	0.03351
2	4	0.15657
3	8	0.28326
4	12	0.42251
5	16	0.58350

Table 5.8-4 Calibration set 2 for the determination of the limit of detection for the photometric determination of phosphorus

Standard	$V_{SSL,2}$ in mL	A
1	0.4	0.00134
2	0.6	0.00228
3	0.8	0.00305
4	1.0	0.00365
5	1.2	0.00419
6	1.4	0.00537

$V_{SSL,2}$ is the volume of stock solution 2 which was pipetted into 25 mL volumetric flasks, A is the measured absorbance

ISO 6878 [33]. The method is based on the measurement of the absorbance at $\lambda = 710$ nm of phosphorus molybdenum blue produced by the reduction of phosphorus molybdenum heteropolyacid by ascorbic acid.

(a) The determination of the limit of detection was begun with calibration data set 1, which was prepared as follows:

A stock solution 1 (SSL 1) was prepared by dilution of 5 mL of a commercial solution with phosphate content 0.1 g L^{-1} into 100 mL.

The volumes of the stock solution $V_{SSL,1}$ given in Table 5.8-3 were pipetted into five 25 mL volumetric flasks, 2 mL molybdenum reagent and 1 mL 10% (m/V) ascorbic acid were added to each, the flasks were filled with water and, after 20 min, the absorbance A was measured. The results are presented in Table 5.8-3.

A second calibration data set was prepared with the same procedure using a stock solution 2 (SSL 2) with phosphate content 0.5 mg L^{-1}. The preparation of the calibration standard solutions and the measured values of the responses are given in Table 5.8-4.

What value does x_{LD} have in the photometric determination of phosphorus?

Estimate the limit of detection by the quick method. Which highest calibration standard should be used for the determination of x_{LD}?

(b) Next, the analytical result (mean value with the confidence interval) of a sample with small amounts of phosphate must be calculated from the

(continued)

Table 5.8-5 Calibration data set for the photometric determination of phosphorus

Standard	c_i in mg L^{-1}	A_i
1	0.20	0.03986
2	0.25	0.04763
3	0.30	0.05897
4	0.35	0.06702
5	0.40	0.07505
6	0.45	0.08752
7	0.50	0.09487

measurement of two replicates: $A_1 = 0.02218$ and $A_2 = 0.02368$. The data set of the calibration is given in Table 5.8-5. Because the predicted value $\hat{\hat{x}}$ must be greater than the limit of quantification, the value of x_{LQ} has to be determined using the data set of calibration standards given in Table 5.8-4.

Determine the analytical result $\hat{\hat{x}} \pm \Delta\hat{\hat{x}}$ mg L^{-1} phosphate and check whether the analytical result is greater than the limit of quantification for a 33% uncertainty according to the requirement

$$x_{LQ} < \hat{\hat{x}} - \mathrm{CI}(\hat{\hat{x}}).$$

Solution to Challenge 5.8-2

(a) The concentration of the stock solution SSL 1 is

$$c_{SSL,1} = \frac{0.1\,\mathrm{g\,L^{-1}} \cdot 5\,\mathrm{mL}}{100\ \mathrm{mL}} \cdot 1{,}000 = 5\ \mathrm{mg\,L^{-1}}. \tag{5.8-12}$$

The concentrations of the calibration solutions calculated by

$$c_i = \frac{V_{SSL,1}\,\mathrm{mL} \cdot 5\ \mathrm{mg\,L^{-1}}}{25\ \mathrm{mL}} \tag{5.8-13}$$

with the volumes of SSL 1 $V_{SSL,1}$ given in Table 5.8-3 are listed in Table 5.8-6.

The regression parameters obtained by appropriate Excel functions are presented in Table 5.8-7.

The detection limit calculated by (5.8-7) is $x_{LD} = 0.177$ mg L^{-1}, but the highest concentration level $c_5 = 3.2$ mg L^{-1} is greater than $10 \cdot x_{LD} = 1.77$ mg L^{-1}. Because the working range exceeds the required limit, the determination must be repeated with a new calibration data set, which is given in Table 5.8-4.

(continued)

Table 5.8-6 Calibration data set 1 for the determination of the limit of detection

Standard	c_i in mg L^{-1}	A_i
1	0.2	0.03351
2	0.8	0.15657
3	1.6	0.28326
4	2.4	0.42251
5	3.2	0.58350

Table 5.8-7 Regression parameters obtained from calibration data set 1

a_0	0.00132	a_1 in $L\,mg^{-1}$	0.17960
$s_{y,x}$	0.01049	$s_{0.x}$ in $mg\,L^{-1}$	0.05838
$\bar{x}$ in $mg\,L^{-1}$	1.64	SS_{xx} in $mg^{-2}\,L^{-2}$	5.7920
n	5	df	3
$s_r\%$	3.56	$t(\bar{P} = 95\%, \mathrm{df})$	2.353

Table 5.8-8 Calibration data set 2 for the determination of the limit of detection

Standard	c_i in mg L^{-1}	A_i
1	0.008	0.00134
2	0.012	0.00228
3	0.016	0.00305
4	0.020	0.00365
5	0.024	0.00419
6	0.028	0.00537

The concentrations calculated according to (5.8-13), but with the new stock solution SSL 2 with $c_{SSL,2} = 0.5$ mg L^{-1}, are given in Table 5.8-8.

The regression parameters obtained by appropriate Excel functions are summarized in Table 5.8-9.

The detection limit $x_{LD} = 0.0031$ mg L^{-1} can be accepted because the highest concentration level $c_5 = 0.028$ mg L^{-1} is smaller than the limit value $10 \cdot x_{LD} = 0.030$ mg L^{-1}.

The limit of detection estimated according to (5.8-10) with $s_{x.0} = 0.000938$ mg L^{-1} and the factor $\Phi_{n=6}(P = 95\%) = 2.2$ obtained by Table 5.8-1 is $x_{LD} = 0.0025$ mg L^{-1}, which is within the range of the exact value. Thus, the highest calibration standard for the exact determination of x_{LD} should be 0.025 mg L^{-1}, which is in good agreement with the standard used.

(b) The regression parameters obtained by appropriate Excel functions using the data set in Table 5.8-5 are given in Table 5.8-10.

The predicted value $\hat{x}$ and its confidence interval CI($\hat{x}$) calculated by (4.2-15)–(4.2-17) using the measured mean value $\hat{\bar{A}} = 0.02293$ is 0.1120 ± 0.0204 mg L^{-1}.

(continued)

Table 5.8-9 Regression parameters obtained from calibration data set 2

a_0	−0.000091	a_1 in L mg^{-1}	0.189143
$s_{y,x}$	0.000177	$\bar{x}$ in mg L^{-1}	0.018
SS_{xx} in mg^{-2} L^{-2}	0.00028	df	4
$s_{x.0}$ in mg L^{-1}	0.000938	s_r in %	5.21
$t(P = 95\%, \text{df})$	2.776		
$t(\bar{P} = 95\%, \text{df})$	2.132		

Table 5.8-10 Regression parameters obtained from the calibration data given in Table 5.8-5

a_0	0.00205	a_1 in L mg^{-1}	0.18635
$s_{y,x}$ in	0.00122	df	5
$\bar{x}$ in mg L^{-1}	0.35	SS_{xx} in mg^{-2} L^{-2}	0.07
$s_{0.x}$ in mg L^{-1}	0.00657	s_r in %	1.88
$\bar{y}$	0.06727	$t(P = 95\%, \text{df})$	2.571

The limit of quantification calculated by (5.8-8) at 33% uncertainty ($k = 3.03$) is $x_{LQ} = 0.0857$ mg L^{-1}. The lower value $\hat{x}_{lower} = 0.1120-0.0204$ mg L^{-1} $= 0.0917$ mg L^{-1} is higher than the quantification limit, which means that the analytical result is valid.

Challenge 5.8-3

The chromatogram covering the peak of the analyte As(III) with concentration $c = 0.8$ μg L^{-1} obtained by HPLC-ICP-MS is given in Fig. 5.8-5.

Determine the limit of detection.

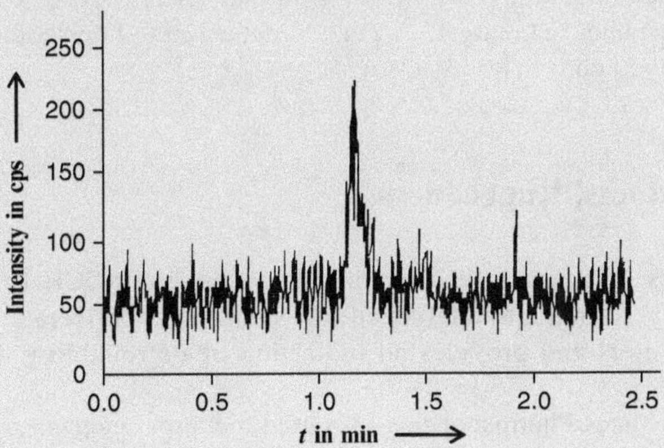

Fig. 5.8-5 Section of the HPLC-ICP-MS chromatogram of a solution with analyte concentration $c = 0.4$ µg L^{-1}As(III)

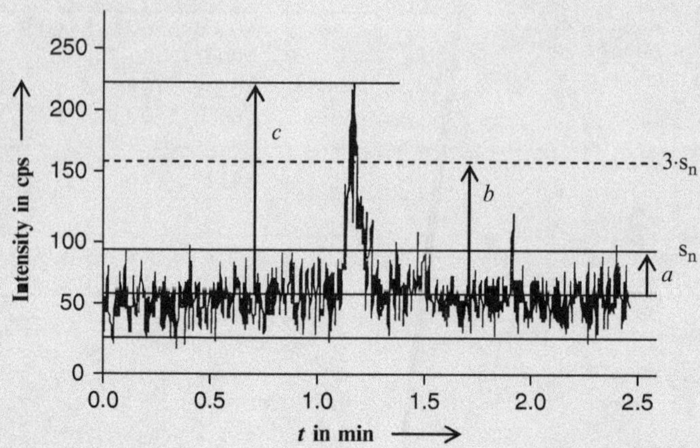

Fig. 5.8-6 Standard deviation of the noise s_n and the critical value $3s_n$ represented as a dashed line in the HPLC-ICP-MS chromatogram

Solution to Challenge 5.8-3

The limit of detection is determined by the signal-to-noise procedure according to (5.8-6) and (5.8-5).

Although in practice the value of *S/N* is given by the software package of the equipment, the standard deviation may also be obtained by the graphical method shown in Fig. 5.8-6.

The standard deviation of the noise obtained from the distance a in Fig. 5.8-5 is 29 cps, therefore, the critical value (distance b) is 87 cps.

The sensitivity used for calculating the limit of detection according to (5.8-5) is determined by the height c of the analyte peak, $c = 163$ cps, which corresponds to 0.8 μg L^{-1}. Therefore, the limit of detection calculated by the *S/N* method is $x_{LD} = 0.43\ \mu g\,L^{-1}$.

5.9 Robustness, Ruggedness

The *robustness* of an analytical method is, according to ICH [3], "a measure of its capacity to remain unaffected by small, but deliberate variations in method parameters and provides an indication of its reliability during normal usage".

The United States Pharmacopoeia [33] used the term *ruggedness* which is "the degree of reproducibility of test results obtained by the analysis of the same sample under a variety of conditions such as different laboratories, different analysis, different instruments... Ruggedness is a measure of reproducibility of test results under the variation in conditions normally expected from laboratory to laboratory and analyst to analyst". Thus, the ICH definition of robustness is related to *intra*-laboratory influences whereas the ruggedness refers to *inter*-laboratory studies. However, in general, ruggedness is also used for intra-laboratory studies.

Robustness is not given explicitly in the list of required validation parameters, but it is recommended as part of method development to establish the critical measurement parameters. It should show the reliability of an analysis with respect to deliberate variations in method parameters [3]. The aim of the robustness test is to find the method parameters that might lead to variations in the results when measurements are carried out under (small) different conditions such as different times or different laboratories.

Common critical parameters can be caused by sample preparation and by measurement conditions. Thus, critical parameters of HPLC relate to the column type (for example, RP-18 available from various producers; altering, etc.), mobile phase (percent organic component in the RP method, pH, and additives), and instrument parameters (dwell volume, flow rate, column temperature, etc.).

A check for ruggedness is made to verify that the method performance is not affected by typical changes in normal experiments or, if influences on the parameters exist, which parameters are critical. If this is the case, a precautionary statement needs to be included in the procedure to ensure that these parameters are tightly controlled between experiments. The parameters used for testing should reflect typical day-to-day variations. Guidance for robustness/ruggedness tests is given in [34].

Principally, ruggedness can be tested by considering each effect separately, but this procedure would require a large number of experiments [35, 36]. Therefore, ruggedness is tested by using a factor design in which several parameters are varied at the same time. Because the total number of experiments $n = 2^k$ for a two-level design strongly increases with the factor k, a fractional factor design is used.

The fractional factor design according to Plackett–Burman considers only the main effects and neglects all interactions between the effects. Plackett–Burman designs are used for screening experiments because main effects are, in general, heavily confounded with two-factor interactions.

The first step in the Plackett–Burman experimental design [36] is the choice of the parameters (factors) to be studied. Each parameter is assigned to one of two levels. The first level is identical with the "normal" parameters which are optimized in the course of method development. The second-level parameters are different from the first, which may be higher or lower.

An alternative approach is to take the extremes of the range over which the normal parameters can possibly vary. The lower limit is indicated by "−" and the upper one by "+". The experimental design for k factors, most simply obtained from software packages, is ordered in a matrix. Note that the number of experiments must be divisible by four.

To investigate the effect of a parameter, the difference Δ between the averages of the results obtained with the parameter is calculated at both alternative levels or the normal level and the alternative level, respectively. Next, the absolute values of the differences are ordered according to their size. The greater the value, the larger the effect of this parameter. However, to decide whether the parameter indeed has an effect, a statistical test is necessary.

The test value is the critical difference calculated by 5.9-1:

$$\hat{\Delta}_{\text{crit}} = \frac{s \cdot t(P, \text{df})}{\sqrt{2}}. \tag{5.9-1}$$

The expected precision of the method is given as the standard deviation s obtained by previous experiments in the course of method development with the degrees of freedom df. The effect of a parameter is statistically significant at the significance level P if its difference Δ exceeds the test value $\hat{\Delta}_{\text{crit}}$.

As example, let us suppose that we have developed a HPLC method for the determination of the assay of a drug. The performance of the HPLC method fulfils the conditions of separating the compounds by a sufficient resolution to enable the determination of a correct assay. But the question is, do small variations of the parameters, e.g. the pH of the mobile phase or the temperature of the column, significantly diminish the performance of the method such that correct analytical results cannot no longer be obtained. This question must be answered by the test for robustness, for which the following steps should be carried out:

- Choose the critical parameters which are to be tested (pH, additives, percentage of the organic component of the mobile phase, flow rate, column temperature, etc.)
- Specify the upper limit values ("+" level) and the lower limit values ("−" level)
- Construct the array of a Plackett–Burman design, e.g. four factors and eight experiments
- Carry out the experiments according to the design
- Calculate the factors having effects, e.g. the resolution of some peaks obtained from the HPLC chromatogram
- Calculate the differences between the averages of the result obtained with the parameter at both alternative levels or the normal level and the alternative level, respectively
- Evaluate the effects statistically.

A test for robustness of a HPLC method is presented in the next Challenge.

Challenge 5.9-1

For the determination of assay and purity of tamoxiphen-dihydrogene-citrate in tablets, a HPLC method was developed which sufficiently separates the API *Z*-isomer (Z) from the byproducts *E*-isomer (E), *bis*-tamoxifen (B), and *des*-methyl-tamoxiphen (D) under the conditions [37] given in Table 5.9-1.

(continued)

Table 5.9-1 Optimal conditions for the HPLC separation of *Z*-tamoxifen from its byproducts

Column	UltraSep ES Pharm RP8
Mobile phase	Water/acetonitrile (ACN), 88% (v/v) ACN
pH of the mobile phase	7.3
Column temperature	35°C
Flow rate	1 mL min^{-1}

Fig. 5.9-1 HPLC chromatogram of the API *Z*-tamoxifene (Z) and the byproducts, *E*-tamoxifene (E), *bis*-methyltamoxifene (B), and *des*-methyltamoxifene (D) obtained under the optimized ("normal") experimental condition given in Table 5.9-1[37]

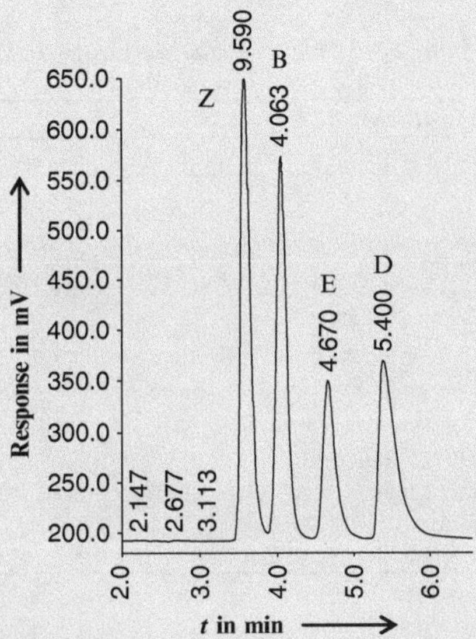

A HPLC chromatogram obtained under the optimal conditions is shown in Fig. 5.9-1.

The chosen limit values of the various parameters are given in Table 5.9-2 and the experimental design (obtained by the software package *Statistica*®) is presented in Table 5.9-3.

The HPLC chromatograms obtained under the varied conditions are shown in Fig. 5.9-2 and the resolutions R_s, the critical value for correct results, are given in Table 5.9-4. The resolution of the pair of the API (Z-tamoxifene) and the adjacent peak (*bis*-methyltamoxifene) R_s(Z/B) (bold type) is the most important parameter for determining the assay of the drug. Therefore, only this parameter will be considered in the following discussion.

(continued)

Table 5.9-2 Optimized (normal) parameters for the HPLC analysis of the API Z-tamoxifene and its byproducts as well as the changed lower (−) and upper (+) limits

Chosen parameters	Optimized	(−) limit	(+) limit
pH value of the eluent	7.3	7.0	7.6
% (v/v) ACN of the mobile phase	88	86	89
Temperature of the column in °C	35	30	40
Flow rate in mL min^{-1}	1.0	0.8	1.2

Table 5.9-3 Experimental design for four factors and eight experiments as well as the normal and varied HPLC conditions

Experimental design				
No.	pH value	% (v/v) ACN	Temperature	Flow rate
1	−1	−1	−1	−1
2	+1	−1	−1	+1
3	−1	+1	−1	+1
4	+1	+1	−1	−1
5	−1	−1	+1	+1
6	+1	−1	+1	−1
7	−1	+1	+1	−1
8	+1	+1	+1	+1
HPLC conditions				
(−) Limit	7.0	86	30	0.8
Normal	7.3	88	35	1.0
(−) Limit	7.6	89	40	1.2

Note that the first experiment, for example, was carried out under the conditions pH = 7.0, 86% (v/v) ACN, temperature = 30°C, and flow rate = 0.8 mL min^{-1}, the second experiment with pH = 7.6, 86% (v/v) ACN, temperature = 30°C, and flow rate = 1.2 mL min^{-1}, and so on.

Evaluate the influence of the HPLC parameter chosen on the resolution of the peaks of the pair of substances Z-isomer, which is the API, and its adjacent peak *bis*-methyltamoxifene (B).

The values of the resolution R_s(Z/B) determined in previous experiments are:

2.41	2.27	2.29	2.30	2.41	2.25	2.24	2.31	2.22

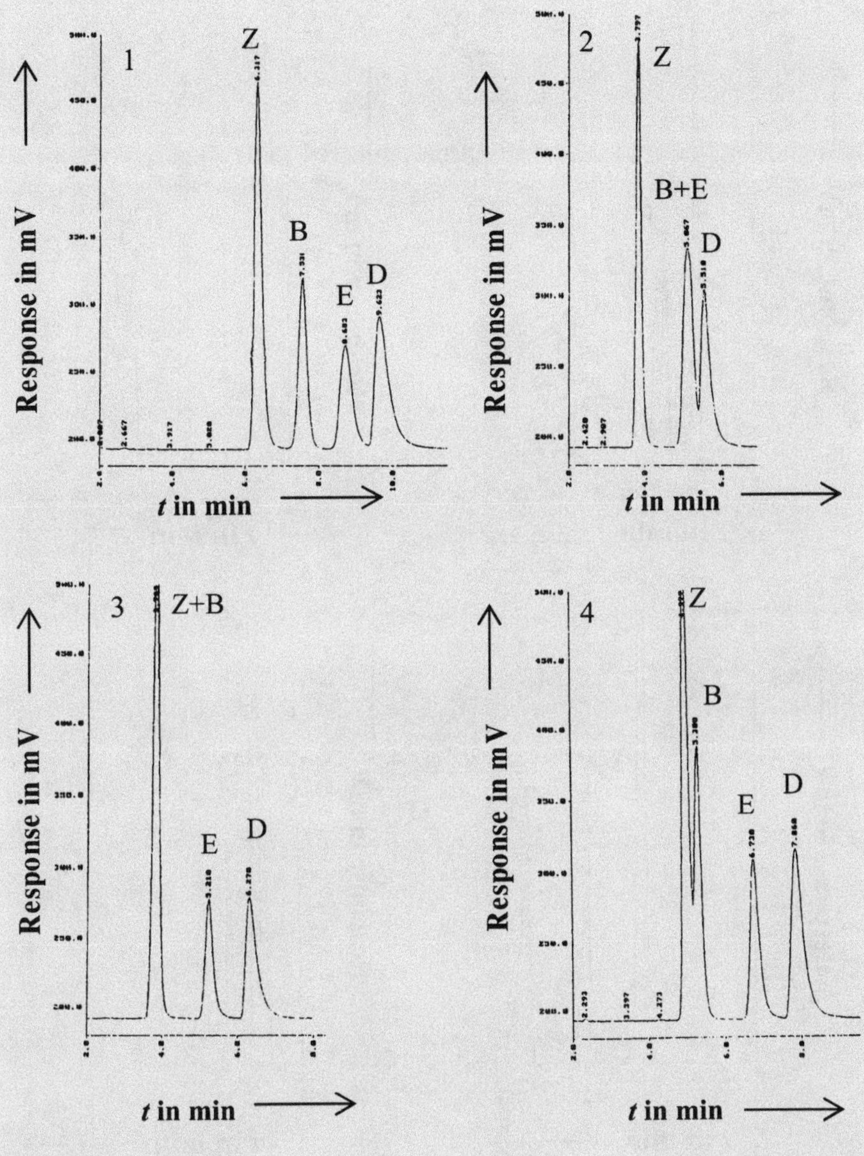

Fig. 5.9-2 (*continued*)

Solution to Challenge 5.9-1

The resolution of the pair of peaks *Z*-tamoxifen and the byproduct *bis*-methyltamoxifene – R_s(Z/B) – is the critical parameter for determining correct results in the assay of the API.

(continued)

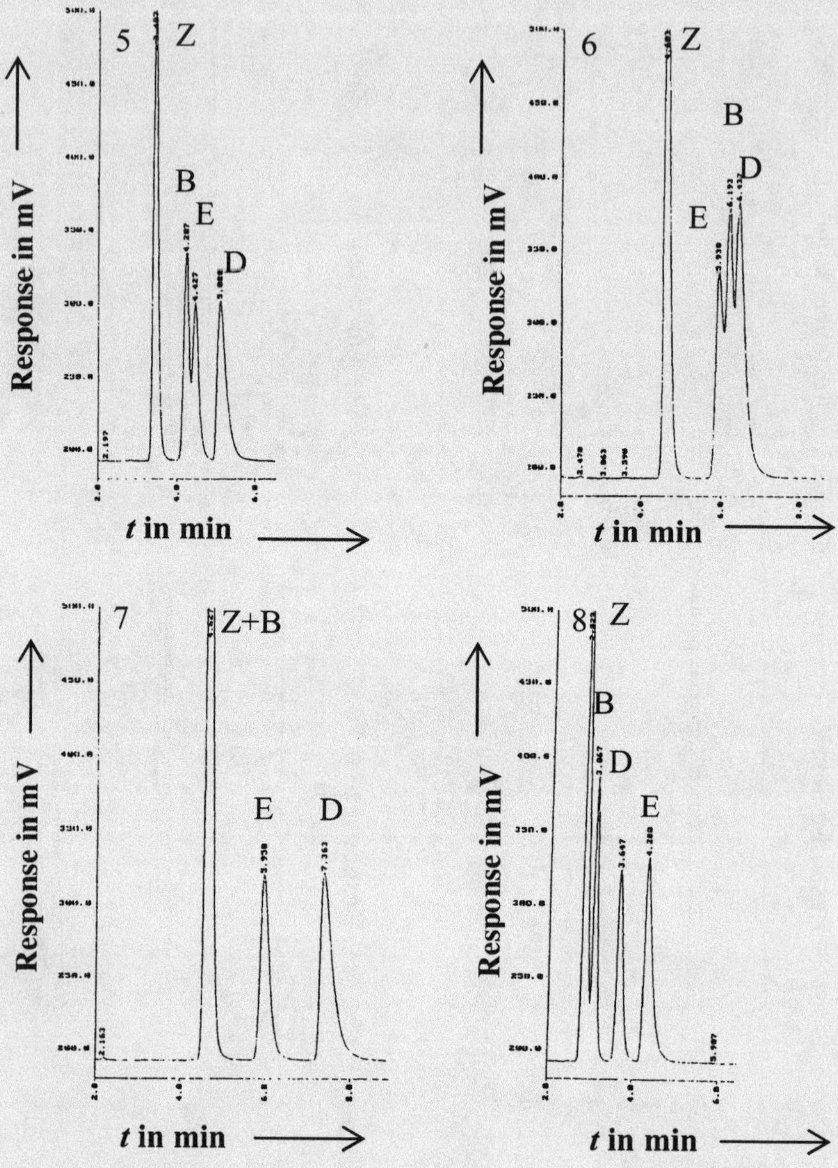

Fig. 5.9-2 HPLC chromatograms obtained by the factor experiments according to the experiments 1–8 given in Table 5.9-4

Z – *Z*-isomer, E – *E*-isomer, B – *bis*-methyltamoxifene, and D – *des*-methyltamoxifene

The factor design, i.e. the changed parameters of the HPLC analysis, are given in Table 5.9-3.

(continued)

Table 5.9-4 Resolution R_s for adjacent pairs of substances obtained by the software package of the HPLC instrument

Experiment no.	R_s(Z/B)	R_s(B/E)	R_s(E/D)
1	2.96	2.19	1.43
2	3.40	0	0.95
3	0	4.33	2.32
4	0.98	3.69	2.18
5	3.28	0.23	1.71
6	4.4	0.72	0.49
7	0	3.78	3.07
8	1.30	2.81	2.12

Table 5.9-5 Factors influencing the resolution R_s of adjacent pairs of substances

Parameter	R_s(Z/B)	R_s(B/E)	R_s(E/D)
pH	0.96	−0.85	−0.70
% (v/v) ACN of the mobile phase	−2.94	2.89	1.23
Temperature of the column	0.41	−0.65	0.13
Flow rate	−0.09	0.78	−0.02

The factors affecting R_s(Z/B) caused by pH are calculated according to the factor design by (5.9-2),

$$\Delta_{\mathrm{pH}} = \frac{1}{4}(3.4 + 0.98 + 4.4 + 1.3) - \frac{1}{4}(2.96 + 0 + 3.28 + 0) = 0.96 \quad (5.9\text{-}2)$$

and those caused by the percentage ACN by (5.9-3):

$$\Delta_{\mathrm{ACN}} = \frac{1}{4}(0 + 0.98 + 0 + 1.3) - \frac{1}{4}(2.96 + 3.4 + 3.28 + 4.4) = -2.94. \quad (5.9\text{-}3)$$

All factors influencing the resolution due to changing each parameter calculated by the same procedure are listed in Table 5.9-5.

The standard deviation obtained with df = 8 experiments is $s = 0.069$ and the statistical two-sided t-factor is $t(P = 95\%, \mathrm{df} = 8) = 2.306$. Thus, the critical difference calculated according to (5.9-1) is $\Delta_{\mathrm{crit}} = 0.112$.

As Table 5.9-5 shows, the (absolute) values of the differences in the parameters pH, % (v/v) ACN of the mobile phase, and the temperature of the column have a significant effect on the resolution R_s(Z/B), whereas the flow rate does not have a significant effect.

(continued)

The composition of the mobile phase (given in % (v/v) ACN) and the pH of the eluent have the greatest influence on the resolution of the critical peak pair API and *bis*-methyltamoxifene (B). Thus, the composition as well as the pH of the eluent must be carefully controlled during routine analysis.

5.10 Application of Method Validation

Now that all required validation parameters, except for "selectivity" which is a topic of method development (see Chap. 6), have been explained, let us now apply method validation to a problem of analytical practice.

Challenge 5.10-1

In an analytical laboratory, a method for the routine analysis of nitrite-N in industrial waste water is to be introduced in order to monitor routinely the limit value of $L_0 = 0.163\ \mathrm{mg\,L^{-1}}$ N. For this purpose, the inexpensive photometric method for determination of nitrite-N in surface water was chosen, for which there is a EURO-Norm DIN EN 26777 [4].

The principle of the method is based upon the fact that in acidic conditions, nitrite ions bind aminobenzene-sulfamide equimolarly to form diazocompounds. Coupling with *N*-(1-naphthyl-)-ethylene-diamine forms a red dye with an absorption maximum $\lambda_{max} = 540$ nm,which is the wavelength used for measuring the absorbance. All experimental conditions conforming to the DIN [4] such as reagents, pH of the solution, and equilibrium time have been previously validated and can, therefore, be applied unchanged.

From pre-tests of waste water, a relative high iron content was determined at an average of 4 mg $\mathrm{L^{-1}}$ Fe. As given in the EURO-Norm [4], this high iron content could lead to interference in the method resulting in false analytical results. Therefore, it should be determined whether the method validated for *surface water* may be applied to the industrial waste water, i.e. whether the matrix does causes a systematic error. A *re-validation* of the method is thus carried out for the changed conditions.

The quality of the substances used for the preparation of the solutions required for the determination of the regression parameters in iron-free solutions, as well as tests on the homogeneity of variances and trueness, are the same as given in DIN EN 26777 [4]. The reagent solution (RS) was prepared by dissolving 20 g aminobenzene-sulfamide and 1 g *N*-(1-naphthyl-)-ethylene-diamine dihydrochloride in a mixture of 250 mL water and 50 mL phosphoric acid ($\rho = 1.71$ g $\mathrm{mL^{-1}}$). The solution was then made up to 500 mL with water.

(continued)

(a) The calibration solutions were prepared as follows: the volumes V_{st} of stock solution 1 with $c_{st,1} = 0.22 \text{ mg L}^{-1}\text{N}$ given in Table 5.10-1 and 1 mL of the reagent solution (RS) described above were each pipetted into nine 25 mL volumetric flasks. The flasks were then filled with water and after 30 min the absorbance was measured at $\lambda = 540$ nm.

Determine the regression parameters using the data set given in Table 5.10-1.

Check the linearity of the calibration function and check the suspect residuals for outliers.

(b) Check the homogeneity of variances in the working range using the data set given in Table 5.10-2. Remember that the calculation of standard deviations requires data sets which must be normally distributed and free of outliers.

(c) Does the iron-containing matrix cause a systematic error?

First, apply the test using the recovery function with simulated matrices.

All calibration solutions for the test using the *recovery function* were prepared as given above, but each calibration solution was spiked by 1 mL

(continued)

Table 5.10-1 Calibration data for determination of the regression parameters

Level i	V_{st} in mL	$y_i(A_i)$
1	4	0.10473
2	6	0.15284
3	8	0.20413
4	10	0.25017
5	12	0.30352
6	14	0.35414
7	16	0.40256
8	18	0.45325
9	20	0.49754

Table 5.10-2 Response values $y_{i,j}(A_{i,j})$ obtained by solutions of calibration levels 1 and 9, respectively

Calibration level	
$i = 1$	$i = 9$
$y_{1,j}(A_{1,j})$	$y_{9,j}(A_{9,j})$
0.10418	0.49954
0.10457	0.49605
0.10463	0.49803
0.10455	0.49648
0.10482	0.49542
0.10447	0.49838
0.10469	0.50649
0.10371	0.49613
0.10489	0.49982
0.10448	0.49963

of an iron-containing solution with $c_{Fe} = 125\ \mathrm{mg\,L^{-1}}$. Thus, the concentration of each iron-containing solution was $c_{Fe} = 5\ \mathrm{mg\,L^{-1}}$, somewhat higher than the average iron content of the waste water.

The results are summarized in Table 5.10-3.

The calibration solutions for the test using the *standard addition method* were prepared as follows: into each of eight 25 mL volumetric flasks was pipetted 10 mL of the stock solution 2 with $c_{st,2} = 0.4\ \mathrm{mg\,L^{-1}}$, 1 mL of the iron-containing stock solution with $c_{Fe} = 125\ \mathrm{mg\,L^{-1}}$, the volumes of the stock solution 3 with $c_{st,3} = 0.155\ \mathrm{mg\,L^{-1}N}$ given in Table 5.10-4, and 1 mL reagent solution RS. The flasks were then filled up with water and after 30 min the absorbance was measured at $\lambda = 540$ nm.The results are given in Table 5.10-4.

(d) Does the matrix significantly affect the precision of the analytical method? Check it on the basis of the recovery function as well as by the standard addition method.

(e) When the matrix significantly affects the regression coefficients, the tested method cannot be applied. As an alternative to the photometric method according to DIN EN, standard addition should be used. The experimental conditions can also be applied to this procedure. In order to validate the modified method, the linearity of the regression line must still be checked, which can be performed using the data set for the trueness test in c.

(continued)

Table 5.10-3 Data set for checking trueness using the recovery function

Level i	V_{st} in mL	$y_i(A_i)$
1	4	0.12538
2	6	0.18331
3	8	0.24404
4	10	0.30132
5	12	0.36426
6	14	0.42211
7	16	0.48745
8	18	0.54329
9	20	0.59301

Table 5.10-4 Data for checking trueness by the standard addition method

Level i	$V_{st,2}$ in mL	$y_i(A_i)$
1	0	0.2240
2	1	0.2452
3	2	0.2634
4	3	0.2801
5	4	0.2982
6	5	0.3146
7	6	0.3365
8	7	0.3558

The standard addition method must be described by a standard operation procedure (SOP), according to regulatory requirements for the creation of SOPs valid for the analytical laboratory. One point of the SOP will concern the "validation of the method" which must be checked by means of a suitable test specified in the SOP. This test must be carried out routinely in order to demonstrate that the method is still valid.

Let us assume the "validation of the method" is specified by the trueness test as follows: a synthetic iron-containing solution with a known amount of nitrite-N is analyzed by the standard addition method. The analytical result is correct and, thus, the method is still valid if the 'true value' lies within the range of the confidence interval of the predicted value $\hat{x}$ in $\mathrm{mg\,L^{-1}}$ obtained by the standard addition method.

Check the "validation of the method" using the following data: 18 mL aliquots of a validation solution $\mu_{val} = 0.07\ \mathrm{mg\,L^{-1}N}$ which contains 5 mg $\mathrm{L^{-1}}$ Fe is pipetted into five 25 mL volumetric flasks. After addition of the volumes of the stock solution 4 with $c_{st,4} = 0.20\ \mathrm{mg\,L^{-1}N}$ given in Table 5.10-5 and 1 mL reagent solution RS, the flasks are filled with water and the absorbance is measured after 30 min. The measured mean value of the absorbance of the blank is $y_{bl} = 0.0004$.

(f) Remember that the analytical method was introduced in order to monitor threshold values of the waste water. The preparation of the spiked solutions was carried out by the same procedure as given above. The volume of each waste water sample is $V_{sample} = 20$ mL and the concentration of stock solution 5 is $c_{st,5} = 0.75\ \mathrm{mg\,L^{-1}}$ N.

The preparation of the spiked solution and the measured response is given in Table 5.10-6. The absorbance of the blank measured using a solution which was prepared only with the reagents is $y_{bl} = 0.0006$.

Check whether the limit value given above is exceeded.

Has the calibration range been properly chosen?

Table 5.10-5 Data set for validation of the standard addition method

Level	1	2	3	4	5
$V_{st,val}$ in $\mathrm{mg\,L^{-1}N}$	0	1.5	3.0	4.5	6.0
$y_i(A_i)$	0.1422	0.1767	0.2069	0.2436	0.2791

Table 5.10-6 Calibration data set for the determination of nitrite-N in a waste water sample according to the standard addition method

Level	1	2	3	4	5
V_{st} in mL	0	1	2	3	4
$y_i(A_i)$	0.3555	0.4418	0.5173	0.6091	0.6978

Solution to Challenge 5.10-1

Note that all intermediate quantities and results are calculated by appropriate Excel functions explained above.

(a) The concentration of the calibration standard solutions calculated according to (4.5-2) as and the residuals calculated by (4.2-3) with the linear regression coefficients $a_0 = 0.005212$ and $a_1 = 2.815625 \text{ L mg}^{-1}$ are listed in Table 5.10-7. The residuals e_i are shown in Fig. 5.10-1. Further regression parameters are: $s_{a_0} = 0.001751$, $s_{a_1} = 0.015229 \text{ L mg}^{-1}$, $s_{y.x} = 0.0020762$, $s_{x.0} = 0.000737 \text{ mg L}^{-1}$, $s_r\% = 0.70$.

Test of linearity

Linearity can be checked by the Mandel test or by the test of the significance of the quadratic regression coefficient a_2. Both methods are applied.

Mandel test

(continued)

Table 5.10-7 Data and residuals e_i for the calibration of the photometric determination of nitrite-N according to DIN EN 26777 0493 [4]

Level	c in mg L^{-1}	$y_i(A_i)$	$\hat{y}_i$ in mg L^{-1}	1000 e_i
1	0.0352	0.10473	0.10432	0.41
2	0.0528	0.15284	0.15388	−1.04
3	0.0704	0.20413	0.20343	0.70
4	0.0880	0.25017	0.25299	−2.82
5	0.1056	0.30352	0.30254	0.98
6	0.1232	0.35414	0.35210	2.04
7	0.1408	0.40256	0.40165	0.91
8	0.1584	0.45325	0.45121	2.04
9	0.1760	0.49754	0.50076	−3.22

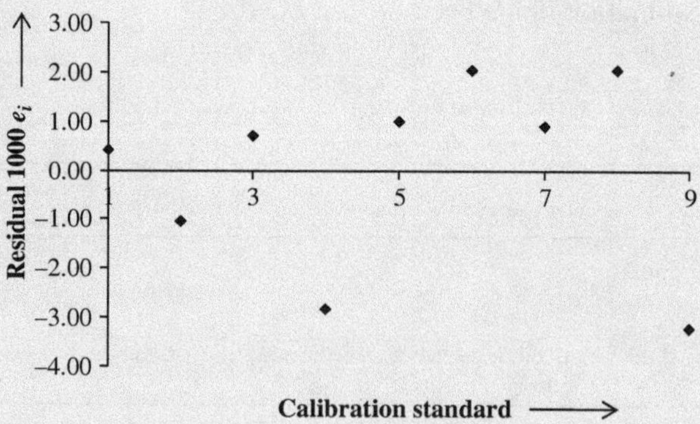

Fig. 5.10-1 Residual plot obtained by the calibration of the determination of nitrite-N according to DIN EN 26777 [4]

The test value $\hat{F}$ calculated according to (5.3.4-1) using the residual error of the linear regression model $s_{y.x,1} = 0.0020762$ and the quadratic model $s_{y.x,2} = 0.0021287$ and the degrees of freedom $\mathrm{df}_1 = 7$ and $\mathrm{df}_2 = 6$, respectively, is $\hat{F} = 0.659$. The critical value $F(P = 99\%, \mathrm{df}_1 = 1, \mathrm{df}_2 = 6) = 13.745$ is much greater than the test value $\hat{F}$, and therefore the linearity given by the randomized residuals around zero is confirmed.

Significance test of the quadratic regression coefficient a_2

The regression coefficient of the second degree equation is $a_2 = -0.317939\ \mathrm{L}^2\,\mathrm{mg}^{-2}$ and its standard deviation is $s_{a_2} = 0.391566\mathrm{L}^2\,\mathrm{mg}^{-2}$. The test value calculated according to (5.3.6-2) is $\hat{t} = 0.812$, which is smaller than the critical value $t(P = 95\%, \mathrm{df} = n - 3 = 6) = 2.447$, and thus the linearity is valid. The same result yields the confidence interval of the quadratic regression parameter a_2 calculated by (5.3.6-1). The value zero is included in the range of $\mathrm{CI}(a_2) = -0.31794 \pm 0.95813$, and thus a_2 cannot be distinguished from zero at the significance level $P = 95\%$.

Outlier test in the regression

Inspection of the residuals in Fig. 5.10-1 shows that the (absolute) greatest value is obtained by level 9. The value $y_9 = 0.49754$ should be tested as to whether it is an outlier in the regression.

First, let us apply the F-test. The test value is $\hat{F} = 7.423$ calculated according to (5.4-1) with the re-calculated residual standard deviation after rejection of the x_9, y_9-values from the data set $s_{y.x,\mathrm{OL}} = 0.0014993$, the degrees of freedom $\mathrm{df}_{\mathrm{OL}} = 6$ and the degrees of freedom $\mathrm{df}_1 = 7$ of the whole data set given above. The test value $\hat{F}$ is compared with the critical value $F(P = 99\%, \mathrm{df}_1 = 1, \mathrm{df}_2 = n - 3 = 6) = 13.745$. Because the critical value is smaller than the test value, the suspect y_9- value is not confirmed to be an outlier and must be included in the data set again.

The t-test is used to check whether the suspect outlier value $y_{\mathrm{OL}} = y_9$ lies within the confidence interval $\mathrm{CI}(\hat{y}_{\mathrm{OL}})$, which is calculated by (5.4-2) and (5.4-3) without the suspect x_9, y_9-values. The confidence interval is calculated with $a_{0,\mathrm{OL}} = 0.0037162$, $a_{1,\mathrm{OL}} = 2.83524\mathrm{L\,mg}^{-1}$, $x_{\mathrm{OL}} = x_9 = 0.176\ \mathrm{mg\,L}^{-1}$, $n_{\mathrm{OL}} = 8$, $\bar{x}_{\mathrm{OL}} = 0.0968\ \mathrm{mg\,L}^{-1}$, $\mathrm{SS}_{xx} = 0.0130099\ \mathrm{mg}^{-2}\mathrm{L}^{-2}$, and $t(P_{\mathrm{two-sided}} = 99\%, \mathrm{df} = 6) = 3.707$.

The test value $y_{\mathrm{OL}} = y_9 = 0.49754$ is included within $\mathrm{CI}(\hat{y}_{\mathrm{OL}}) = 0.5027 \pm 0.00705$, i.e. the range 0.49567–0.50977, and therefore the tested response is not confirmed to be an outlier in the regression.

(b) Both data sets are normally distributed as checked by the David test. The test values are $\hat{q}_{r,1} = 3.47$ and $\hat{q}_{r,9} = 3.44$ calculated according to (3.2.1-1) with $x_{\max,1} = 0.10489$, $x_{\min,1} = 0.10371$, $x_{\max,9} = 0.50649$, $x_{\min,9} = 0.49542$, $s_1 = 0.000340$, and $s_9 = 0.00322$ for the lowest and the highest calibration levels, respectively. The test values lie within the

(continued)

critical lower (2.67) and upper (3.685) limits at the significance level $P = 95\%$.

The highest value in the upper range $y_{9,7} = 0.50649$ should be checked as a suspected outlier. After ranking the data set, the test values must be calculated for $n = 10$ by (5.10-1):

$$\hat{Q} = \frac{|x_1^* - x_2|}{|x_1^* - x_{n-1}|} = \frac{0.50649 - 0.49982}{0.50649 - 0.4960} = 0.639. \qquad (5.10\text{-}1)$$

Because the critical value $Q(P = 95\%, n = 10) = 0.477$ is smaller than $\hat{Q}$, the measured absorbance $y_{9,7} = 0.50649$ must be rejected from the data set.

The test value $\hat{F} = 25.945$ calculated by (4.3-1) with $s_1 = s_9 = 0.001734$ and $s_2 = s_1 = 0.000340$ exceeds the critical value $F(P = 95\%, \mathrm{df}_1 = 8, \mathrm{df}_2 = 9) = 3.230$, which means that the variances are inhomogeneous within the working range.

Note that this result was to be expected for a range whose highest value is about five times the lowest one. If the calibration method according to DIN EN 26777 can be applied, weighted regression should be used.

(c) Test of trueness according to the recovery function explained in Sect. 5.7.5

The concentrations $c_{m,i}$ of the matrix solutions calculated with the regression coefficients obtained by the matrix-free solutions are listed in Table 5.10-8. The regression parameters of the recovery function are $a_{0,m} = 0.000592\,\mathrm{mg\,L^{-1}}$, $s_{a_{0,m}} = 0.001141\,\mathrm{mg\,L^{-1}}$, $a_{1,m} = 1.196674$, and $s_{a_{1,m}} = 0.009922$. The recovery function is shown in Fig. 5.10-2.

The *linearity* of the recovery function is checked by the Mandel test using the residual errors $s_{y,x,m} = 0.001353\ \mathrm{mg\,L^{-1}}$, $s_{x.y,m,2} = 0.001311\ \mathrm{mg\,L^{-1}}$, and the degrees of freedom $\mathrm{df}_m = 7$, and $\mathrm{df}_{m,2} = 6$, where the index 2 refers to the data calculated for the quadratic regression line. The test value $\hat{F} = 1.454$ is much smaller than the critical value $F(P = 99\%, \mathrm{df}_1 = 1, \mathrm{df}_2 = 6) = 13.745$, and thus the linearity of the regression function is valid.

(continued)

Table 5.10-8 Calibration data for the determination of the recovery function and the predicted values $\hat{x}_{m,i}$ $(= \hat{c}_{m,i})$ calculated by the regression coefficients of the recovery function

Level i	c_i in $\mathrm{mg\,L^{-1}}$	$y_i(A_i)$	$\hat{c}_{m,i}$ in $\mathrm{mg\,L^{-1}}$
1	0.0352	0.12538	0.04268
2	0.0528	0.18331	0.06325
3	0.0704	0.24404	0.08482
4	0.0880	0.30132	0.10517
5	0.1056	0.36426	0.12752
6	0.1232	0.42211	0.14807
7	0.1408	0.48745	0.17127
8	0.1584	0.54329	0.19110
9	0.1760	0.59301	0.20876

Fig. 5.10-2 Recovery function

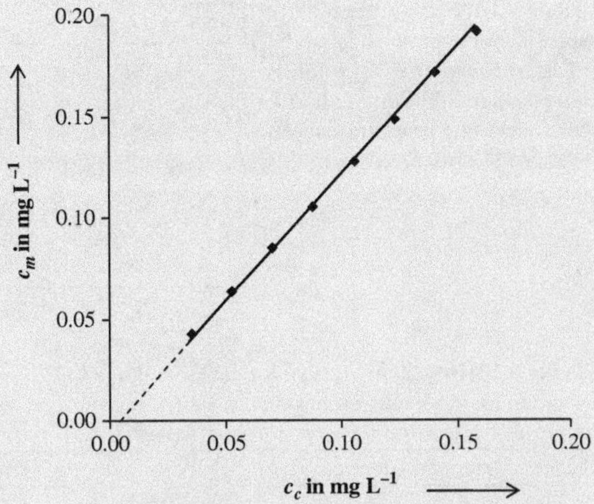

Note that the check of the quadratic regression parameter a_2 yields the same results: $\hat{t} = 1.206$, $t(P = 95\%, \mathrm{df} = 6) = 2.447$, $\mathrm{CI}(P = 95\%, \mathrm{df} = 6) = -0.2907 \pm 0.5900$.

Check for *proportional* systematic error according to the recovery function

The confidence interval of the slope of the recovery function $\mathrm{CI}(a_{1,m}) = 1.19667 \pm 0.02346$ calculated by (5.7.5-6) with the data given above does not include the value 1, and therefore the matrix causes a proportional systematic error at the significance level $P = 95\%$.

Check for *constant* systematic error

The confidence interval of the intercept of the recovery function $\mathrm{CI}(a_{0,m}) = 0.0005918 \pm 0.0026973\ \mathrm{mg\,L^{-1}}$ calculated by (5.7.5-5) with the data given above includes zero, and therefore the matrix does not cause a constant systematic error at the significance level $P = 95\%$.

Check for *proportional* systematic error by the standard addition method

The spiked concentrations calculated from the concentration of the added volumes of the stock solution given in Table 5.10-4 and the measured calibration data are given in Table 5.10-9.

The parameters of the linear and quadratic regression functions of the standard addition method and further data used for tests are summarized in Table 5.10-10.

The test value calculated by (5.3.4-1) of the Mandel test is $\hat{F} = 0.417$, which does not exceed the critical value $F(P = 99\%, \mathrm{df}_1 = 1, \mathrm{df}_2 = 5) = 16.258$. Thus, the regression line of the standard addition calibration is linear at the significance level $P = 99\%$.

(continued)

Table 5.10-9 Calculated spiked concentrations c_{sp} and measured responses $y_i(A_i)$of the standard addition method for checking a proportional systematic error

Level	c_{sp} in $mg\,L^{-1}$	$y_i(A_i)$
1	0	0.2240
2	0.0062	0.2452
3	0.0124	0.2634
4	0.0186	0.2801
5	0.0248	0.2982
6	0.0310	0.3146
7	0.0372	0.3365
8	0.0434	0.3558

Table 5.10-10 Parameters of the linear and quadratic regression functions of the standard addition method, and further data used for tests of linearity by Mandel and by the significance of the quadratic regression coefficient a_2

Linear regression function			
$a_{0,sp}$	0.22511	$a_{1,sp}$ in $L\,mg^{-1}$	2.97773
$s_{a_{0,sp}}$	0.00108	$s_{a_{1,sp}}$ in $L\,mg^{-1}$	0.04181
$s_{y.x,sp}$	0.00168	$s_{x.0,sp}$ in $mg\,L^{-1}$	0.00056
$\bar{x}$ in $mg\,L^{-1}$	0.0217	$s_r\%$	2.60
df	6	s_p	0.030524
$t(P = 95\%, df = df_c + df_{sp} = 7 + 6 = 13)$			2.160
Quadratic regression line			
$a_{2,sp}$ in $L^2\,mg^{-2}$ in $L^2\,mg^{-2}$	2.29178	$s_{a_{2,sp}}$ in $L^2\,mg^{-2}$	3.54862
$s_{y.x,sp,2}$	0.001768	df	5

The same result yields the significance checks of the regression parameter a_2. The test value $\hat{t} = 0.646$ does not exceed the critical value $t(P = 95\%, df = 5) = 2.571$, and the confidence interval of the quadratic regression parameter $CI(a_2) = 2.29178 \pm 9.122$ cannot be distinguished from zero at the significance level $P = 95\%$.

Testing the trueness is carried out by comparison of the slopes obtained by the calibration method $a_{1,c} = 2.815625\,L\,mg^{-1}$ and $a_{1,sp} = 2.97773\,L\,mg^{-1}$. A proportional systematic error is caused by the matrix if the test value $\hat{t}$ calculated by (5.7.6-1) and (5.7.6-2) is greater than the critical value $t(P, df_{tot} = df_c + df_{st})$ at the chosen significance level P. Because the test value $\hat{t} = 10.929$ calculated with the data given in Table 5.10-10 is much greater than the critical value $t(P = 95\%, df = 13) = 2.160$, the matrix is confirmed to have a significant influence on the regression parameters, resulting in false results.

(d) The tests of trueness presented above are only allowed when the iron-containing matrix does not affect the precision of the method. This is checked by an F-test.

(continued)

The test value for the recovery method is $\hat{F} = 3.365$ calculated by (5.7.5-7) with $s_{y.x,m} = 0.001353\ \mathrm{mg\,L^{-1}}$ and $s_{x.0,c} = 0.000737\ \mathrm{mg\,L^{-1}}$. The test value does not exceed the critical value $F(P = 99\%, \mathrm{df}_1 = \mathrm{df}_2 = 7) = 6.993$, and thus the matrix does not affect the precision of the analytical method.

The same result is obtained by the standard addition method but, because the calibration error of the standard addition method $s_{y.x,\mathrm{sp}} = 0.00168$ is smaller than that of the calibration method $s_{y.x,c} = 0.002076$, the test value must not be calculated.

(e) The regression parameters and further data for the calculation of the confidence interval calculated with the calibration data set given in Table 5.10-11 are listed in Table 5.10-12.

The predicted value $\hat{x}$ and its confidence interval $\mathrm{CI}(\hat{x})$ calculated according to (5.7.8-3) and (5.7.8-4) are:

$$\hat{x}_{val} = \frac{0.14156 - 0.0004}{2.839167\ \mathrm{L\,mg^{-1}}} \cdot \frac{25\ \mathrm{mL}}{18\ \mathrm{mL}} = 0.069\ \mathrm{mg\,L^{-1}} \qquad (5.10\text{-}2)$$

and

$$\begin{aligned}\mathrm{CI}(\hat{x}_{\mathrm{val}}) &= \frac{0.001916}{2.839167\ \mathrm{L\,mg^{-1}}} \cdot 3.182 \\ &\quad \cdot \sqrt{1 + \frac{1}{5} + \frac{(0.14156 - 0.2097)^2}{2.839167^2\ \mathrm{L^2\,mg^{-2}} \cdot 0.00144\ \mathrm{mg^2\,L^{-2}}}} \\ &\quad \cdot \frac{25\ \mathrm{mL}}{18\ \mathrm{mL}} \\ &= 0.004\,\mathrm{mg\,L^{-1}}. \end{aligned} \qquad (5.10\text{-}3)$$

(continued)

Table 5.10-11 Calculated spiked concentration $c_{\mathrm{sp,val}}$ and the measured response y_i for checking the standard addition method

Level	$c_{\mathrm{sp,val}}$ in $\mathrm{mg\,L^{-1}}$	$y_i(A_i)$
1	0	0.1422
2	0.012	0.1767
3	0.024	0.2069
4	0.036	0.2436
5	0.048	0.2791

Table 5.10-12 Parameters of the linear regression of the standard addition method and further data necessary for the calculation of the confidence interval

$a_{0,\mathrm{val}} = \hat{y}_{0,\mathrm{val}}$	0.14156	$a_{1,\mathrm{val}}$ in $\mathrm{L\,mg^{-1}}$	2.839167
$s_{y.x,\mathrm{val}}$	0.001916	$\bar{y}_{\mathrm{val}}$	0.2097
$\mathrm{SS}_{xx,\mathrm{val}}$ in $\mathrm{mg^2\,L^{-2}}$	0.00144	$t(P = 95\%, \mathrm{df} = 3)$	3.182

Thus, the true value $\mu_{val} = 0.07\ \text{mg L}^{-1}$ lies within the range of the confidence interval 0.065–0.073 mg L^{-1}, which means that the analytical method is valid.

(f) The regression parameters calculated with the calibration data given in Table 5.10-13 and further parameters necessary for the calculation of the confidence interval are listed in Table 5.10-14. Note that the test of limit values is a one-sided problem; therefore, the one-sided *t*-value must be used in order to calculate the confidence interval.

The predicted value $\hat{x}$ and its confidence interval $\text{CI}(\hat{x})$ calculated by (5.7.8-2) and (5.7.8-4) are

$$\hat{x}_a = 0.124\ \text{mg L}^{-1}, \tag{5.10-4}$$

$$\text{CI}(\hat{x}_a) = 0.0050\ \text{mg L}^{-1}. \tag{5.10-5}$$

Considering the volume factor

$$f_V = \frac{V_{\text{flask}}}{V_{\text{sample}}} = \frac{25\ \text{mL}}{20\ \text{mL}} = 1.25 \tag{5.10-6}$$

the concentration of the waste water sample is $c_{\text{sample}} = 0.1555 \pm 0.0063\ \text{mg L}^{-1}$ N. Thus, the upper analytical result is $c_{\text{sample}} + \text{CI}(c_{\text{sample}}) = 0.162\ \text{mg L}^{-1}$ N which does not exceed the threshold value $L_0 = 0.163\ \text{mg L}^{-1}$.

(continued)

Table 5.10-13 Calculated spiked concentrations for the analysis $c_{sp,a}$ and measured responses $y_i(A_i)$ by the standard addition method

Level	$c_{sp,a}$ in mg L^{-1}	$y_i(A_i)$
1	0	0.3555
2	0.03	0.4418
3	0.06	0.5173
4	0.09	0.6091
5	0.12	0.6978

Table 5.10-14 Parameters of the linear regression of the standard addition method and further data necessary for calculation of the confidence interval

$a_{0,a} = \hat{y}_{0,\text{val}}$	0.35392	$a_{1,a}$ in L mg^{-1}	2.839667
$s_{y.x,a}$	0.004783	$\bar{y}_a$	0.5243
$\text{SS}_{xx,a}$ in $\text{mg}^2\,\text{L}^{-2}$	0.009	n_a	5
$t_{\text{one-sided}}(\bar{P} = 95\%, \text{df})$	2.353	df_a	3

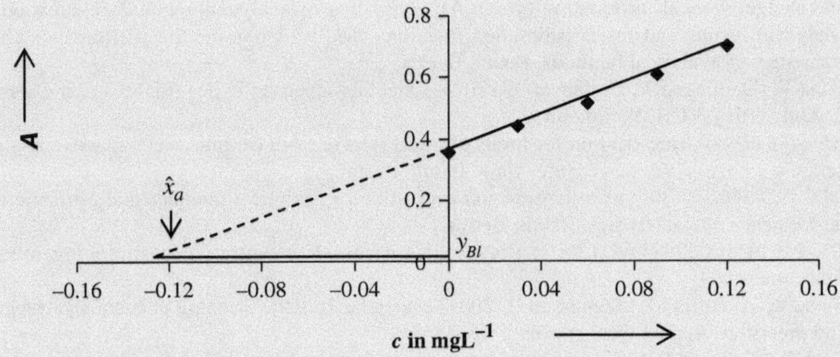

Fig. 5.10-3 Determination of the predicted value $\hat{x}$ by the standard addition method

Note that using the confidence interval calculated by the two-sided t-factor $t(P = 95, \mathrm{df} = 3) = 3.182$, the upper analytical result is $\hat{x} = 0.164\ \mathrm{mg\,L^{-1}}$, resulting in a false decision as the limit value is exceeded.

The graphical representation of the determination of the predicted value $c_{\text{stocked sample}} = \hat{x}_a$ is shown in Fig. 5.10-3. Note that the dilution factor must still be taken into account in the analytical result.

References

1. DIN EN ISO/IEC 17025 (1999) General requirements for the competence of testing and calibration laboratories. Beuth, Berlin
2. ISO 3534-2 (2006) Statistics – vocabulary and symbols – part 2: applied statistics. International Organization for Standardization, Geneva
3. ICH Harmonised Tripartite Guideline (2005) Validation of analytical procedures: text and methodology Q2(R1). http://www.ICH.org
4. DIN EN 26777 (1993) Water quality; determination of nitrite; molecular absorption spectrometric method. Beuth, Berlin
5. Joint Committee for Guides in Metrology (JCGM) (2008) International vocabulary of metrology – basis and general concepts and associated terms (VIM). International Organization for Standardization, Geneva
6. Pharm. EUR. 6. Ausgabe, hppt://www.pharmeur.org
7. OECD-Dokumente zur Guten Laborpraxis, hppt://bfr.bund.de/cd/473
8. Good Laboratory Practice, hppt://de.wikipedia.org/wiki/GLP
9. Albert R, Horowitz W (1997) A heuristic derivation of the Horowitz curve. Anal Chem 69:789–790
10. Horowitz W, Kamps LR, Boyer KW (1980) Quality assurance in the analysis of of foods for trace constituents. J Assoc Off Anal Chem 63:1344
11. Massart DL, Vandeginste BGM, Buydens LMC, De Jong S, Lewi PJ, Smeyers-Verbeke J (1997) Handbook of chemometrics and qualimetrics, part A. Elsevier, Amsterdam

12. DIN 38402–51 (1986) German standard methods for the examination of water, waste water and sludge; general information (group A); calibration of analytical methods, evaluation of analytical results and linear calibration functions used to determine the performance characteristics of analytical methods. Beuth, Berlin
13. Funk W, Dammann V, Donnevert G (2005) Qualitätssicherung in der Analytischen Chemie, 2. Aufl. Wiley-VCH, Weinheim
14. BS 6748 (1986) Specification for limits of metal release from ceramic ware, glassware, glass ceramic ware and vireous enamel ware. Beuth, Berlin
15. DIN ISO 8466-2 (2004) Calibration and evaluation of analytical methods and estimation of performance characteristics. Beuth, Berlin
16. DE-Patent 10042498A1 (2002) Polyasparaginsäure Konzentrationsbestimmung mittels Fluorometrie
17. Menditto A, Patria M, Magnusson B (2007) Understanding the meaning of accuracy, trueness and precision. Accred Qual Assur 12:45–47
18. DIN 38402-71 (2002) German standard methods for the examination of water, waste water and sludge – general information- part 71: Procedure for verifying the equivalence of two analytical methods involving quantitative characteristics with a continous set of values. Beuth, Berlin
19. DIN EN ISO 10304-1 (2009) Water quality – determination of dissolved anions by liquid chromatography of ions – part 1, determination of bromide, chloride, fluoride, nitrate, nitrite, phosphate and sulfate. Beuth, Berlin
20. Thurman EM, Mills MS (1998) Solid-phase-extraction – principles and practice, chemical analysis: A series of monographs on analytical chemistry and its applications Vol 147. Wiley-VCH, Weinheim
21. Pawliszyn J (1996) Solid-phase-microextraction: theory and practice. Wiley-VCH, Weinheim
22. Kolb B, Ettre LS (2006) Static headspace-gas chromatography – theory and practice, 2nd edn. Wiley-Interscience, New York
23. DIN32633 (1998) Chemical analysis – methods of standard addition – procedure, evaluation. Beuth, Berlin
24. Ehrlich G, Danzer K (2007) Nachweisvermögen von Analysenverfahren – Objektive Bewertung und Ergebnisinterpretation. Springer, Berlin, Heidelberg
25. Danzer K (2007) Analytical chemistry – theoretical and metrological fundamentals. Springer, Berlin, Heidelberg
26. DIN 32645 (1994) Nachweis-, Erfassungs- und Bestimmungsgrenze. Ermittlung unter Wiederholbedingungen. Begriffe Verfahren Auswertung, Beuth, Berlin
27. ISO11843–2 (2000) Capability of detection, part 2: methodology in the linear calibration case. International Organization for Standardization, Geneva
28. Currie LA (1995) Nomenclature in evaluation of analytical methods including detection and quantification capabilities. Pure Appl Chem 67:1699–1723
29. Kleiner J (2000), Software SQS2000. http://www.kleiner-j.de
30. Geiß S, Einax JW (2001) Comparison of detection limits in environmental analysis – is it possible? An approach on quality assurance in the lower working range by verification. Fresenius J Anal Chem 370:675–678
31. ICH Harmonised Tripartite Guideline (2005) Validation of analytical methods: Methodology, Q2A http://www.ich.org
32. DIN EN ISO 6878 Water quality – Determination of phosphorus – Ammonium molybdate spectrometirc method, Beuth, Berlin
33. The United States pharmacopoeia, USP24, The National formulary NF19
34. Heyden YV, Nijhuis A, Smeyers-Verbeke J, Vandeginste BGM, Massart DL (2001) Guidance for robustness/ruggedness tests in method validation. J Pharmaceutical and Biomedical Analysis 24:723–53
35. Ellison SLR, Berwick VJ, Duguid Farrant TJ (2009) Practical statistics for the analytical scientist, 2nd edn. RSC, Cambridge

36. Doerffel K (1990) Statistik in der analytischen Chemie 5. Aufl. Deutscher Verlag für Grundstoffindustrie, Leipzig
37. Gliesing S, García DeLa Calle, Reichenbächer M, Böse M (1997) HPLC determination of tamoxifendihydrogencitrate with the base selective column UltraSep ES Parm RP8 in tablets. Pharmazie 52:3–4

Chapter 6
Aspects of Method Development

6.1 General Remarks

Chronologically, the selection and development of an appropriate analytical method for a specific analytical purpose is the first stage in the validation of a method which should be capable of producing results that are fit for a particular purpose. However, method development is a wide field with specific investigations for each method, and therefore in this book only some aspects of selected analytical methods – in particular, applications of chromatography – will be given; for detailed information see the corresponding literature, for example [1, 2].

In general, one of the chromatographic methods (GC, HPLC, IC) is chosen for the analysis of organic compounds. The next stage is the planning and carrying out of test experiments using starting conditions with the chosen phase system, e.g. the combination of the stationary and mobile phases, the detector, and other parameters such as column temperature, flow rate, etc. The purpose of these experiments is the optimization of the chromatographic conditions for a sufficient separation of all components of the sample which have to be determined, as demonstrated, for example, by the HPLC chromatogram in Fig. 5.9-1 and the validity of some performance parameters.

Such performance parameters, given for the simple chromatogram in Fig. 6.1-1, are:

- *Retention factor* of the component i, k_i'

  The retention factor describes the migration rate of an analyte on a column:

$$k_i' = \frac{t_{r,i} - t_M}{t_M}. \tag{6.1-1}$$

$t_{r,i}$ is the retention time of the component i of a solute which is taken as the elapsed time between the time of injection of a solute and the time of elution of the peak maximum of that solute, and t_M is the time taken for the mobile phase to pass through the column, also called dead time.

M. Reichenbächer and J.W. Einax, *Challenges in Analytical Quality Assurance*,
DOI 10.1007/978-3-642-16595-5_6, © Springer-Verlag Berlin Heidelberg 2011

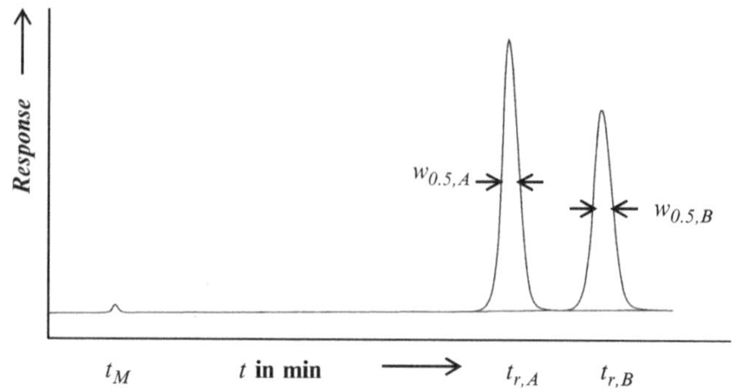

Fig. 6.1-1 Chromatogram of a sample with the components A and B

- *Selectivity factor α*

The selectivity factor describes the separation of two peaks:

$$\alpha = \frac{k'_{\mathrm{B}}}{k'_{\mathrm{A}}}. \tag{6.1-2}$$

The selectivity factor must be greater than 1.0 in order to separate two compounds, but whether the separation of two compounds is sufficient for a quantitative determination of both compounds is not determined by the selectivity α. The selectivity describes the separation based on the peak centres but does not take into account peak widths.

- *Peak symmetry*

Peak shape is an important factor in obtaining correct counts of the peak areas by means of correct integration.

The asymmetry factor As is calculated by (6.1-3)

$$\mathrm{As} = \frac{w_{5\%}}{2d}, \tag{6.1-3}$$

where d is the distance between the perpendicular dropped from the peak maximum to the leading edge of the peak at 5% of the peak height and $w_{5\%}$ is the width of the peak at 5% of peak height.

Unless otherwise stated in the regulatory documents, the values of As should fall between 0.8 and 1.6. Note that As = 1.0 corresponds to ideal symmetry.

- *Resolution R_s*

The resolution of two compounds, A and B, is defined as [3]

$$R_s = 1.18 \cdot \frac{t_{\mathrm{r,B}} - t_{\mathrm{r,A}}}{w_{0.5,\mathrm{A}} + w_{0.5,\mathrm{B}}}, \tag{6.1-4}$$

where $w_{0.5,\mathrm{A}}$ and $w_{0.5,\mathrm{B}}$ are the peak widths at half height of A and B, respectively.

$R_s \geqslant 1.2$ is, in general, sufficient for quantitative analysis and baseline resolution is achieved when $R_s \geqslant 1.5$.

The resolution is determined by three terms, according to (6.1-5):

$$R_s = 0.25 \cdot \underset{\mathrm{I}}{\sqrt{N}} \cdot \underset{\mathrm{II}}{\left(\frac{1+k'_B}{k'_B}\right)} \cdot \underset{\mathrm{III}}{\left(\frac{\alpha-1}{\alpha}\right)}, \quad (6.1\text{-}5)$$

To obtain high resolution, these three terms must be optimized. An increase in N, the number of theoretical plates (term I), by lengthening the column leads to an increase in retention time (and, therefore, the analysis time!) and increased band broadening – which is not desirable. Instead of increasing the number of plates, the height equivalent of a theoretical plate can be reduced by reducing the size of the stationary phase particles.

It is often found that separations can be considerably improved by controlling the capacity factor k' (term II). This can be achieved by changing the temperature (in GC) or the composition of the mobile phase (in HPLC).

Optimization of the selectivity factor α (term III) is the best way to improve separations. In general, k' is optimized first and then α is increased, for example by

- Changing the mobile phase composition (percentage or change of the organic component, pH, additives, for example)
- Changing the stationary phase (a different polarity of the column, for example)
- Changing the temperature of the column in HPLC or the temperature program in GC.

Although the performance parameters are obtained by the software package of the instrument, one should also be able to determine these parameters from a chromatogram.

The validation of HPLC methods in pharmaceutical analysis according to various regulatory requirements are compared in [4].

Challenge 6.1-1

Figure 6.1-2 shows a section of a GC chromatogram of the separation of some aromatics using the column SPB 5 ($d_f = 0.25$ µm, $L = 25$ m). Determine the selectivity factor α and the resolution $R_s(A/B)$ for the peak pair A/B as well as the peak asymmetry As at the peak C. The dead time is $t_M = 0.62$ min.

Is the resolution sufficient for quantitative analysis of the components A and B?

Solution to Challenge 6.1-1

The peak widths of A and B obtained using the scale 19 mm = 0.1 min are $w_{0.5,A} = 0.02368$ min and $w_{0.5,B} = 0.02632$ min (Fig. 6.1-3)

(continued)

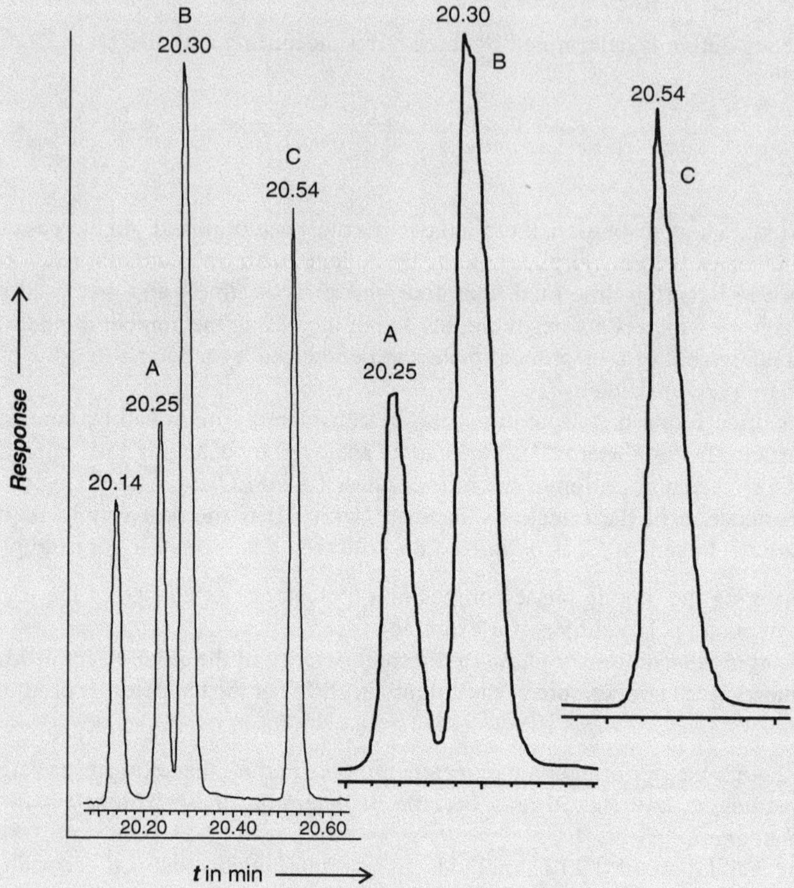

Fig. 6.1-2 Section of a GC chromatogram of aromatics

The results are summarized in Table 6.1-1.

Note that chromatographic methods are not stable in time, usually due to alterations in the column, as shown in Fig. 6.1-4 by the two chromatograms of a test mixture which were obtained with the same sample using the same column but at different times.

Testing must therefore be used to confirm that the system will function correctly whenever the method is applied. Chromatographic systems must have a system suitability requirement which has to be specified in the course of the method development. System suitability parameters are needed to ensure the quality of separation. Acceptance criteria should be established based on data observed during method development. Whenever the method is applied, it must be checked by a

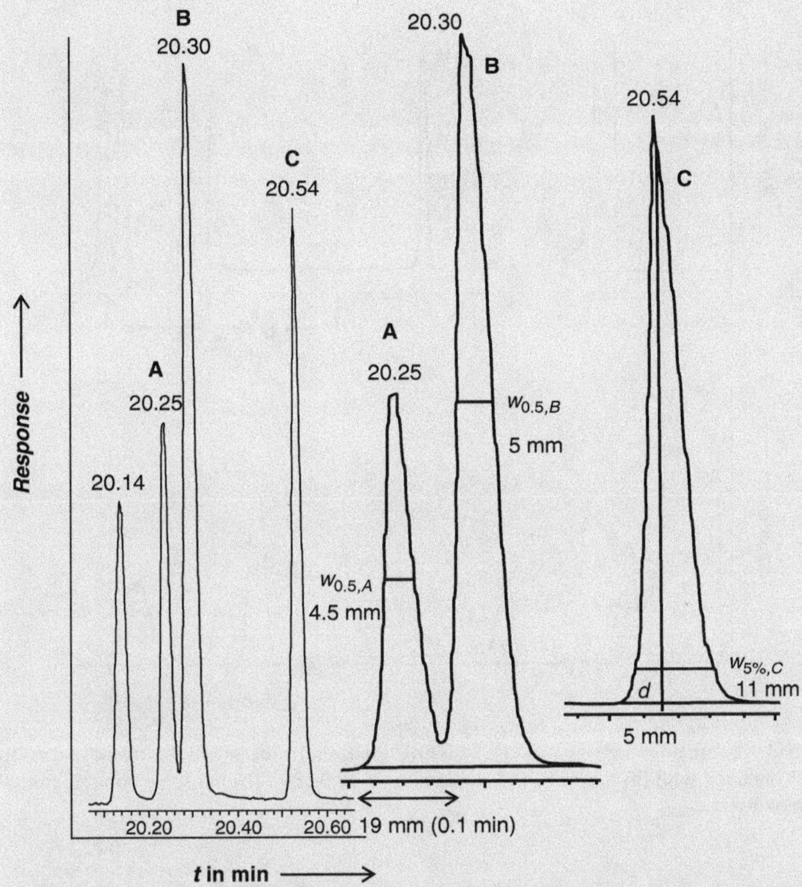

Fig. 6.1-3 Delineated peak widths at half height ($w_{0.5}$) and 5% height ($w_{5\%}$), respectively, for the calculation of performance parameters of the chromatographic system

Table 6.1-1 Performance parameters of the chromatographic system calculated by the parameters given in Fig. 6.1-3

Parameter	Equation	Value
Retention factor k'_A	(6.1-1)	31.66
Retention factor k'_B	(6.1-1)	31.74
Selectivity factor α	(6.1-2)	1.0025
Resolution $R_s(A/B)$	(6.1-4)	1.18
Asymmetry factor As(C)	(6.1-3)	1.10

so-called *system suitability test* (SST) to determine whether these parameters are still met.

As an example, the suitability parameters given by the results obtained during the method development of the determination of the assay of the *Z*-isomer of

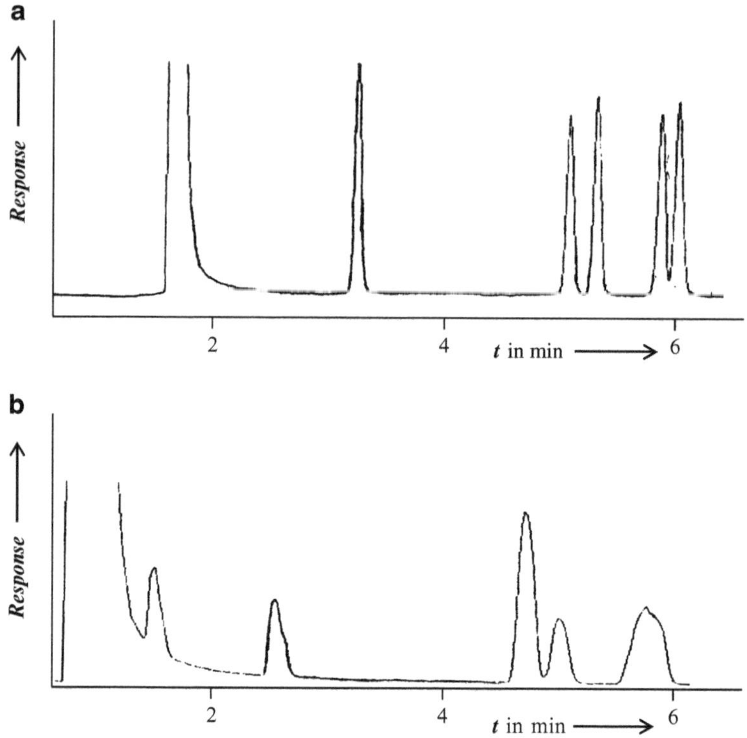

Fig. 6.1-4 Chromatograms of a mixture of BTXE (benzene, toluene, ethyl benzene, *p*-, *m*- and *o*-xylene) obtained with (**a**) a new SPB 5 column, $L = 20$ m, and (**b**) the same column after about three months' use

Table 6.1-2 Parameters of the HPLC chromatogram of Fig. 5.9-1 obtained under optimized conditions by means of the SST software package of the instrument

RT in min	Compounds	k'	Plate in m^{-1}	As	α	R_s
4.367	*Z*-isomer	2.485	32,556	1.56		
5.030	*Bis*-tamoxifen	3.015	34,742	1.16	1.213	2.30
5.887	*E*-isomer	3.699	22,399	1.43	1.227	2.29
6.603	Desmethyl-tamoxifen	4.271	39,391	1.77	1.155	1.75

tamoxifene by HPLC, presented in Fig. 5.9-1, are summarized in Table 6.1-2. The requirements of the SST are based on these data.

However, there is another important parameter, the *peak purity*, which describes the co-elution of peaks. Is the peak area of the analyte generated by the analyte alone, or do other components interfere with the peak? If the latter is the case, then false peak areas are obtained which yield incorrect analytical results. The peak purity is related to the validation parameters selectivity and specificity which are, because of their relevance to the trueness of analytical results, explicitly listed as validation parameters, but these parameters must already be checked at the stage of method development.

6.2 Selectivity and Specificity

Selectivity and specificity are measures that assess the reliability of measurements in the presence of interferences. The *selectivity* refers to the extent to which particular analyte(s) in a complex mixture can be detected without interferences from the other components in the mixture. A method that is completely *selective* for *one individual* analyte in the mixture is said to be specific for that substance.

There are some procedures to establishing the selectivity of the method; for example:

1. Confirmation of the analyte identity and ability to measure the analyte in isolation from the interferences by measurement of the sample and corresponding reference materials.

   Let us return to the GC chromatogram of the analytes BTXE shown in Fig. 6.1-4a. The retention times of peaks obtained by the CRS will be identical to that of the mixture. But the injection of a solution of CRS (*m*-xylene) yields a chromatogram with retention time identical to *p*-xylene using a unpolar column SPB 5, whereas the *ortho*-isomer gives a chromatogram with greater retention time. Thus, the comparison of the retention times is not fit for testing the selectivity of *p*-xylene in GC under the conditions given in Fig. 6.1-2. All other BTXE compounds can be detected with CRS.
2. Comparison of spectral data obtained at various positions of the peak with those of the library of the instrument or with spectra obtained by CRS, for example, spectral properties (selective UV-wavelength, fluorescence, IR spectra), mass spectrometry (MS) including fragmentation, or selective reactions (sensor).

   Let us consider further the problem in Fig. 6.1-2. The mass spectra of the three xylene isomers as well as ethyl benzene cannot be distinguished either according to their molecular peaks or the fragmentations. MS of all four compounds gives the same molecular ion peak, $m/z = 91$ amu.
3. Changing the phase should always be included in the checks for selectivity. This may be done by using another mobile phase (changing the organic component or its proportion in the mobile phase, pH, additives) as demonstrated by Fig. 6.2-1, applying another stationary phase, changing the temperature of the column, etc.

   Note that the selectivity of optical isomers can be checked only by optically active phases.

   Let us return to the problem of the separation of the isomers *m*- and *p*-xylene. The change from the unpolar stationary phase, giving the chromatogram in

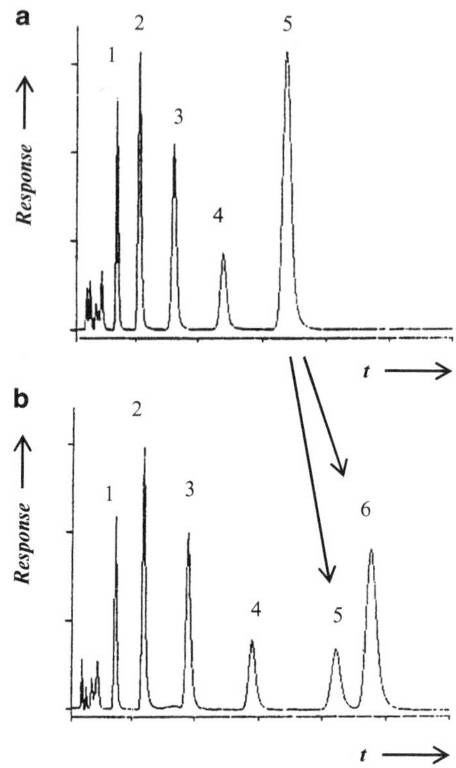

Fig. 6.2-1 HPLC chromatograms of aromatics obtained by two different compositions of the mobile phase: (**a**) methanol/water/THF: 45/42/13% (v/v/v), (**b**) methanol/water/THF: 50/42/8% (v/v/v)

Fig. 6.1-4a, to a strongly polar phase will cause the separation of the isomers *m*- und *p*-xylene enabling us to estimate the peak purity. Now, the separation takes place not only on the basis of the boiling points, which are nearly equal for both isomers, but preferentially on the basis of the polarity, which is greater for the *m*-isomer because of its dipole moment. Therefore, the retention time of *m*-xylene will increase and it is thus separated from the analyte *p*-xylene if present.

4. Changing the detector in GC analysis, for example ECD instead of FID, can also be used in order to check if the analyte peak has interferences superimposed on it. Note that the estimation of peak symmetry, sometimes recommended in checking the selectivity, can give rise to errors because the asymmetry can also be caused by an unsuitable phase system: for example, separation of unpolar compounds on a strongly polar solid phase.

Challenge 6.2-1

(a) Figure 6.2-2 shows a section of the HPLC chromatogram of a vitamin D_3 assay and the diode array detection (DAD) spectra obtained at the ascending and the descending positions of the analyte peak.
Estimate the selectivity of the peak at the retention time $t_r = 5.302$ min.

(continued)

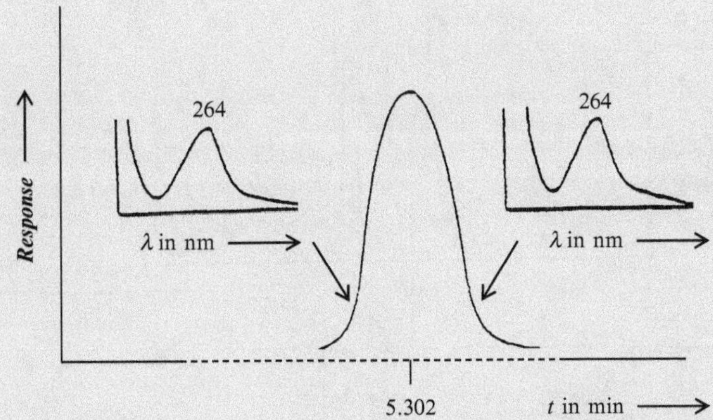

Fig. 6.2-2 A section of the HPLC chromatogram of a vitamin D_3 assay and the DAD spectra obtained at the ascending and the descending positions of the analyte peak (marked by *arrows*)

(b) A section of the SPME-GC chromatogram of a white wine is given in Fig. 6.2-3, together with the mass spectra obtained at three positions, marked by arrows, of the two peaks **A** and **B**. The symmetry of peak **A** is $As = 1.05$ whereas peak **B** shows a pronounced shoulder.
Estimate the selectivity of both peaks.

Solution to Challenge 6.2-1

(a) The spectra obtained at two different positions of the analyte peak are identical, and therefore the peak at the retention time $t_r = 5.302$ min does not have other interferences superimposed on it, which means that selectivity is adequate. Furthermore, the maxima of both UV spectra coincide with the known spectra of vitamin D_3.

(b) The highly symmetrical peak **A** shows at its descending position a significantly different mass spectrum than is obtained at the ascending position. Thus, for example, the intense peaks at $m/z = 41$ and 59 amu disappear and new peaks arise ($m/z = 42, 55, 86$ amu). It is obvious that another compound is superimposed on the peak; therefore, the selectivity is not adequate although it is a highly symmetrical peak.

The fairly unsymmetrical peak **B**, however, shows identical mass spectra at all three positions with the base peak at $m/z = 60$ amu, which is caused by a McLafferty rearrangement of a aliphatic carboxylic acid. The asymmetry of this peak is caused by the large difference between the polarity

(continued)

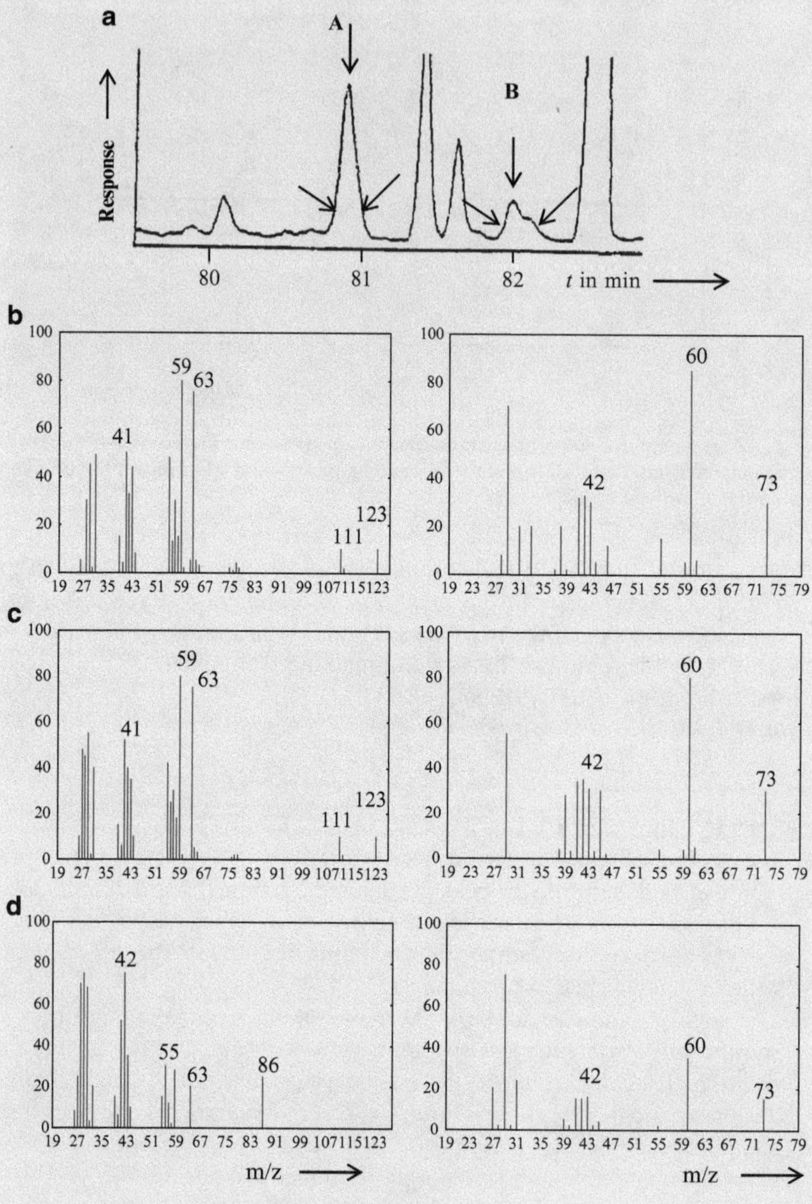

Fig. 6.2-3 Section of the SPME-GC chromatogram using the nonpolar column SPB 5 (**a**) together with the mass spectra obtained at the ascending (**b**), the maximum (**c**) and the descending positions (**d**) of each peak **A** and **B** [6]

of the analyte and the low polarity of the column which disturbs the interactions between analyte and stationary phase along the column.
Now, after we have learned the most important stages of the development of chromatographic methods, as well as the subsequent method validation procedure, we will apply this knowledge to an example, the development and validation of analysis by headspace gas chromatography.

6.3 Method Development of Headspace Gas Chromatography

Headspace gas chromatography (HS-GC) is one of the most important analytical methods for the determination of organic compounds in liquid and solid samples [5]. Examples of its application are:

- Liquid samples
  Organic compounds in drinking or waste water
  Trace components in beverages
  Aromatics in exhausted mineral oil
  Determination of the alcohol content in blood
- Soluble samples
  Content of monomers in polymers
  Residual solvents in pharmaceuticals
- Insoluble samples
  Quality control of foodstuffs and semi-luxury foods
  Volatile compounds in soil

After thermal equilibration of the sample in a tightly closed headspace vial, a partial amount of the analyte in the headspace is transferred to the injection system of the gas chromatograph. This procedure is called *static* headspace GC. Therefore, headspace GC, in contrast to normal GC, is an *indirect* and a *partial* method; *indirect*, because the analyte is not directly injected as in normal GC, and *partial*, because only a small part of the sample in the headspace is transferred to the injector.

Because the analytes are separated from the matrix, HS-GC is called a matrix-free analytical method. An advantage of the HS-GC method is the omission of time-consuming sample extraction steps, but there are some problems in the quantitative analysis. Remember that only a part of the analyte, defined by Henry's law, is transferred into the headspace, only a small part of this amount is transferred to the injector, only a part of this, determined by the split relation, finally arrives at the column, and furthermore the peak area of the chromatogram is caused by the detector response. Calibration is therefore required for quantitative analysis, which can be performed, in general, without difficulty for liquid samples, for example by means of the known standard addition method. But calibration for solid samples is not possible because, in general, no standards are available. However, before we turn to quantitative determination by HS-GC let us examine factors which determine the *sensitivity* of HS-GC, an important parameter for its application in trace analysis.

The sensitivity expressed as the peak area A of the chromatogram is determined by the concentration $c_{0,i}$ of the analyte i in the sample, the partition constant K of the analyte i and the phase relation β:

$$A = \mathrm{Rf} \cdot \frac{c_{0,i}}{K + \beta} \qquad (6.3\text{-}1)$$

with

$$K = \frac{c_i^l}{c_i^{\mathrm{HS}}} \qquad (6.3\text{-}2)$$

and

$$\beta = \frac{V^{\mathrm{HS}}}{V^l}. \qquad (6.3\text{-}3)$$

Rf is the response factor, c_i^l,c_i^{HS} are the concentration of the analyte in the liquid and the headspace, respectively, and V^l, V^{HS} are the volumes of the liquid and the headspace, respectively.

As the equations show, for highly volatile compounds having very small partition constants (close to zero at the temperature of the equilibration), the peak area is mainly determined by the phase relation β. According to (6.3-3), the sensitivity increases with the volume of the sample V^l, whereas for water-soluble compounds the sensitivity is hardly influenced by the sample volume. Decreasing the constant K by enhancement of the temperature of the equilibration is limited by the boiling point of water; thus 80°C is, in general, the highest temperature.

A further enhancement in sensitivity can be achieved by salting the sample, which diminishes the partition constant K. Finally, the sensitivity can be influenced by the response factor Rf in (6.3-1) using a detector of a higher sensitivity, for example ECD instead of FID.

Next in the method development of a HS-GC method is finding the optimal *time of the equilibration* which should not be markedly greater than necessary to reach equilibrium, because longer time can lead to loss of analytes by diffusion into the septum. The optimal time has to be determined experimentally for all analytes.

Finally, the *septa* used as closure of the headspace flasks must be checked as to whether they are appropriate. Testing whether parts of the analyte will diffuse into the septa during the equilibration is made by repeated headspace analysis of the septa used. The chromatogram must be free of peaks in the range of the analyte.

The *development* of the headspace conditions is finished after optimization of the following parameters:

- Equilibration temperature
- Equilibrium time
- Appropriate septa used for the headspace flasks
- The volume of the sample used for the headspace flasks
- The headspace volume which is transferred into the GC injector
- GC conditions (column, flow rate, detector, split rate)

Next, the *validation* steps can begin.

The *precision* of the injection must be determined with a test sample. This is performed using replicates. Note that in contrast to the normal GC method, the replicates must be carried out with each *new* prepared headspace sample. If the relative standard deviation of the injection precision is greater than a given limit, e.g. $s_r\% \geqslant 2$, then it can be tested whether the *internal standard* method will improve the injection precision. In this case, $s_r\%$ is calculated by relative peak areas obtained by the proportion of the peak areas of the analyte and those of the standard added to each sample in the same amount. The internal standard used should be chosen from the same class of compounds, it should lie roughly in the middle of the chromatogram and it must not interfere with other peaks.

After explaining the steps of method development, let us turn to the problem of the *quantitative analysis* carried out by HS-GC methods.

Remember that in normal GC the peak area A is proportional to the analyte i concentration c_0:

$$A = \mathrm{Rf} \cdot c_{0,i} \tag{6.3-4}$$

but in static HS-GC analysis the peak area is caused by the partial vapor pressure of the component i:

$$A = \mathrm{Rf}' \cdot p_i. \tag{6.3-5}$$

According to Raoult's law, the partial vapor pressure is given by the vapor pressure of the pure component i p_i^0, the mole fraction x_i, and the activity coefficient γ_i:

$$p_i = p_i^0 \cdot x_i \cdot \gamma_i. \tag{6.3-6}$$

The influences on the peak areas must be evaluated by calibration.

The method of calibration is preferably given by the estimation of the activity coefficient γ in (6.3-6) of the calibration solution sample relative to the sample.

Three cases may be distinguished using the activity coefficient:

1. The activity coefficients of the matrix of the sample and the calibration solutions are equal, $\gamma_{\text{sample}} = \gamma_{\text{calibration}}$.

   For example, the calibration standards are prepared with unused mineral oil for the determination of aromatics in exhausted mineral oil.

2. The activity coefficient of the calibration solution can be simulated, $\gamma_{\text{sample}} \approx \gamma_{\text{calibration}}$.
   For example, the calibration standards used for the determination of aromatic compounds in white wine are prepared using 10% (v/v) alcoholic solutions instead of pure water.
3. The activity coefficient of the sample is unknown and it cannot be simulated, $\gamma_{\text{sample}} \neq \gamma_{\text{calibration}}$.
   In this case, a matrix-independent method must be used for the calibration.

Calibration solutions for the determination of analytes in samples of cases 1 and 2 are prepared as in the examples above, and the quantitative analysis is carried out as described in Chap. 5.

Let us turn to the samples of case 3, to which, for example, the solid samples belong. If the solid sample can be dissolved in water or in an organic solvent, then the determination of the analytes can be carried out analogously to samples of case 2.

For example, the determination of styrene in polystyrene can be achieved by dissolving the polymer in dimethyl-formamide (DMF), and the calibration solutions can also be prepared by dissolving styrene in DMF. Apart from the fact that the solvent DMF can damage the GC column, dissolving the sample will strongly enhance the partition coefficient of styrene and, therefore, diminish the sensitivity so that the analyte cannot be determined in the given concentration range.

A better analytical method for solid samples, and also for insoluble samples, is multiple headspace extraction (MHE). After the extraction of the total amount of the sample by multiple extraction steps n, the sum of peak areas $\sum A_n$ is proportional to the concentration of the analyte c_0 in the sample:

$$\sum A_n = \text{Rf} \cdot c_0. \tag{6.3-7}$$

The determination of the response factor Rf will be explained later. First, we will learn how to determine the sum of peak areas $\sum A_n$.

The peak areas decrease exponentially with the number of extraction steps n, which is shown in Fig. 6.3-1 for the example of MHE of benzene in a soil sample.

In general, the graph

$$\ln A = f(n) \tag{6.3-8}$$

yields a straight line as shown in Fig. 6.3-2.

If the linearity of the plot $\ln A = f(n)$ is confirmed (and only under this condition!), the sum of areas $\sum A_n$ according to (6.3-9) can be derived:

$$\sum A_n = \frac{A_1}{1 - e^{-k}}, \tag{6.3-9}$$

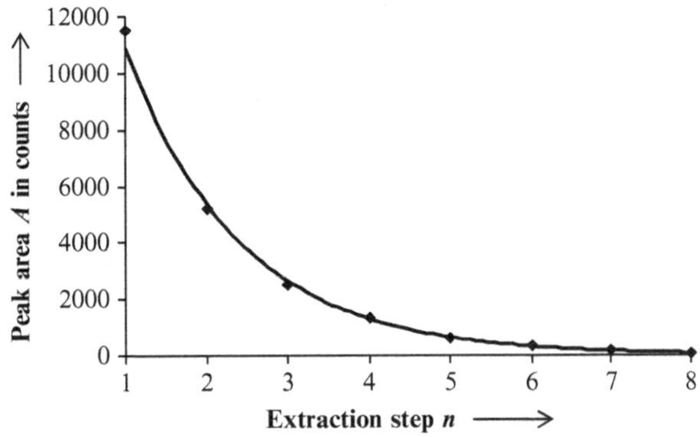

Fig. 6.3-1 Decrease of the peak areas A in the multiple headspace extraction GC chromatograms for benzene in a soil sample

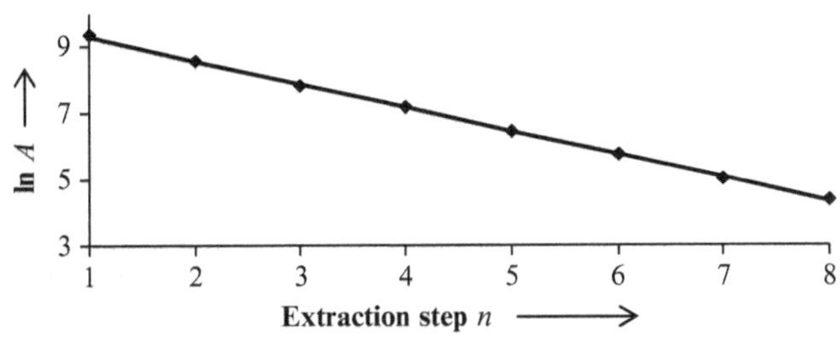

Fig. 6.3-2 The linear plot $\ln A = f(n)$ for the MHE given in Fig. 6.3-1

with

$$k = \ln\frac{A_1}{A_2}, \tag{6.3-10}$$

and

$$\sum A_n = \frac{A_1^2}{A_1 - A_2}. \tag{6.3-11}$$

A_1 and A_2 are the peak areas of the first and second extraction step, respectively.

Thus, according to (6.3-11) the sum of areas $\sum A_n$ is obtained by only two extraction steps, provided the linearity of (6.3-8) was confirmed in the method validation. If linearity is not present the MHE method cannot be applied at all.

Although the sum of areas can be obtained by two extraction steps, if more accurate results are desirable the value of A_1 should be determined by the intercept of the function $\ln A = f(n)$ verified with more than two extraction steps. The decisive quantity A_1 required for the determination of the sum of areas according to (6.3-11) is then confirmed by more than two values.

According to (6.3-9), if k is known, the determination of the sum of areas $\sum A_n$ can be reduced to *one* extraction. The constant k can be determined by samples with known analyte content, provided that the matrix does not alter the thermal equilibrium. This must be checked if this simplification of MHE is applied, for example, in the course of the quality control of batches all produced by the same method.

In order to calculate the analytical result c_0 according to (6.3-7) the response factor Rf must still be determined, which can be done in two ways:

1. The sum of areas is determined under the same HS and GC conditions using a reference sample with known content of the analyte(s):

$$m_{i,\text{sample}} = \frac{\sum A_{n.i,\text{sample}}}{\sum A_{n,i,\text{ref}}} \cdot m_{\text{ref}}. \tag{6.3-12}$$

   The mass of the analyte is then calculated by (6.3-12) in which the index *ref* *r*efers to the reference sample.
2. Organic reference materials of solid samples are usually not available; the response factor is then determined by injecting a known amount of the analyte into the headspace vial and carrying out the analysis under the same headspace and GC conditions. The headspace vial is filled with glass pearls so that the headspace volumes of the sample and that of the calibration are not different. The measured peak area relates to the amount of the analyte in the headspace. The mass of the analyte in the sample m_{sample} is obtained by the known mass m_{cal} which was injected into the headspace, the sum of the peak areas of the sample $\sum A_{n,\text{sample}}$ and the peak area of the calibration run A_{cal}:

$$m_{\text{sample}} = \frac{\sum A_{n,\text{sample}}}{A_{\text{cal}}} \cdot m_{\text{cal}}. \tag{6.3-13}$$

Instead of using MHE, liquid inhomogenous or highly viscous samples or solid samples which do not provide clear homogeneous solutions can be analyzed by the stock method explained in Sect. 5.7.6.

If the preparation of a spiked solution of water-insoluble organic compounds is required, the analytes must be included by means of a water-soluble modifier such as acetone, as described in Challenge 4.5-3. The analytical results can be obtained by a calibration line according to (5.7.8-2) or with only one calibration solution whose concentration is approximately that of the sample. Note that the so-called *single point method* places the calibration line through the zero point, which is not correct. But the error may be neglected if both the concentrations are almost the same.

If the single point method is used, the concentration of the sample c_{sample} is calculated by (6.3-14) for the external method

$$c_{\text{sample}} = \frac{A_{\text{sample}}}{A_{\text{spiked}}} \cdot c_{\text{spiked}}, \tag{6.3-14}$$

where c_{spiked} is the concentration of the spiked sample, and A_{sample} and A_{spiked} are the peak areas obtained from the headspace GC of the sample and spiked sample, respectively.

Of course, the internal standard method can also be applied.

If the headspace analysis is carried out by an HS autosampler, then the amount of the headspace volume which is transferred to the injector of the gas chromatograph is determined by the *injection time,* which is the duration of opening of the outlet valve of the headspace vial. However, transfer of the headspace volumes can also be done using a gastight syringe. The data given in the following Challenges were obtained by the headspace autosampler HS 40 from Perkin-Elmer®.

Challenge 6.3-1

In the course of method development of HS-GC analysis of benzene in waste water, the following tests were carried out using a test sample of 10 ppm (w/w) benzene in water which was dissolved using the modifier acetone. The volume of the headspace vials used was 21 mL closed by butyl rubber septa. The sample volumes used are given in the following Challenges.

GC-conditions:

Carrier gas	H_2
Column	SPB 5, 0.32 mm, 0.25 µm, 15 m
Split	10:1

(a) Determination of the equilibration time
The peak areas *A* obtained by various equilibration times at 80°C are given in Table 6.3-1. The sample volumes were each 4 mL.
Which equilibrium time should be chosen for the headspace analysis of benzene in water?

(b) Salting
To improve the sensitivity, the salting effect was tested using NaCl and Na_2SO_4. Two grams of each salt were added to 5 mL sample solutions prepared as given above. The equilibration conditions were 25 min at 80°C. The results are listed in Table 6.3-2.
Check whether salting with NaCl and Na_2SO_4, respectively, significantly improves the sensitivity.

(continued)

Table 6.3-1 Determination of the thermal equilibrium

t in min	Peak area *A* in counts
1	36,820
3	105,342
6	158,562
12	197,384
20	208,967
25	210,655
30	210,393

Table 6.3-2 The peak areas A in counts obtained by salting with NaCl and Na_2SO_4 and without salting

No salting	NaCl	Na_2SO_4
200,352	204,684	287,463
204,492	213,823	278,469
202,076	211,084	298,357
194,867	202,593	270,641
198,563	200,832	286,023
206,572	218,431	297,367

Table 6.3-3 Peak areas of the benzene analyte A_{bz} and the internal standard A_{IS} obtained by six replicates

A_{bz}	A_{IS}
22,200	24,995
21,507	24,114
26,889	25,138
21,895	24,726
23,793	26,524
22,456	25,273

(c) Injection precision
The injection precision should be $s_r\% \leqslant 2$. Check whether the required precision is achieved by the *external* standard procedure or whether the *internal* standard procedure must be applied on the basis of the results given in Table 6.3-3. The test solution described above was spiked with 2 μL internal standard solution prepared using 50 μL n-octane in 1 mL acetone. The headspace conditions are the same as given above.

Solution to Challenge 6.3-1

(a) As the function $A = f(t)$ presented in Fig. 6.3-3 shows, thermal equilibrium is achieved after about 25 min. This time can be regarded as the equilibrium time for the headspace analysis of benzene.

(b) The significance of the influence of salting on the sensitivity must be checked by a mean value t-test according to (3.5-5). The intermediate quantities and results of the tests for outliers, homogeneity of variances and the t-test are summarized in Table 6.3-4.

The test values of Dixon's test calculated by (3.2.3-1) with $b = 2$ and $k = n$ do not exceed the critical value $Q(P = 95\%, n = 6) = 0.560$, which means that no data need be rejected.

The precision is not significantly influenced by salting, which is checked by the Cochran test. The test value $\hat{C} = 0.6054$ calculated according (3.4-1) does not exceed the critical value $C(P = 95\%, k = 3, \text{df} = 5) = 0.7071$. Therefore, the mean value t-test can be carried out

(continued)

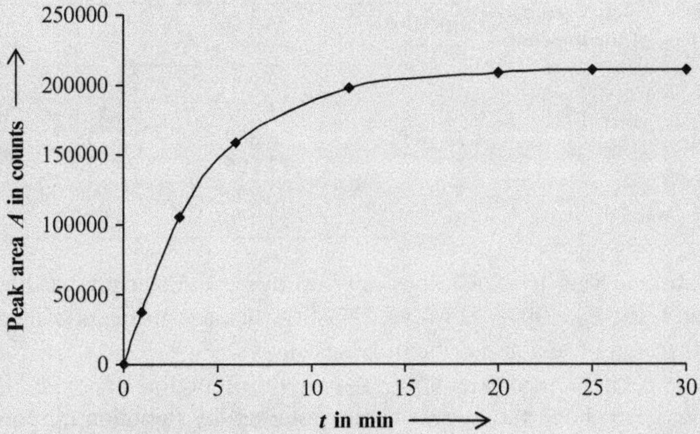

Fig. 6.3-3 The function $A = f(t)$ at 80°C for the headspace analysis of benzene

Table 6.3-4 Intermediate quantities and results of the significance test of salting estimated by the peak areas of benzene A_{bz}

Dixon outlier test			
	Without Salt	NaCl	Na_2SO_4
$x_1^* = x_{max}$	206,572	218,431	298,357
x_2	204,492	213,823	297,367
$\hat{Q}_{x_{max}}$	0.141	0.262	0.036
$x_1^* = x_{min}$	191,867	200,832	270,641
x_2	198,563	202,593	278,469
$\hat{Q}_{x_{min}}$	0.455	0.100	0.282
Cochran test for homogeneity of variances			
$s^2_{A,bz}$	26,689,981	48,371,115	115,152,562
$s^2_{A,bz,max}$	115,152,562	$\sum s^2_{A,bz}$	190,213,658
$\hat{C}$	0.6054		
t-test			
$s_{A,bz}$	5,166	6,955	10,731
$\bar{x}_{A,bz}$	200,653.7	208,574.5	286,386.7
Comparison "without salt" and "salting"			
NaCl			Na_2SO_4
s_p	6,126.22	s_p	8,421.48
$\hat{t}_{NaCl}$	2.239	$\hat{t}_{Na_2SO_4}$	17.633

in order to decide whether salting will improve the sensitivity and which of the salts tested should be used.

As Table 6.3-4 shows, the test value calculated for salting by NaCl is smaller than the critical value $t(P = 99\%, df = 10) = 3.169$, and

(continued)

Table 6.3-5 Relative peak areas A_{bz}/A_{IS} of the injection precision

Injection number	A_{bz}	A_{IS}	A_{bz}/A_{IS}
1	22,200	24,995	0.88818
2	21,507	24,114	0.89189
3	26,889	25,138	1.06966
4	21,895	24,726	0.88551
5	23,793	26,524	0.89704
6	22,456	25,273	0.88854

therefore salting by NaCl does not improve the sensitivity at the significance level $P = 99\%$. However, Na_2SO_4 has a significant effect, as the comparison of test and critical values shows.

(c) As the relative peak areas A_{bz}/A_{IS} of the injection precision listed in Table 6.3-5 show, the largest value, obtained by injection number three, must be checked as an outlier. The test value calculated by the Dixon test with $x_1 = 1.06966$, $x_2 = 0.89704$, and $x_n = 0.88551$ is $\hat{Q}_{x_{max}} = 0.937$. Because the test value exceeds the critical value $Q(P = 95\%, n = 6) = 0.560$, the peak areas of injection number 3 must be rejected.

The relative standard deviations of the injection precision calculated by the outlier-free data sets are $s_r\%_{A_{bz}} = 3.89$, which corresponds to the external standard procedure, and $s_r\%_{A_{bz/IS}} = 0.50$, which is the precision of the internal standard procedure. Thus, the internal standard procedure fulfills the requirement $s_r\% \leqslant 2$.

According to the results obtained by the checks on the method development given in (a)–(c), the headspace analysis of benzene should be carried out under the following conditions:

- Using the internal standard procedure with the internal standard *n*-octane
- The time for the equilibration at 80°C should be 25 min
- Salting with Na_2SO_4

Challenge 6.3-2

The validation of the headspace analysis of benzene in waste water must be carried out for the working range 5–20 ppm (w/w). The preparation of the calibration solution was made as described in Challenge 4.5-3 but 2 g Na_2SO_4 was added to each sample ($V = 10$ mL) and each sample was spiked with 5 μL internal standard before the headspace vials were closed.

The calibration data obtained by the headspace analysis of benzene with the optimized conditions are listed in Table 6.3-6.

(a) Check the regression function for linearity.
(b) Examine the calibration line for the presence of an outlier.
(c) Determine the calibration function and the relative standard deviation of the method.

(continued)

Table 6.3-6 Calibration data of the headspace analysis of benzene using the internal standard method

Level	1	2	3	4	5
c_{bz} in ppm (w/w)	4.4	8.8	13.2	17.6	22.0
A_{bz} in counts	133,983	191,693	277,492	366,251	456,295
A_{IS} in counts	235,206	228,972	238,221	240,792	249,547

Table 6.3-7 Peak areas obtained by a test sample and a spiked sample whose concentration is $c_{spiked} = 5.5$ ppm (w/w)

Analyte	Sample	Spiked sample
A_{bz} in counts	324,583	242,149
A_{IS} in counts	228,764	243,059

(d) The trueness of the headspace analysis is checked by the recovery rate of two spiked waste water samples, with the results given in Table 6.3-7. The concentration of the spiked sample is $c_{spiked} = 5.5$ ppm (w/w). Check whether the result is true.

Remember that in order to evaluate the relative standard deviation for trueness its upper and lower limit values must be calculated.

(e) A waste water sample analyzed by the validated HS-GC method, the following peak areas are obtained from the chromatograms: $A_{bz} = 325{,}824$ counts and $A_{IS} = 220{,}835$ counts

Calculate the predicted value $\hat{x}_{bz}$ and its confidence interval $CI(\hat{x})$ in ppm (w/w).

Solution to Challenge 6.3-2

(a and c) Application of the internal standard procedure means that the relative peak areas A_{bz}/A_{IS} are used for the calculation of the regression parameters, which are given in Table 6.3-8.

The regression coefficients obtained by Excel functions are $a_0 = 0.223778$ and $a_1 = 0.0727623$ ppm^{-1}. The calibration line with the confidence intervals is shown in Fig. 6.3-4. The linearity of the regression function should be confirmed by visual inspection.

(continued)

Table 6.3-8 Data set for the calculation of the regression parameters for the determination of benzene by the HS-GC method

Level	1	2	3	4	5
c_{bz} in ppm (w/w)	4.4	8.8	13.2	17.6	22.0
A_{bz}/A_{IS}	0.570	0.837	1.165	1.521	1.828

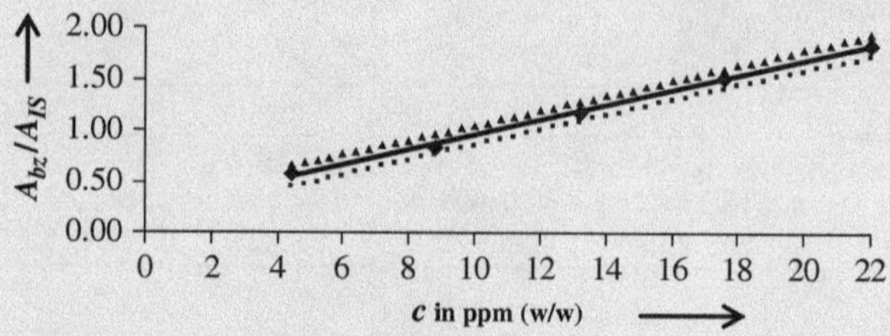

Fig. 6.3-4 Calibration line with confidence intervals for the HC-GC determination of benzene in water

The Mandel linearity test requires at least seven levels, and therefore the significance of the quadratic regression coefficient a_2 is used for testing the linearity of the calibration function.

The test value $\hat{t} = 1.175$ calculated by (5.3.6-2) with $a_2 = 0.00039976$ and $s_{a_2} = 0.00034011$ obtained by Excel function LINEST is smaller than the critical value $t(P = 95\%, \mathrm{df} = 2) = 4.303$. Thus, the null hypothesis H_0: $a_2 = 0$ is valid, which means that the calibration line is indeed linear.

(b) The outlier test according to (5.4-1) does not make sense because the small data set gives a very high critical value $F(P = 99\%, \mathrm{df}_1 = 1, \mathrm{df}_2 = 2) = 98.20$. However, the inspection of Fig. 6.3-4 shows that no measured y-values lies outside the confidence interval, and therefore an outlier is not present in the calibration line.

(d) The analytical results of the sample and the spiked sample calculated by (6.3-14)

$$\hat{x} = \frac{A_{\mathrm{bz}}/A_{\mathrm{IS}} - 0.223778}{0.0727623\ \mathrm{ppm}^{-1}} \tag{6.3-14}$$

are $\hat{x}_{\mathrm{sample}} = 10.6$ ppm and $\hat{x}_{\mathrm{spiked\ sample}} = 16.4$ ppm, obtained with the relative peak areas 0.9963 and 1.4189, respectively.

The recovery rate obtained according to (5.7.3-1) is Rr% = 105.6.

The confidence interval $\mathrm{CI}(\hat{x} = 16.4) = \pm 1.28$ ppm, calculated according to (4.2-17) with $s_{y.x} = 0.02616$, $\mathrm{SS}_{xx} = 193.6\ \mathrm{ppm}^2$, $\bar{y} = 1.184$, $t(P = 95\%, \mathrm{df} = 3) = 3.182$, as well as the other values given above. The lower limit is 15.1 ppm and 92.2%, respectively, and the upper limit is 17.7 ppm and 107.8%, respectively. The experimentally determined recovery rate lies within the range, which means that the result is true.

(e) After checking linearity and trueness, the calibration function can be used for the analysis.

(continued)

The predicted value calculated by (6.3-14) with the relative peak area $A_{bz}/A_{IS} = 1.47542$ is $\hat{x} = 17.2$ ppm (w/w) and the confidence interval calculated as described above is $CI(\hat{x}) = 1.3$ ppm (w/w) at the significance level $P = 95\%$. Thus, the result is 17.2 ± 1.3 ppm (w/w) benzene.

Challenge 6.3-3

The MHE-GC method was chosen for the determination of benzene in soil samples. The time of equilibration is 60 min at 85°C. The sample amount is 2 g. The peak areas obtained by the chromatograms of seven MHE replicates are given in Table 6.3-9.

(a) Present the measured peak areas of benzene A_{bz} as the function of the extraction steps n $A_{bz} = f(n)$ as well as the function $\ln A_{bz} = f(n)$, and confirm that the determination of the sum of areas according to (6.3-11) is allowed.
(b) Calculate the sum of areas for the MHE of benzene.
(c) Determine the concentration of benzene in the soil sample in ppm (w/w) by the following calibration:
A headspace flask of the same volume was filled with glass pearls so that the headspace volume of the sample and that of the calibration were the same. Then 10 μL of a benzene solution in acetone with concentration 1.0 μg mL^{-1} was injected into the headspace vial and the vial was quickly closed. The peak area of benzene after HS-GC analysis under the same conditions is $A_{bz,cal} = 83,294$ counts.
(d) Because of the quadratic dependence of the sum of peak areas $\sum A_n$ on the peak area of the first extraction step A_1 [see (6.3-11)], the error of the analytical result is mainly determined by A_1. For more precise results the peak area for the first extraction step A_1^* has to be calculated from the linear regression function with the whole data set obtained by MHE. Calculate the concentration of benzene in the soil sample using the value A_1^* and compare the result with that calculated using A_1.

Table 6.3-9 Peak areas of benzene A_{bz} in counts obtained by the MHE analysis of benzene in a soil sample

Extraction step	A_{bz} in counts
1	2,786,634
2	1,333,514
3	838,188
4	428,373
5	238,250
6	130,675
7	70,378

Solution to Challenge 6.3-3

(a) Figures 6.3-5 and 6.3-6 show the plot of the experimental decrease of the peak areas of benzene A_{bz} with the number of extraction steps in MHE-GC and the linear plot of $\ln A_{bz} = f(n)$, respectively.

As the function $\ln A_{bz} = f(n)$ shows, there is no indication of an outlier in the regression line because no measured y-values lie outside of the confidence interval, and therefore a check for outliers will not be done. But the check for linearity is necessary because linearity is a precondition for applying (6.3-11) to the calculation of the sum of peak areas. Linearity is highly probable because the residuals of the function $\ln A = f(n)$ are randomly distributed around zero, as Fig. 6.3-7 shows. However, a statistical test is still necessary. With seven levels, the Mandel test can be applied. The test value calculated according to (5.3.4-1) is $\hat{F} = 0.023$. The standard

(continued)

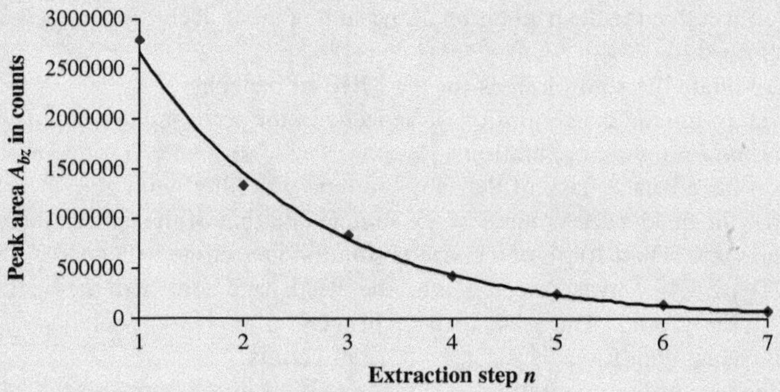

Fig. 6.3-5 Decrease of peak areas A of benzene in MHE-GC

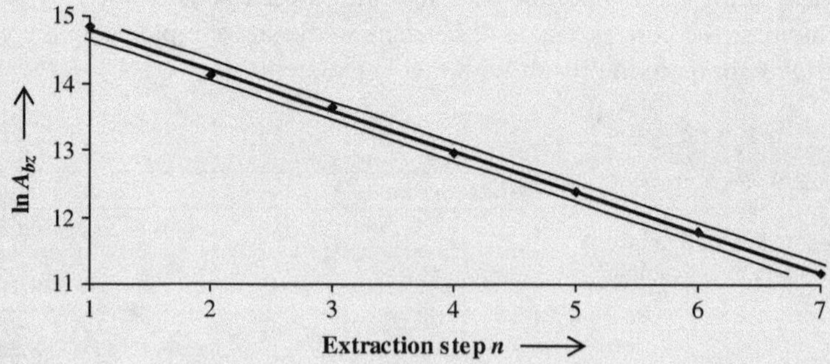

Fig. 6.3-6 Linear plot of the decrease in peak areas of benzene in MHE-GC

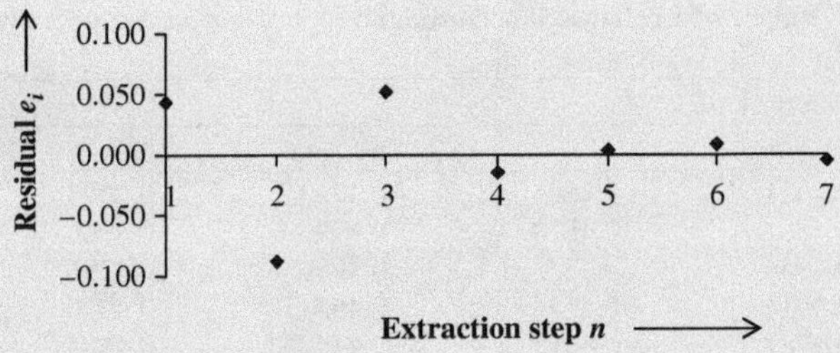

Fig. 6.3-7 Residual plot for the function $\ln A_{bz} = f(n)$ of the MHE-GC

deviations for the linear and quadratic regression functions $s_{y.x,1} = 0.05053$ and $s_{y.x,2} = 0.05633$, respectively, are obtained by Excel functions. The test value is much smaller than the critical value $F = (P = 99\%, \mathrm{df}_1 = 1, \mathrm{df}_2 = 4) = 21.198$, and therefore quadratic regression is not a better regression model.

(b and d) Because of the linearity of the regression function $\ln A = f(n)$, the sum of peak areas can be calculated according to (6.3-11):

$$\sum A_n = \frac{2,786,634^2}{2,786,634 - 1,333,514} = 5,343,901. \qquad (6.3\text{-}15)$$

The value of the first extraction step A_1^*, which is the intercept of the regression function $\ln A = f(n)$, is $\ln A_1^* = 15.4019$ and $A_1^* = 4,886,123$, which differs from the value in (6.3-15) by 8.5%.

(c) Under the same conditions, 10 ng benzene was analyzed by the injection of 10 μL of a benzene solution of concentration 1 μg μL^{-1}, giving a peak area of 83,294 counts. Therefore, 2 g of the soil sample contains 641.6 ng benzene. The content of benzene is thus 321 ppb (w/w). Using the value A_1^* for the calculation of the sum of areas, the sample content is 293 ppb (w/w) benzene.

Challenge 6.3-4

According to (6.3-9), only one extraction step is required in order to determine the sum of peak areas $\sum A_n$ if k is a constant for all samples. This can be the case, for example, in the quality control of residual monomer in polymer batches which were obtained by the same technological procedure. In the method development, a set of representative batches are analyzed under the same HS-GC conditions by replicates and k is calculated for all runs using (6.3-10). If k is equivalent for all batches, (6.3-9) with only *one* extraction step may be used in quality control for the determination of $\sum A_n$ with the

(continued)

Table 6.3-10 Values of k calculated according to (6.3-10) of $m = 5$ polystyrene samples with $n_i = 6$ replicates

n_i	Samples m				
	1	2	3	4	5
1	0.4821	0.4645	0.4733	0.5003	0.4536
2	0.4654	0.4870	0.4793	0.4966	0.4693
3	0.4954	0.4735	0.4950	0.4794	0.4622
4	0.4918	0.4674	0.4621	0.4940	0.4694
5	0.4781	0.4693	0.4788	0.4683	0.4644
6	0.5025	0.4963	0.4838	0.4933	0.4599

mean value of the experimentally determined constant k. This procedure would halve the analysis time!

To determine whether the simple (6.3-9) can be used for the determination of the sum of peak areas, five representative polystyrene samples were analyzed with six replicates each under the same HS-GC conditions. The k values of the five polystyrene samples calculated according to (6.3-10) are given in Table 6.3-10.

Is k equivalent in all batches, so that (6.3-9) can be applied in quality control?

Solution to Challenge 6.3-4

Because n_j replicate measurements are performed for each sample m, ANOVA must be used to check whether k is a constant for all samples. The ANOVA procedure has already been discussed in Sect. 3.6. The maximum and the minimum values of k for each sample are checked for an outlier with the Dixon test; see Sect. 3.2.3. The results are listed in Table 6.3-11. All test

(continued)

Table 6.3-11 Intermediate quantities and results for the Dixon and Cochran tests

Samples m	1	2	3	4	5
Dixon outlier test according to (3.2.3-1), with $b = 2,\ k = n$					
$x^* = x_{max}$	0.5025	0.4963	0.4950	0.5003	0.4694
x_2	0.4954	0.4870	0.4838	0.4966	0.4693
$\hat{Q}_{max}$	0.1914	0.2933	0.3405	0.1139	0.0084
$x^* = x_{min\,x}$	0.4654	0.4645	0.4621	0.4683	0.4536
x_2	0.4781	0.4674	0.4733	0.4794	0.4599
$\hat{Q}_{min}$	0.3421	0.0894	0.3405	0.3472	0.3978
Cochran test for homogeneity of variances according to (3.4-1)					
s_j^2	0.000179	0.000158	0.000119	0.000149	0.000036
s_{max}^2	0.000179	$\sum s_i^2$	0.000642	$\hat{C}$	0.2795

Table 6.3-12 Intermediate quantities and results of ANOVA

Samples m					
$\bar{x}_j$	0.4859	0.4763	0.4787	0.4887	0.4631
$\bar{\bar{x}}$	0.4785				
$1{,}000\, n_j(\bar{x}_j - \bar{\bar{x}})^2$	0.32349	0.02972	0.000174	0.61305	1.4221
$SS_{bw} = \sum n_j(\bar{x}_j - \bar{\bar{x}})^2$	0.002389				
$df_{bw} = m - 1$	4				
$s^2_{bw} = \frac{SS_{bw}}{df_{bw}}$	0.000597				
$1{,}000\, SS_j$	0.89649	0.78877	0.59428	0.74658	0.18172
$SS_{in} = \sum SS_j$	0.003208				
$df_{in} = n - m$	25				
$s^2_{in} = \frac{SS_{in}}{df_{in}}$	0.000128				
$\hat{F} = \frac{s^2_{bw}}{s^2_{in}}$	4.654	$F(P = 99\%, df_{bw} = 4, df_{in} = 25)$			4.177

values for the maximum k values $\hat{Q}_{max,k}$ and the minimum k values $\hat{Q}_{min,k}$ are smaller than the critical Q $(P = 95\%,\ n = 6) = 0.560$.

Because no outlier could be detected, all groups have the same number of k values, and therefore the Cochran test of homogeneity described in Sect. 3.4 is suitable for multiple comparison of variances. The test value $\hat{C} = 0.2795$ is smaller than the critical value of $C(\chi^2, P = 95\%, k = 5, df = 5) = 0.5065$, so the variances of the k values are homogeneous.

The results of ANOVA for the MHE problem for $m = 5$ samples with $n_j = 6$ replicates are presented in Table 6.3-12.

$\bar{x}_j$ is the mean k value of each sample, $\bar{\bar{x}}$ is the total of the mean k values, SS_{bw} and s^2_{bw} are the sum of squares and the variance, respectively, *between* the k values, SS_{in} and s^2_{in} are the sum of squares and the variance, respectively, *within* the groups of the k values, and df_{bw} and df_{in} are degrees of freedom between and within the groups of k values, respectively. The test value $\hat{F} = 4.654$ is higher than the critical F-value at the significance level $P = 99\%$. This means that the k values are not homogeneous or, in other words, k is not a constant for all samples. The simplification of the determination of the sum of peak areas cannot be applied.

References

1. Chan CC, Lee YC, Lam H, Zhang XM (2009) Analytical method validation and instrument performance verification. Wiley-Interscience, New York
2. Ermer J, Miller JHMcB (2005) Method validation in pharmaceutical analysis – a guide to best practice. Wiley-VCH, Weinheim

3. Pharm Eur (2009) 6. Ausgabe, 2.2.46, http://www.pheur.org
4. Shabir GA (2003) Validation of high-performance liquid chromatography methods for pharmaceutical analysis: understanding the differences and similarities of the US food and drug administration, the pharmacopeia and the International Conference on Harmonization. J Chromatogr A 987:57–66
5. Kolb B, Ettre LS (2006) Static headspace-gas chromatography – theory and practice, 2nd edn. Wiley-Interscience, New York
6. De la Calle D, Magnaghi S, Reichenbächer M, Danzer K (1996) Systematic optimization of the analysis of wine bouquet components by solid-phase microextraction. J High Resolut Chromatogr 19:257–261

Chapter 7
Performance Verification of Analytical Instruments and Tools: Selected Examples

7.1 General Remarks on Qualification and Performance Verification of Laboratory Instruments

Besides the validity of the analytical methods, controlled by internal and external tests, as well as proper training of the analysts, the reliability of all the instruments used for experiments and measurements provides the fundamentals of analytical quality assurance. There are therefore regulatory agency requirements for the qualification, calibration, and verification of analytical instruments.

The requirements in EN ISO/IC 17025, the "General Requirements for the Competence of Testing and Calibration Laboratories" are stated in its Sect. 4.5.2 [1]:

"Equipment and its software used for testing, calibration and sampling shall be capable of achieving the accuracy required and shall comply with specifications relevant to the tests and/or calibration concerned. Calibration programmes shall be established for the key quantities or values of the instruments where these properties have a significant effect on the results." Similar requirements are also stated in ICH Guideline Q7 [2]. The life cycle of an instrument starts from planning to bring a new instrument into the laboratory and ends with the decommissioning of the instrument. It involves, in general, three phases [3–5]:

1. Prepurchase planning phase
2. Postpurchase phase
3. Routine operation phase

The last phase includes all activities which have to be performed by the users (not by the instrument providers) to document that the instrument is fit-for-purpose.

If the instrument is qualified by the provider it can be used to generate analytical data. A standard operation procedure (SOP) must be written for the new instrument which must include, besides the important operational instructions, all activities for the maintenance, calibration, and performance verification. These are described in the following section for the example of an UV–Vis spectrometer and an HPLC instrument.

Note that not only the analytical instruments (spectrometers, chromatographic instruments, etc.) must be checked for performance verification and calibration, but all other

M. Reichenbächer and J.W. Einax, *Challenges in Analytical Quality Assurance*, DOI 10.1007/978-3-642-16595-5_7, © Springer-Verlag Berlin Heidelberg 2011

measurement devices, such as volumetric flasks, thermometers, pH-meters, balances, etc. must also be checked as fit-for-purpose. This will be described for the example of the balance. Note that the results must be documented in the so-called *log-book*.

7.2 UV–Vis Spectrophotometers

Because of its ease of use and speed of analysis, UV–Vis spectrophotometry is often used for qualitative and quantitative analyses. Therefore, the user must document on the basis of appropriate experiments that the UV–Vis spectrophotometer is fit-for-purpose. The performance requirements of spectrophotometers vary according to the nature of the measurements and the design of the instrument, e.g. whether it is a scanning spectrometer with a single beam or double beam design or whether it is a diode instrument.

The regulatory requirements for pharmaceutical analysis include tests for the attributes *wavelength accuracy, stray light, resolution*, and *photometric accuracy*. The following discussion focuses on these four parameters. Further characteristics, such as noise, baseline flatness, and stability can usually be checked by the software which is integrated in modern instruments [3–5].

The acceptance criteria are focused on the regulatory requirements in pharmaceutical analysis but they should be generally applicable for any spectrophotometric analysis.

7.2.1 *Wavelength Accuracy*

Wavelength accuracy is defined as the deviation of the measured wavelength from the "true" value of the absorption band. Wavelength deviations can cause errors in qualitative and quantitative analysis. For example, confirmation of identity of a pharmaceutical steroid can be determined by the established value λ_{max} of the absorption band, say $\lambda_{max} = 265 \pm 1$ nm. If the value obtained by the spectrometer is 267 nm, then the identity of the test sample is not confirmed.

In addition to the qualitative problem, wavelength deviation also affects quantitative analysis. Usually, the absorbance is measured at the maximum of the absorption band (λ_{max}) because of the highest sensitivity and the lowest effect of the measurand on the absorbance resulting from the natural slight shift in wavelength at this location. But it is sometimes necessary to use a measurement wavelength in the upslope or downslope of the absorption band. Then, a small deviation in wavelength will cause a large effect on the absorbance, as shown for the absorption band in Fig. 7.2.1-1.

Wavelength accuracy verification is checked by the comparison of the measured wavelength obtained by a reference standard with the wavelength listed in the certificate. There are many standards which can be used and all of these are commercially available, for example emission lines of D_2 or the mercury lamp,

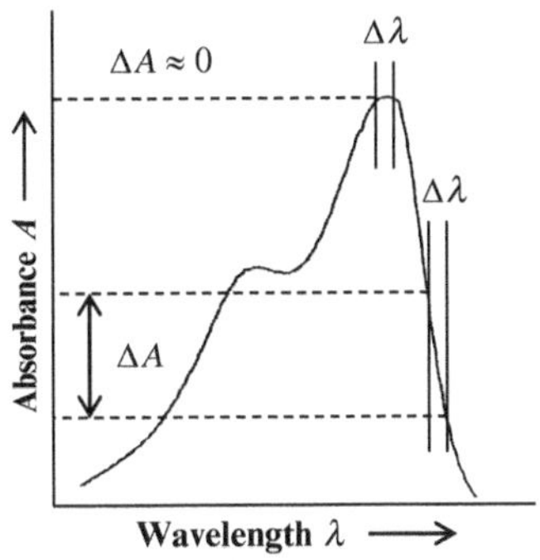

Fig. 7.2.1-1 Effects of the position (at λ_{max} and at a steep downslope of the absorption profile, respectively) of the wavelength shift $\Delta\lambda$ on the absorbance A

Table 7.2.1-1 Reference values of the wavelength of a solution of holmium oxide in perchloric acid

Spectral bandwidth 0.5 nm λ_{max} in nm	1.0 nm λ_{max} in nm	2.0 nm λ_{max} in nm
241.01	241.13	241.08
249.79	249.87	249.98
278.13	278.10	278.03
287.01	287.18	287.47
333.34	333.44	333.40
345.52	345.47	345.49
361.33	361.31	361.16
385.50	385.66	385.86
416.09	416.28	416.62
467.80	467.83	467.94
485.27	485.29	485.33
536.54	536.64	536.97
640.49	640.52	640.84

or a solution of 4% holmium oxide in 10% perchloric acid. The latter is presented in Table 7.2.1-1.

The *acceptance criteria* are:

- The deviation of the measured values from the reference value may not greater than ±1 nm in the UV range (200–380 nm) and ±3 nm in the visible range.
- Three repeated scans of the same method should be within ± 0.5 nm.

7.2.2 *Stray Light*

Stray light is false light caused by scattering or by higher-order diffraction of the monochromator which decreases the absorbance and reduces the linearity of the spectrophotometer.

The relation between absorbance A and transmission T is given by the definition:

$$A = -\log(T). \tag{7.2.2-1}$$

The (pure) transmission T is the relation of the transmitted light intensity I and the incident intensity I_0:

$$T = \frac{I}{I_0}. \tag{7.2.2-2}$$

In the presence of stray light, the stray light intensity I_s is included in (7.2.2-2), and the transmission T^* is given by:

$$T^* = \frac{I + I_s}{I_0 + I_s}. \tag{7.2.2-3}$$

According to (7.2.2-3) the effect of stray light increases with the decrease in the transmitted intensity. Therefore, the effect of stray light must be considered for highly absorbent samples because it causes deviation of the linearity of the absorbance (see the results in Challenge 7.2-1). Besides being the minimum of the relative error of the absorbance measurement as shown in Fig. 2.2.5-1, the stray light effect is a further reason for the best range of the absorbance being 0.3–1.0.

The stray light can be estimated by various cut-off filter aqueous solutions: KCl (12 g L^{-1}) for the measurement wavelength 200 nm, NaI (10 g L^{-1}) for 220 nm, and $NaNO_2$ (50 g L^{-1}) for 340 nm. The cut-off filter solutions block the light at the measuring wavelength, and thus the measured absorbance at the wavelengths given above is caused by stray light.

The *acceptance criterion is*:

The values of the transmission measured in a 1 cm cell against water should be less than 0.01 which means, according to (7.2.2-1), that the value of the absorbance should be greater than 2.0.

7.2.3 Resolution

The resolution describes the separation of a peak pair. In spectrophotometry the resolution is related to the spectral bandwidth: the smaller the spectral bandwidth the higher the resolution. The resolution is determined by the slit width and the dispersive power of the monochromator and by the number of diodes in the array in the diode array instrument. Simple spectrometers, mostly used in routine analysis, are equipped by a fixed slit width.

Insufficient resolution decreases the absorbance, as shown in Fig. 7.2.3-1 for the absorption band of toluene around 269 nm measured with two different slit width of the monochromator.

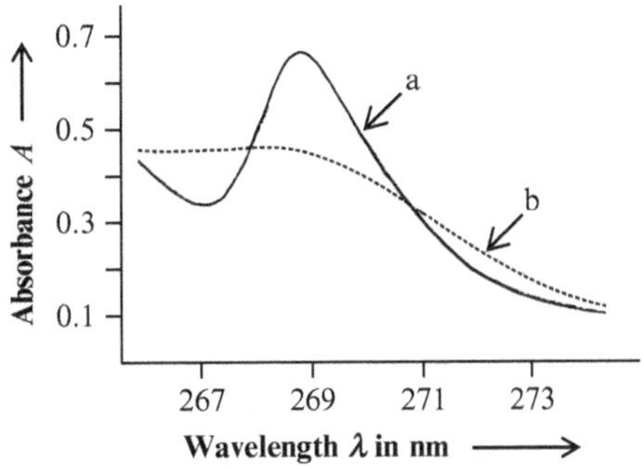

Fig. 7.2.3-1 Part of the absorption band of toluene measured with two various slit widths; a: 0.5 nm, b: 4 nm

The resolution of a UV–Vis spectrometer is estimated by the ratio of the absorbances at $\lambda_{max} = 269\,\text{nm}$ and $\lambda_{min} = 267\,\text{nm}$ measured with a solution of 0.02% (v/v) toluene in *n*-hexane (UV-grade).

The *acceptance criterion* is:

The resolution is sufficient if the ratio is greater than 1.5.

7.2.4 Photometric Accuracy

Photometric accuracy concerns the measurement of the absorbance as the parameter for quantitative analysis on the basis of the Lambert–Beer law. As long as there is linearity over the range, the photometric accuracy is not critical but the photometric accuracy is important, for example, for the determination of the extinction coefficient as a specific parameter characterizing an analyte.

Photometric accuracy is determined by comparing the measured absorbance or transmission of commercial standard filters or standard solutions with the specified values of the standards.

Because certified glass filters are expensive and not stable over a long period, the simple potassium dichromate method can be used to check the photometric accuracy. The procedure given in the European Pharmacopeia [6] is as follows:

- Potassium dichromate is dried to constant weight at 130°C.
- An accurately weighed sample $m_{K_2Cr_2O_7}$ in the range 5.7–6.3 mg is dissolved in 100 mL 0.01 N sulfuric acid.
- The absorption spectrum is measured in the range 220–380 nm against 0.01 N sulfuric acid.
- The adjusted values A_{cor} are calculated by (7.2.4-1) at the wavelengths 235, 257, 313, and 350 nm; see Fig. 7.2.4-1:

$$A_{cor} = \frac{m_{K_2Cr_2O_7}\text{in mg}}{6.006\text{ mg}} \cdot A_{measured}. \qquad (7.2.4\text{-}1)$$

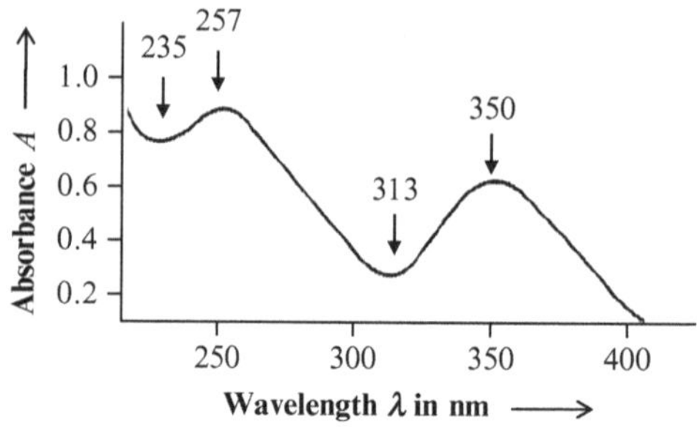

Fig. 7.2.4-1 Absorption spectrum of a solution of potassium dichromate in 0.01 N sulfuric acid for the determination of photometric accuracy

Table 7.2.4-1 Reference values for potassium dichromate solution at selected wavelengths

λ in nm	A_{ref}	α in L mol^{-1} cm^{-1}
235	0.742	3.635
257	0.861	4.217
313	0.291	1.425
350	0.639	3.130

The *acceptance criterion* is:

- The photometric accuracy is sufficient if the difference between the specified absorbance given in Table 7.2.4-1 and the adjusted absorbance is within the limit ± 0.01.
- The relative standard deviation obtained by six replicates is $s_r\% < 0.5\%$.

Challenge 7.2-1

The spectrophotometer Specord M 500 (Carl Zeiss Jena®) used for the determination of the API acetylsalicylic acid in tablets must be checked for wavelength accuracy, stray light, resolution, and photometric accuracy. Noise, baseline flatness, and stability can be checked by the instrument software.

The following results are obtained:

(a) *Wavelength accuracy*

The wavelength accuracy was checked by the holmium perchlorate method. Because the analyte absorbs in the UV range, the commercial holmium perchlorate solution was measured only in the range 200–380 nm. The spectrum is shown in Fig. 7.2.4-2 and the values of the test wavelength obtained by the instrument software are given in Table 7.2.4-2 for all replicates.

(continued)

Fig. 7.2.4-2 UV-Absorption spectrum of a commercial solution of holmium perchlorate

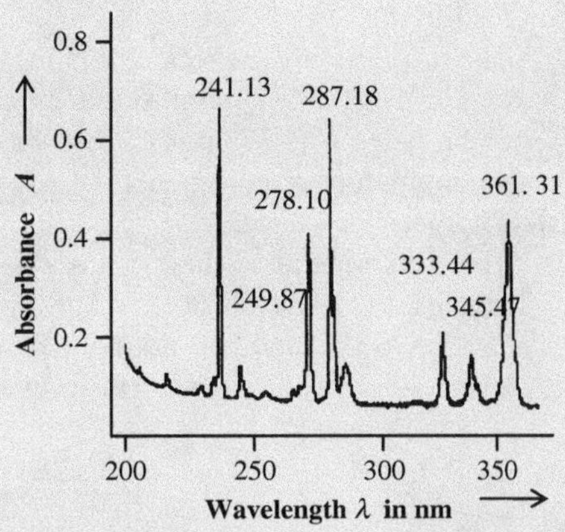

Table 7.2.4-2 Values of the wavelengths of holmium perchlorate in the UV range obtained by three replicates

Wavelength in nm			
Reference values	Measured values (obtained by the instrument software)		
241.13	241.61	241.45	241.67
249.87	249.21	249.32	249.15
278.10	278.33	278.24	278.39
287.18	287.02	287.16	287.05
333.44	333.58	333.68	333.45
345.47	345.63	345.69	345.58
361.31	361.42	361.57	361.47

Check whether the wavelength accuracy fulfills the regulatory requirements.

(b) *Stray light*

Stray light was estimated at a wavelength of 220 nm using the cut-off filter solution NaI. The measured absorbance was $A_{220\text{nm}} = 2.055$.

1. Check whether the wavelength accuracy fulfills the regulatory requirements.
2. Calculate the percentage deviation from pure transmission caused by stray light for the absorbances 0.25, 0.5, 1.0, 1.5, 2.0, 2.5, 3.0, and examine the results.
3. Compile the graph $A_{\text{with stray light}} = f(A_{\text{without stray light}})$ in the absorbance range 0–3.0 with a stray light transmission of 1% and show the deviation from linearity.

(continued)

(c) *Resolution*
The resolution was checked for slit width 1 nm. The absorption spectrum of a 0.02% (v/v) solution of toluene in *n*-hexane is shown in Fig. 7.2.4-3 in the range 266–273 nm. Check whether the wavelength accuracy fulfills the regulatory requirements.

(d) *Photometric accuracy*
The photometric accuracy was checked by the potassium dichromate method as described above. The measured values of the absorbance at 235, 257, 313, and 350 nm with six replicates are listed in Table 7.2.4-3. Check whether the wavelength accuracy fulfills the regulatory requirements.

(continued)

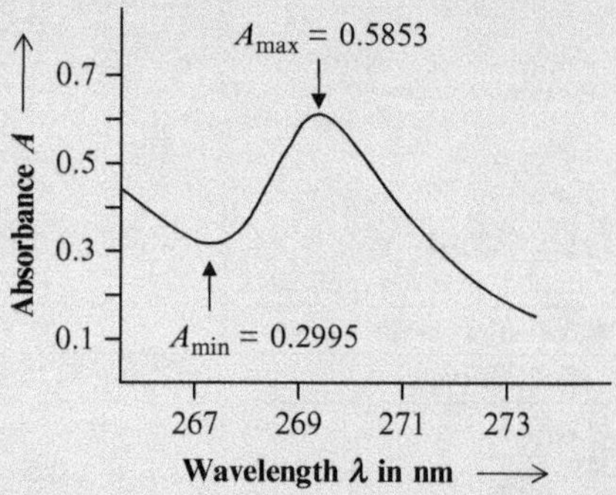

Fig. 7.2.4-3 Absorption spectrum of toluene in *n*-hexane

Table 7.2.4-3 Values of the absorbance of potassium dichromate at selected wavelengths and the concentration of the potassium dichromate solutions

λ in nm	Concentration of potassium dichromate solutions in mg L^{-1}					
	60.015	60.005	60.024	60.018	60.011	60.008
	Measured absorbances with six replicates					
	1	2	3	4	5	6
235	0.7425	0.7408	0.7421	0.7456	0.7448	0.7467
257	0.8607	0.8632	0.8672	0.8645	0.8653	0.8639
313	0.2926	0.2916	0.2946	0.2922	0.2928	0.2948
350	0.6391	0.6412	0.6389	0.6395	0.6405	0.6401

Solution to Challenge 7.2-1

(a) Intermediate quantities and results of the wavelength accuracy check are given in Table 7.2.4-4:
λ_{ref} is the reference value of the wavelength according to Table 7.2.4-2.

(continued)

Table 7.2.4-4 Intermediate quantities and results of the wavelength accuracy check

λ_{ref} in nm	$\bar{\lambda}_{exp}$ in nm	$\lvert\lambda_{ref} - \bar{\lambda}_{exp}\rvert$ in nm	λ_{max} in nm	λ_{min} in nm	$\lambda_{max} - \lambda_{min}$ in nm
241.13	241.58	0.45	241.67	241.45	0.22
249.87	249.23	0.64	249.32	249.15	0.17
278.10	278.32	0.22	278.39	278.24	0.15
287.18	287.08	0.10	287.16	287.02	0.14
333.44	333.57	0.13	333.68	333.45	0.23
345.47	345.63	0.16	345.69	345.58	0.11
361.31	361.49	0.18	361.57	361.42	0.15

Table 7.2.4-5 Effect of stray light observed at 220 nm on the absorbance

A	T ($T = 10^{-A}$)	T^* (7.2.2-3)	A^* (7.2.2-1)	Bias $A - A^*$	Bias in %
0.25	0.5623	0.5662	0.2471	0.0029	0.29
0.50	0.3162	0.3222	0.4919	0.0081	0.81
1.00	0.1000	0.1079	0.9671	0.0329	3.29
1.50	0.0316	0.0401	1.3971	0.1029	10.29
2.00	0.0100	0.0186	1.7294	0.2706	27.06
2.50	0.0032	0.0119	1.9256	0.5744	57.44
3.00	0.0010	0.0097	2.0121	0.9879	98.79

$\bar{\lambda}_{exp}$ is the mean value of the observed wavelengths given in Table 7.2.4-2 obtained by three replicates.

$\lvert\lambda_{ref} - \bar{\lambda}_{exp}\rvert$ is the difference between the reference and the mean values of the wavelengths, which is the first criterion of acceptance.

$\lambda_{max} - \lambda_{min}$ is the difference between the greatest and the smallest observed values of the wavelengths, which is the second criterion of acceptance. Thus, the result is tested by both criteria of acceptance.

As Table 7.2.4-4 shows, both acceptance criteria are fulfilled by all wavelengths checked.

(b) 1. The observed absorbance $A = 2.055$ and the transmission $T = 0.0088$ do not exceed the limit values of the acceptance $T_{lim} = 0.01$ and $A_{lim} = 2.0$, respectively, and therefore the acceptance criterion is fulfilled.
2. The intermediate quantities and results are given in Table 7.2.4-5 calculated with stray light observed at 220 nm ($T_s = 0.0088$). According to the results (bias in %), the working range chosen should not exceed the absorbance 1.0.
3. The graph $A_{\text{with stray light}} = f(A_{\text{without stray light}})$ based on the value of absorbance calculated by (7.2.2-1) and (7.2.2-3) is shown in

(continued)

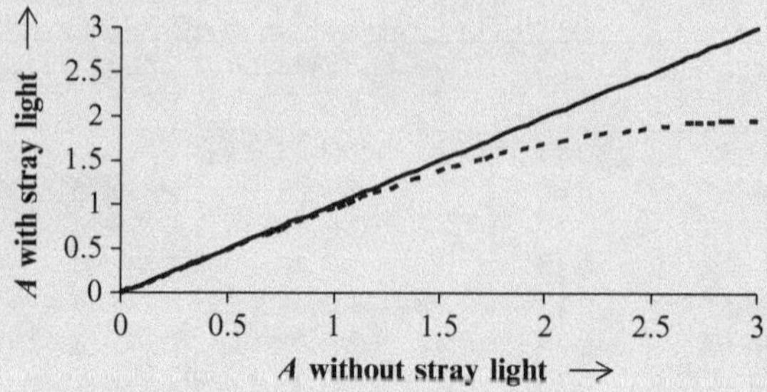

Fig. 7.2.4-4 Deviation (*dotted line*) from linearity as a result of stray light intensity $I_s = 1\%$

Table 7.2.4-6 Intermediate quantities and results for checking the photometric accuracy

Adjusted absorbances at $\lambda = 235$ nm ($A_{ref} = 0.742$)					
0.7419	0.7401	0.7417	0.7451	0.7442	0.7461
$\bar{A}_{235}$	0.7432	$\|A_{ref} - \bar{A}_{235}\|$	0.0012	s	0.0023
$s_r\%$	0.31	Result	$s_r\%_{observed} < s_r\%_{ref}$		
Adjusted absorbances at $\lambda = 257$ nm ($A_{ref} = 0.861$)					
0.8601	0.8624	0.8667	0.8639	0.8646	0.8632
$\bar{A}_{257}$	0.8635	$\|A_{ref} - \bar{A}_{257}\|$	0.0025	s	0.0022
$s_r\%$	0.26	Result	$s_r\%_{observed} < s_r\%_{ref}$		
Adjusted absorbances at $\lambda = 313$ nm ($A_{ref} = 0.291$)					
0.2924	0.2913	0.2944	0.2920	0.2926	0.2945
$\bar{A}_{313}$	0.2929	$\|A_{ref} - \bar{A}_{313}\|$	0.0019	s	0.0013
$s_r\%$	0.45	Result	$s_r\%_{observed} < s_r\%_{ref}$		
Adjusted absorbances at $\lambda = 350$ nm ($A_{ref} = 0.639$)					
0.6386	0.6406	0.6385	0.6391	0.6400	0.6395
$\bar{A}_{350}$	0.6394	$\|A_{ref} - \bar{A}_{350}\|$	0.0004	s	0.0008
$s_r\%$	0.13	Result	$s_r\%_{observed} < s_r\%_{ref}$		

Fig. 7.2.4-4. As the graph shows, the deviation as a consequence of 1% stray light is appreciably greater than the absorbance of about 1.0.

(c) The ratio of the observed absorbances at $\lambda_{max} = 269$ nm and $\lambda_{min} = 267$ nm is 1.95 and thus greater than the acceptance value 1.5. The resolution can be accepted.

(d) The adjusted absorbance values calculated by (7.2.4-1) and the intermediate quantities for the test value $s_r\%$ are summarized in Table 7.2.4-6. As the values for the relative standard deviation of the absorbances obtained by six replicates do not exceed $s_r\% = 0.5$ for each test wavelength, the photometric accuracy can be accepted.

7.3 HPLC Instruments

HPLC is one of the most important techniques used for the analysis of organic compounds. Numerous analytical methods have been developed for pharmaceutical, chemical, food, and environmental applications. In order to provide reliable results, the performance of the HPLC system must be checked at specified intervals.

The performance of an HPLC system can be evaluated by examining the modules of the instrument without the column (also called *operational qualification*) and by holistic testing (also called *performance qualification*) which can be verified according to the parameters of the system suitability test (*SST*) described in Sect. 6.1.

The performance verification of the HPLC system includes the following modules:

1. Pump
   Performance attributes of the pump are flow rate accuracy, gradient accuracy, and pressure stability.
   The *flow rate accuracy* can simply be checked by measuring the collected volume of the mobile phase for a specified time at different flow rates. The deviations should not exceed $\pm$ 2% of the set flow rate.
   The *gradient accuracy* crucial for proper chromatographic separation and reproducibility can be indirectly checked for a binary system by monitoring the absorbance change obtained by altering the composition of the mobile phase according to a given program. Channel A, for example, is filled with a pure solvent and channel B is filled with a solvent containing an UV-absorbing substance, e.g. caffeine. The gradient accuracy and linearity is checked by the step-like chromatogram obtained by the gradient program which changes from 100% A to 100% B. Details of the procedure are given in [3–5].
   Pressure testing involves the checking for leaks within the pump system. It is verified by testing the pressure decay after plugging the outlet of the pump with a dead-nut. The general expectation of pressure decay is <520 kPa min^{-1}.
2. Injector
   The volume *precision* of the injector is critical if various amounts of standard and sample solutions are to be injected. It can be checked by making at least six replicate injections of a sample solution. The relative standard deviation should not be greater than 1%.
   The *linearity* of the injected volume by automated injectors is especially important if various volumes have to be injected, for example during the quantitative determination of impurities present in different concentrations. The linearity is checked by making injections over a wide range, for example 5–100 μL. The relationship between the response and the injection volume is checked for linearity by known methods (see Sect. 5.3).
   A further check concerns the problem of *carryover* which will affect the accurate quantitative determination of the analyte, especially when a dilute sample is injected after a concentrated sample. The carryover can be checked by injecting a blank after a highly concentrated test sample. The response of the test sample obtained in the

blank expressed as the percentage of the response of the concentrated sample is a measurand of the level of carryover, which should be smaller than 1%.

3. Detector

   There are different detectors in HPLC, such as UV–Vis, diode array detector (DAD), fluorescence detector, refractive detector (RI), mass-selective detector (MSD), light scattering (ELSD) detector, electrochemical detector (EC), etc. Each of them requires specific test procedures which cannot be described within the scope of this book. The UV–Vis detector is mostly used in routine analysis for AQA and the checking of this detector for wavelength accuracy and linearity of response is substantially to the same as the procedures given above. The linearity of the detector important for quantitative analysis, for example, can be checked by filling the flow cell with a series of test solutions of various concentration of the test sample.

4. Column temperature

   The temperature of the HPLC column affects its efficiency because of the dependence of the capacity factor k' on temperature. Generally, the retention time drops by 1–3% for each increase of 1°C. Maintenance of a constant and accurate column temperature is important in order to achieve stable retention time and resolution of the analytes, and can be achieved by a column heater. The temperature accuracy of the column heater is evaluated by a calibrated thermometer placed in the column heater. The deviation between the measured and the set temperature should not be greater than ±2°C.

5. Dead volume

   The dead volume of the HPLC instrument is the volume between the injector and the detector cell, and is measured *without* the column. The dead volume affects the sharpness as well as the shape of the peak and, therefore, the separation performance of the HPLC instrument. The higher the dead volume the broader the peak, especially at early eluting peaks, and the smaller the resolution of two adjacent peaks.

   The dead volume V_d is estimated by the time t taken by a test substance (for example, acetone) from the injector to the detector at the low rate $\dot{F}$

$$V_d = \dot{F} \cdot t. \tag{7.3-1}$$

A dead volume V_d between 20 and 25 μL is a very acceptable value, but $V_d > 70\mu L$ causes observable peak broadening whereas $25 < V_d < 70$ can still be accepted.

Challenge 7.3-1

(a) The flow rate accuracy of the pump of an isocratic HPLC instrument must be tested.

The measured time for $V = 10$ mL at the set flow rate $\dot{F} = 2$ mL min^{-1} is given below:

t in s	301	304	302	303	306	305

(continued)

Check whether the flow rate fulfills the regulatory requirements.

(b) The injection precision of the automated injector of an HPLC instrument was checked by the injection of 25 μL of the test solution 0.5% (m/V) acetylsalicylic acid in the eluent H_2O/MeOH.
The following HPLC parameters are applied:

Column	ODS 1, 5 μm
Mobile phase	H_2O/MeOH ($V + V = 80 + 20$)
Flow rate	1 mL min^{-1}
Detection	272 nm
Column temperature	25°C

The peak areas A in counts obtained by six replicates are:

157,935	157,032	156,585	157,672	156,472	157,928

Estimate the injection precision of the automated injector.

(c) The performance of the UV detector of a HPLC instrument must be estimated.
The detector cell was filled with a solution of phenanthrene in acetonitrile and the values of λ_{max} and λ_{min} were scanned. The UV spectrum of phenanthrene presented in Fig. 7.3-1 shows a sharp band at $\lambda_{max} = 250$ nm and minima at $\lambda_{min,1} = 225$ nm and $\lambda_{min,2} = 264$ nm. The following scanned values were obtained: $\lambda_{max} = 251$ nm, $\lambda_{min,1} = 224$ nm, $\lambda_{min,2} = 266$ nm
Evaluate the wavelength accuracy.

(d) The choice of an appropriate detection wavelength as well as the importance of wavelength accuracy is preferably set by the analytical problem. Let us assume we have a mixture of the analytes phenanthrene and azobenzene whose UV spectra are given in Fig. 7.3-1.

(continued)

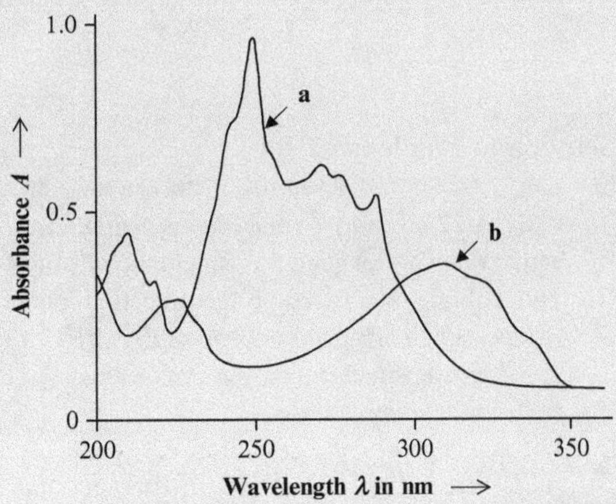

Fig. 7.3-1 UV spectra of (**a**) phenanthrene and (**b**) azobenzene (each 0.003 mol L^{-1}) in acetonitrile

Table 7.3-1 Response values of solutions of phenanthrene in acetonitrile

c in mol L^{-1}	0.001	0.0015	0.002	0.0025	0.003	0.0035
A	58,935	88,028	117,191	146,848	175,896	205,892

Which detection wavelength should be chosen for the following problem?

1. Determination of azobenzene only and vice versa.
2. Determination of both substances.
3. Determination of azobenzene in the presence of biphenyl with the highest sensitivity. Note that biphenyl does not absorb above $\lambda = 310$ nm.
4. Determination of phenanthrene with the highest sensitivity.

(e) The linearity was checked by measuring the response of solutions of various concentrations of phenanthrene in acetonitrile in the detector cell at 250 nm. The results are given in Table 7.3-1.
Evaluate the linearity of the UV detector.

(f) After the injection of 25 μL of a highly concentrated solution of phenanthrene in acetonitrile (0.01 mol L^{-1}) giving a peak area of $A = 615,819$ counts at $\lambda = 250$ nm, the pure eluent was injected. The blank value measured at $\lambda = 250$ nm was $A_{bl} = 1852$ counts.
Evaluate the carryover.
What problems can be caused by the carryover and how can they be avoided?

(g) Evaluate the dead volume of a HPLC instrument on the basis of the following data obtained with the test substance acetone:
Set flow rate: $\dot{F} = 0.1\,\mathrm{mL\,min^{-1}}$
Retention time t_r measured by six replicates:

t_r in s	35	32	34	38	35	33

(h) Which modules can be monitored by the relative standard deviation of the retention time as well as the peak areas of a test substance?

Solution to Challenge 7.3-1

(a) The relative deviation between the measured mean value $\bar{V} = 1.977\,\mathrm{mL\,min^{-1}}$ and the set flow rate $\dot{F} = 2\,\mathrm{mL\,min^{-1}}$ is 1.15%, which does not exceed the acceptance limit of 2%.

(b) The volume precision of the injection obtained by the six replicates is $s_V\% = 0.42$ calculated with $\bar{A} = 157,270.7$ counts and $s = 663.24$ counts. This value is smaller than the limit value $s_{V,\,\mathrm{lim}}\% = 1.0$. Thus, the injection precision is acceptable.

(c) The deviation of the obtained λ_{max}- and λ_{min}-values from the reference values are within the limits of the acceptance criterion. The somewhat greater deviation for the reference value $\lambda_{min,\,2} = 264$ nm may be caused by the broader and flatter valley of the absorption band at this region.

(continued)

(d) 1. With the wavelength $\lambda = 342$ nm, only azobenzene is detected because phenanthrene does not absorb in this range.

There is no wavelength at which azobenzene does not absorb, but at the detection wavelength $\lambda = 250$ nm there is a minimum of the absorbance for azobenzene and the highest absorbance for phenanthrene, and therefore this wavelength should be chosen. Note that wavelength accuracy is necessary for this problem because of the narrow absorption band of phenanthrene at 250 nm.

2. The absorbance in the range $\lambda \approx 290$–300 nm is comparable for both substances, and therefore a detection wavelength in this range yields nearly the same sensitivity for both compounds. Wavelength accuracy is not absolutely necessary for this problem.
3. An appropriate detection limit of azobenzene is $\lambda \approx 315$ nm because at this wavelength the by-product biphenyl does not absorb and azobenzene has the highest sensitivity.
4. The highest sensitivity for the detection of phenanthrene is given at $\lambda = 250$ nm.

(e) According to the plot of the response against the test concentrations shown in Fig. 7.3-2, linearity is present, which is confirmed by the statistical test of the significance of the quadratic regression term a_2 (see Sect. 5.3).

The test value calculated by (5.3.6-2)

$$\hat{t} = \frac{a_2}{s_{a_2}} = \frac{289{,}642{,}857}{131{,}966{,}110} = 2.195 \qquad (7.3\text{-}2)$$

does not exceed the critical value $t(P = 95\%, \mathrm{df} = 3) = 3.182$, and therefore the quadratic term is not significant.

The linearity is confirmed.

(f) The measured peak area of the blank is only 0.30% of the peak area obtained by the injection of a highly concentrated solution. The effect of carryover can be accepted.

Because of the carryover of small sample amounts from the previous to the next injection, carryover will affect the results of quantitative determinations.

(continued)

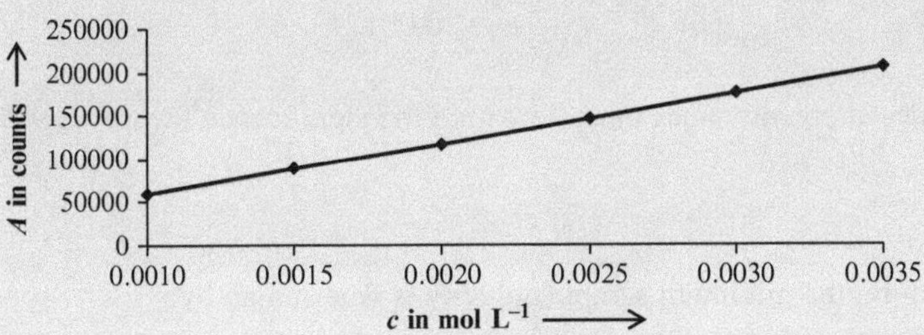

Fig. 7.3-2 Plot of response against concentration of the test sample phenanthrene in acetonitrile

To avoid contamination from the preceding sample injection, all parts of the injector system that come into contact with the sample have to be thoroughly cleaned after the injection.

(g) The dead volume calculated by (7.4-1) using the mean value of the obtained retention times of the test sample acetone, $\bar{t} = 35$ s, is:

$$V_d = 0.1\,\mathrm{mL\,min^{-1}} \cdot \frac{\bar{t}}{60}\,\mathrm{min} = 0.058\,\mathrm{mL}.$$

The dead volume $V_\mathrm{d} = 58\ \mu\mathrm{L}$ can be accepted for most applications.

(h) The relative standard deviation of the retention time of a test substance can control the accuracy of the pump (flow rate, gradient former) and the column heater.

The relative standard deviation of the peak area controls the injection system.

7.4 Balances

"Pharmaceutical testing and assay requires balances that vary in capacity, sensitivity, and reproducibility. Unless otherwise specified, when substances are to be 'accurately weighed' for assay the measurement uncertainty (random plus systematic error) of the weighing device must not exceed 0.1% of the reading. Measurement uncertainty is satisfactory if three times the standard deviation of not less than ten replicated weighings, divided by the amount weighed, does not exceed 0.001" [7].

Requirements according to USP are:

- The measurement uncertainty U may not be greater than 0.1% of the minimal sample quantity ($\mathrm{SQ_{min}}$):

$$U \leq SQ_{\min} \cdot 0.1\% \tag{7.4-1}$$

or

$$\frac{U}{SQ_{\min}} \leq 0.001. \tag{7.4-2}$$

- The measurement values must lie within the significance level $P = 99\%$:

$$U = 3 \cdot \sigma. \tag{7.4-3}$$

- Therefore, the minimum sample quantity is determined by:

$$SQ_{\min} = 3{,}000 \cdot \sigma. \tag{7.4-4}$$

Challenge 7.4-1

(a) The standard deviation of a laboratory balance is specified by the manufacturer as $\sigma = \pm 0.001$ mg. For which minimal sample quantity SQ_{min} can this balance be used?

(b) After optimal siting of a balance (vibration-free site, no direct sunlight, air-conditioned room) the following test parameters were read with a 15 mg weight:

0.015001 g	0.015000 g	0.015002 g	0.015001 g	0.014999 g
0.015000 g	0.015001 g	0.015002 g	0.014999 g	0.015001 g

Can this balance be used for weighing 10 mg according to the USP norm?

Solution to Challenge 7.4-1

(a) The minimal sample quantity is calculated by (7.4-3): $SQ_{min} = 3{,}000 \cdot 0.001$ mg $= 3$ mg.
According to USP the balance can be used for weighing the minimal sample quantity of 3 mg.

(b) The standard deviation is $s = 0.000001075$ g or $s = 0.001075$ mg which gives the values of the uncertainty $U = 0.003225$ mg calculated by (7.4-3).
The minimal sample quantity is $SQ_{min} = 9.675$ mg, calculated according to (7.4-4). This value is smaller than 10 mg, and thus, the USP requirement is fulfilled for this balance.

References

1. DIN EN ISO/IEC 17025 (1999) General requirements for the competence of testing and calibration laboratories. Beuth, Berlin
2. ICH Harmonised Tripartite Guideline (2005) Good manufacturing practice – guide for active pharmaceutical ingredients. http://www.ich.org
3. Chan CC, Lam H, Lee YC, Zhang XM (eds) (2004) Analytical method validation and instrument performance verification, Chap. 10. Wiley, New York
4. Huber L (1999) Validation and qualification in analytical laboratories. Interpharm/CRS, Cambridge
5. Kromidas S (2000) Handbuch Validierung in der Analytik 1 Aufl. Wiley-VCH, Weinheim
6. Pharm. EUP (2009) 6. Ausgabe, http://www.pharmeur.org
7. US-Pharmacopeia, USP Sec. 4e 1. Weights and Balances

Chapter 8
Control Charts in the Analytical Laboratory

8.1 Quality Control

Whereas the term "quality assurance" (QA) involves the overall measures taken by the laboratory to regulate quality, "quality control" (QC) relates to the individual measurements of samples. The two aspects of QC concern *internal* quality control, the subject of this chapter, and proficiency testing, which is a form of the *external* QC described in the next chapter.

Internal laboratory quality control provides evidence of reliability of analytical results. Monitoring of analytical performance on an on-going basis is an important element of quality management in the laboratory. It is documented during the stage of method development and validation that the analytical method applied in routine analysis is fit-for-purpose. But the question is whether the data routinely produced by this method are still fit-for-purpose each time. This is accomplished by analysis of reference materials or control samples under the same conditions, i.e. if, for example, the test material is analyzed using two replicates then the QC material must also be analyzed using two replicates. If there are no significant differences according to tests of trueness and precision, the analytical method is still under statistical control, which means the variation in the variable measured belongs to the same distribution.

The required control material may be a certified reference material, an in-house reference material which can be prepared by the laboratory for the purpose of QC, or it can be excess test materials from earlier batches. There are some requirements for QC materials: they must be stable and available in sufficient quantity, and they should receive the same treatment as the samples.

The data obtained regularly from the QC materials are, in general, evaluated by *control charts*. Control charts are extremely valuable in providing a means of monitoring the total performance of the analyst, the instruments, and the test procedure and can be utilized by any laboratory.

There are a number of different types of control charts but they all illustrate changes over time. In the following, Shewhart charts and CuSum charts will be described. For further information see references [1–4].

M. Reichenbächer and J.W. Einax, *Challenges in Analytical Quality Assurance*,
DOI 10.1007/978-3-642-16595-5_8, © Springer-Verlag Berlin Heidelberg 2011

8.2 Shewhart Charts

The general pattern of a Shewhart chart at the start of routine analysis constructed by the parameters obtained in a pre-period is shown in Fig. 8.2-1. The central line of the control chart is a mean value around which the measured values obtained by observations vary at random. The mean value $\bar{x}$ is the "true value" obtained by measurements of an in-house reference material or given from certified reference materials. Mostly, the assigned value is obtained in the pre-period, or the mean of the most recent observations considered to be under control should be used as the centre line. Measured values which lie on the central line are assumed to be unbiased.

Using the mean μ and the standard deviation s obtained, the upper and lower action limit lines UAL and LAL and the upper and lower warning limit lines UWL and LWL, respectively, are constructed, as in the following equations

Warning limit lines WL:

$$\bar{x} \pm 2 \cdot \Delta. \tag{8.2-1}$$

Action limit lines AL:

$$\bar{x} \pm 3 \cdot s. \tag{8.2-2}$$

Note that the warning limit lines are also called control limit lines CL.

In practice, the standard deviation s will be unknown and will have to be estimated from historical data.

On the assumption that the frequency distribution of the measured values follows a normal distribution, the three-sigma limits include 99.7% of the area

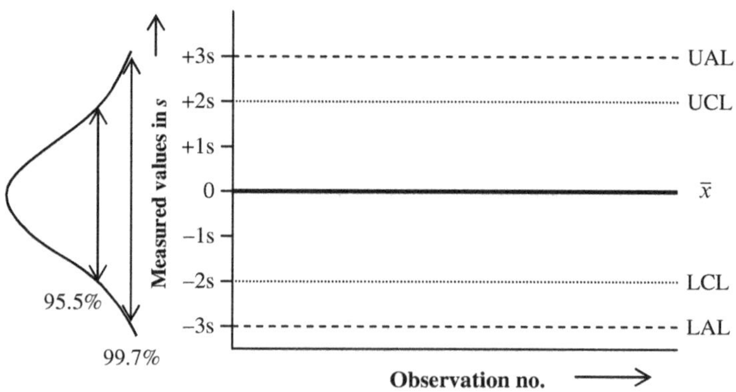

Fig. 8.2-1 The general pattern of a Shewhart chart and the curve of the normal distribution of the analytical results obtained in the pre-period with the "true" value $\bar{x}$ and the limits at the significance levels $P = 95.5\%$ and $P = 99.7\%$, respectively

under a normal curve and the two-sigma lines include 95.5% of the values, as Fig. 8.2-1 shows.

When single QC runs are carried out, the standard deviation s is estimated directly from the standard deviation of single results in different runs, but when the QC results are averaged by replicates per run, the standard deviation s must be calculated from separate estimation of within- and between-run variances according to the rules of ANOVA calculated by

$$s = \sqrt{s_{\text{bw}}^2 + \frac{s_{\text{in}}^2}{n_j}}, \tag{8.2-3}$$

where n_j is the number of the replicates per run.

Finally, the data set used for construction of the control chart has to be inspected to see whether extremely large or small values must be rejected as outliers, because such values will distort the charts and make them less sensitive and, therefore, less useful in detecting problems.

Data obtained by the observations are plotted in chronological order. By comparing current data to the limit lines, one can draw conclusions about whether the process variation is consistent (in control) or is unpredictable (out of control): affected by special causes of variation. If an out-of-control situation is detected, the measurement process should be stopped, the causes of this variation must be sought and eliminated or changed.

Besides the out-of-control rules given in [5], there are some additional rules which are illustrated in Fig. 8.2-2:

1. One measured point lies out of the upper or the lower action line.
2. Nine consecutive measured points lie on one side of the central line.
3. Two consecutive measured points lie outside the warning line.
4. Nine consecutive measured points show an upward trend.
5. Nine consecutive measured points show a downward trend.

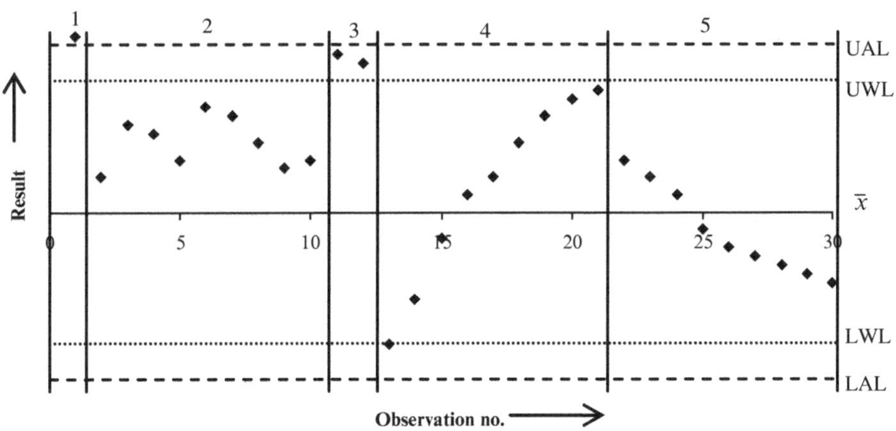

Fig. 8.2-2 Presentation of some out-of-control situations

A Shewhart control chart constructed according to Fig. 8.2-1 can be applied as:

- Mean control chart, preferably, for recognition of the precision or trends of an analytical method.
- Blank control chart, for control of reagents and measurement instruments. Note that blank control charts include not analytical results but measured values.
- Recovery control chart, for control of proportional systematic errors caused by the matrix.

These charts are primarily used for detecting bias in an analytical system. A special chart, the *range chart*, is applied for monitoring the analytical precision. Analytical precision is concerned with variability between repeated measurements of the same analyte, irrespective of presence or absence of bias. The range, i.e. the difference between the largest and smallest values, obtained by replicate measurements within each analytical run is used to control the stability of analytical precision and it thus checks the homogeneity of variances.

The format of a range chart is shown in Fig. 8.2-3.

In order to construct the limits of the range charts, the ranges R_i of all sub-groups must be determined according to

$$R_i = x_{i,\max} - x_{i,\min}, \tag{8.2-4}$$

for which the average range $\bar{R}$ is calculated by (8.2-5)

$$\overline{R} = \frac{\sum R_i}{n}. \tag{8.2-5}$$

The upper action limit line UAL and upper warning (or control) limit line UWL are obtained by multiplying the average range by tabulated multipliers which are given in Table 8.2-1 for various numbers of replicates n_j.

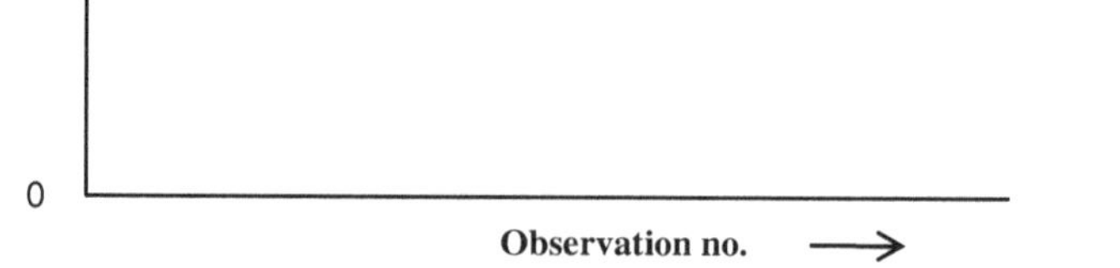

Fig. 8.2-3 Format of a range chart

Table 8.2-1 *D*-factors for the calculation of the limits of range charts for n_j replicates per run

n_j	D_{WL} $P = 95\%$	D_{AL} $P = 99.7\%$
2	2.809	3.267
3	2.176	2.575
4	1.935	2.282
5	1.804	2.115
6	1.721	2.004
7	1.662	1.924
8	1.617	1.864
9	1.583	1.816
10	1.555	1.777

These multipliers D_{WL} and D_{AL} correspond to the two- and three-sigma level, respectively:

Warning limit lines WL:

$$WL = \overline{R} \cdot D_{WL} \tag{8.2-6}$$

Action limit lines AL:

$$AL = \overline{R} \cdot D_{AL} \tag{8.2-7}$$

An out-of-control situation can be detected by the rules given above.

Challenge 8.2-1

The performance of a test method for the determination of copper in soil samples by optical emission spectroscopy with inductively coupled plasma (ICP-OES) was monitored by analyzing a quality control material without replicates. The analytical results obtained in the pre-period are given in Table 8.2-2.

The Cu-containing soil sample was used as "in-house reference material" for quality control in routine analysis. The results for the first 35 control measurements are summarized in Table 8.2-3.

(a) Construct a Shewhart mean value control chart with warning and action limits equivalent to the 95.5% and 99.7% confidence limits on the basis of the data set obtained in the pre-period.
(b) Check whether the method is under statistical control at each control point in routine analysis.

Table 8.2-2 Analytical results of Cu in a soil sample determined in the pre-period by ICP-OES obtained by single observations

Observation no.	c_{Cu} in mg kg^{-1}	Observation no.	c_{Cu} in mg kg^{-1}
1	24.5	16	24.4
2	24.1	17	23.8
3	26.3	18	23.5
4	22.7	19	22.9
5	23.9	20	24.3
6	24.1	21	24.8
7	30.1	22	24.1
8	23.6	23	24.6
9	23.8	24	24.6
10	24.6	25	24.7
11	22.2	26	24.1
12	23.6	27	24.2
13	23.9	28	23.5
14	24.0	29	22.7
15	24.8	30	24.8

Table 8.2-3 Analytical results of Cu determined by ICP-OES in routine analysis

Observation no.	c_{Cu} in mg kg^{-1}	Observation no.	c_{Cu} in mg kg^{-1}
1	23.9	18	24.3
2	23.7	19	21.9
3	24.9	20	22.0
4	21.0	21	22.9
5	24.7	22	23.6
6	25.1	23	25.2
7	23.8	24	25.3
8	23.6	25	26.7
9	23.6	26	24.7
10	24.5	27	24.9
11	23.4	28	24.6
12	22.9	29	25.1
13	22.3	30	25.0
14	26.2	31	24.8
15	25.0	32	24.3
16	23.2	33	23.2
17	25.1	34	23.9

Solution to Challenge 8.2-1

(a) If the whole data set given in Table 8.2-2 is used for the determination of the standard deviation s the following limits are calculated:

$\bar{x} = 24.24\,\text{mg kg}^{-1}$	$s = 1.3594\,\text{mg kg}^{-1}$	$\text{UAL} = 28.32\,\text{mg kg}^{-1}$
$\text{UWL} = 26.96\,\text{mg kg}^{-1}$	$\text{LWL} = 21.52\,\text{mg kg}^{-1}$	$\text{LAL} = 20.16\,\text{mg kg}^{-1}$

(continued)

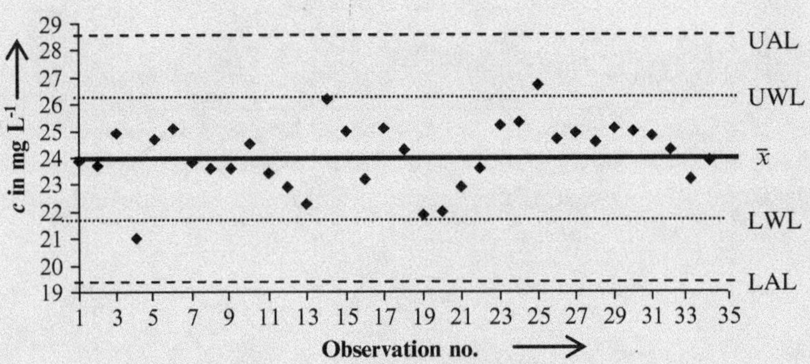

Fig. 8.2-4 Shewhart mean value control chart of the observations given in Table 8.2-3 with the limits calculated by the whole data set from the pre-period listed in Table 8.2-2

The Shewhart mean value control chart constructed with these parameters is shown in Fig. 8.2-4.

As Fig. 8.2-4 shows, no out-of-control situation can be detected. But are the limits used for the construction of the Shewhart chart valid?

Inspection of the data set in Table 8.2-2 shows that the value of observation no. 7 measured in the pre-period is unusually high, and therefore this value must be detected as an outlier. The Grubbs test must be used because $n = 30$.

The test value calculated according to (3.2.3-2) with $x_{\max} = x^* = 30.1$, $x_{\min} = 22.2$, and $s = 1.3594$ is $\hat{r}_m = 4.311$ which is greater than the critical value $r_m(P = 95\%, n = 30) = 2.745$. Thus, the measured value for observation no. 7 must be rejected from the data set.

The recalculated limits on the basis of the outlier-free data set are:

$\bar{x} = 24.04\,\mathrm{mg\,kg^{-1}}$	$s = 0.8033\,\mathrm{mg\,kg^{-1}}$	$\mathrm{UAL} = 26.45\,\mathrm{mg\,kg^{-1}}$
$\mathrm{UWL} = 25.64\,\mathrm{mg\,kg^{-1}}$	$\mathrm{LWL} = 22.43\,\mathrm{mg\,kg^{-1}}$	$\mathrm{LAL} = 21.63\,\mathrm{mg\,kg^{-1}}$

The control chart presented in Fig. 8.2-5 shows three out-of-control situations:

1. The value of observation no. 4 lies outside the action limit.
2. Two successive observations (nos. 19 and 20) lie between the control and the action limits.
3. The value of observation no. 25 lies outside the action limit.

This example demonstrates the importance of the evaluation of data used for the determination of the control limits.

Clearly, the determination of the standard deviation used for the calculation of the control limits requires data sets which are normally distributed, which can be checked by the David test. Strictly speaking, the test value $\hat{q}_r = 5.81$ lies outside from the upper value which is 5.26 at the significance level $P = 99\%$, but the difference is only small.

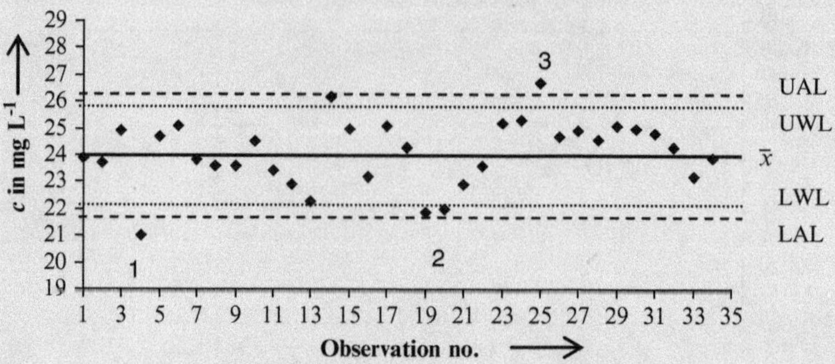

Fig. 8.2-5 Shewhart mean value control chart of the observations given in Table 8.2-3 with the limits calculated by the outlier-free data set from the pre-period listed in Table 8.2-2

Table 8.2-4 Twelve sets of three replicate potency assay measurements obtained from a control material

Observation no.	c in % (w/w)	Observation no.	c in % (w/w)
1	80.37	7	80.99
	80.95		80.51
	80.81		80.83
2	81.05	8	80.92
	80.74		80.85
	80.99		80.91
3	80.85	9	80.87
	81.09		80.64
	80.96		80.58
4	81.12	10	80.88
	81.05		80.99
	80.92		80.81
5	81.05	11	80.96
	80.91		80.81
	81.18		81.04
6	80.83	12	80.95
	80.63		81.41
	80.97		81.09

Challenge 8.2-2

Table 8.2-4 presents the results of the determination of the potency assay of a control material of a pharmaceutical product obtained in the pre-period by each three replicates, and the first nine results in routine analysis are given in Table 8.2-5.

(a) Construct the corresponding chart for controlling the mean values! Determine whether the analytical system is under control!
(b) Construct the corresponding chart for controlling of the precision and determine whether the homogeneity of variances is given!

(continued)

Table 8.2-5 The first nine analytical results of three replicates obtained by the quality control in routine analysis

Observation no.	c in % (w/w)	Observation no.	c in % (w/w)
1	80.82	6	80.97
	80.26		81.01
	80.65		81.13
2	80.27	7	81.65
	81.00		81.75
	81.05		81.97
3	80.99	8	80.53
	80.88		80.77
	81.11		80.97
4	80.28	9	81.02
	80.02		81.01
	80.57		81.03
5	80.65		
	80.62		
	80.78		

Solution to Challenge 8.2-2

First, the mean values $\bar{x}_j$ of the i observations of the measured values in the pre-period listed in Table 8.2-7 are checked for normal distribution and outliers.

The test value calculated according to (3.2.1-1) $\hat{q}_r = 3.30$ lies within the critical limits of the David table for $P = 95\%$ and $n = 12$ which are 2.80 and 3.91, and therefore the data can be assumed to be normally distributed.

The intermediate quantities and results given in Table 8.2-6 show that the data set is free of outliers by the Dixon test for $n = 12$. Thus, the whole data set can used to construct the appropriate control charts.

(a) Because replicates were performed, the standard deviation necessary for the estimation of the control limits according to (8.2-1) and (8.2-2) must be determined by the variance components s^2_{bw} and s^2_{in} according to (8.2-3), which must be obtained by ANOVA. The intermediate quantities and results of ANOVA are listed in Table 8.2-7.

The standard deviation required for the setup of the Shewhart mean control chart is $s = 0.2580\%$ (w/w) calculated according to (8.2-3) using the variances given in Table 8.2-7. The limits of the mean value control charts shown in Fig. 8.2-5 calculated by (8.2-1) and (8.2-2) are:

(continued)

Table 8.2-6 Intermediate quantities and results of the Dixon outlier test on the highest and smallest mean value obtained during the pre-period

Test value	x_1	x_b	x_k	$\hat{Q}$
$x_{\max}$	81.15	81.03	80.71	0.273
$x_{\min}$	80.70	80.78	81.05	0.229

The symbols refer to (3.2.3-1) for $b = 3$ and $k = n–1$. The critical value is $Q(P = 95\%, n = 12) = 0.546$.

Table 8.2-7 Intermediate quantities and results of ANOVA

Observation no.	$\bar{x}_j$	$n_j(\bar{x}_j - \bar{\bar{x}})^2$	SS_j
1	80.71	0.11181	0.18320
2	80.93	0.00167	0.05407
3	80.97	0.01214	0.02887
4	81.03	0.04834	0.02060
5	81.05	0.06187	0.03647
6	80.81	0.02598	0.05840
7	80.78	0.04792	0.11947
8	80.89	0.00028	0.00287
9	80.70	0.12779	0.04687
10	80.89	0.00028	0.01647
11	80.94	0.00339	0.02727
12	81.15	0.18294	0.11120
$\bar{\bar{x}}$	80.90		
$\sum n_j(\bar{x}_j - \bar{\bar{x}})^2 = SS_{bw}$	0.62443	$\sum SS_j = SS_{in}$	0.70573
n_i	12	n_j	3
$df_{bw} = n_i - 1$	11	$df_{in} = n_j \cdot n_i - n_i$	24
s^2_{bw}	0.05677	s^2_{in}	0.02941

$\bar{x} = 80.90\%$ (w/w)	
UAL = 81.68% (w/w)	UWL = 81.42% (w/w)
LWL = 80.39% (w/w)	LAL = 80.13% (w/w)

Figure 8.2-6 shows the mean value charts for controlling the potency assay of a pharmaceutical drug during routine analysis, constructed with the limit values obtained in the pre-period and the mean values given in Table 8.2-8. Inspection of Fig. 8.2-6 shows an out-of-control situation at observation no. 7. After correction of the problem caused by the preparation of the sample, the analytical system is once more under control, as shown by the measured value of the next observation.

(b) The range chart is based on the range values obtained in the pre-period which are given in Table 8.2-9. The limit values of the range chart calculated according to (8.2-6) and (8.2-7) with the mean value $\bar{R}_i = 0.3042\%$ (w/w), and the *D*-factors from Table 8.2-1 for $n_j = 3$ (2.575 and 2.176, respectively) are: UAL = 0.783% (w/w) and UWL = 0.662% (w/w).

The range chart is shown in Fig. 8.2-7 for the first nine observations in routine analysis with the range values listed in Table 8.2-8. Observation no. 2 shows an out-of-control situation, because the range value lies outside the upper action line. After removal of the cause, e.g., exchanging the HPLC injection syringe, the analytical system is again under control. As the results of this Challenge show, the combination of a mean value and a range chart is appropriate for checking large deviations of the mean, the precision, and also trends in the analytical system.

(continued)

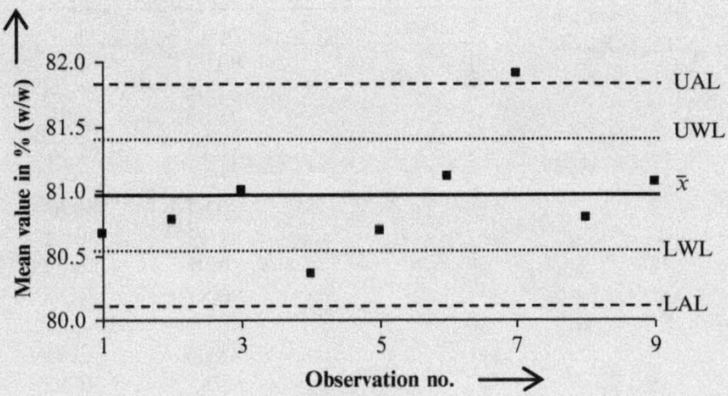

Fig. 8.2-6 Mean value charts for controlling the potency assay of a pharmaceutical drug during routine analysis

Table 8.2-8 The values of the mean and the range of the results given in Table 8.2-5

Observation no.	$\bar{x}_j$	$x_{j,\max}$	$x_{j,\min}$	R_j
1	80.58	80.82	80.26	0.56
2	80.77	81.05	80.27	0.78
3	80.99	81.11	80.88	0.23
4	80.29	80.57	80.02	0.55
5	80.68	80.78	80.62	0.16
6	81.04	81.13	80.97	0.16
7	81.79	81.97	81.65	0.32
8	80.76	80.97	80.53	0.44
9	81.02	81.03	81.01	0.02

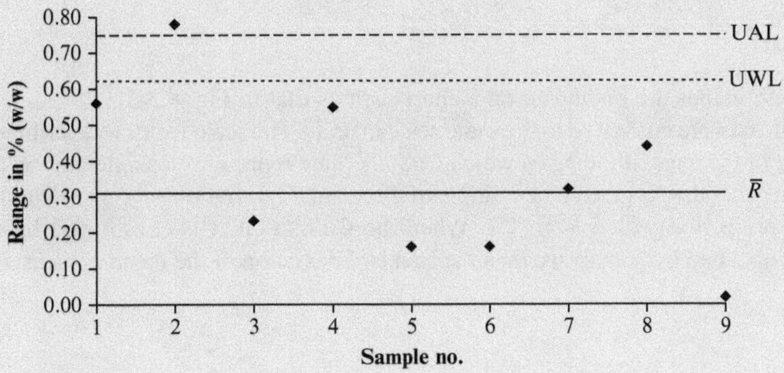

Fig. 8.2-7 Range charts for controlling the potency assay of a pharmaceutical drug during routine analysis

Table 8.2-9 Range values of the data set of Table 8.2-4

Observation no.	x_{max}	x_{min}	R_i
1	80.95	80.37	0.58
2	81.05	80.74	0.31
3	81.09	80.85	0.24
4	81.12	80.92	0.20
5	81.18	80.91	0.27
6	80.97	80.63	0.34
7	80.99	80.51	0.48
8	80.92	80.85	0.07
9	80.87	80.58	0.29
10	80.99	80.81	0.18
11	81.04	80.81	0.23
12	81.41	80.95	0.46

8.3 CuSum Charts

For a series of measurements $x_1, x_2, \ldots, x_n$ the *cumulative sum of differences* (CuSum) between the observed value and the target value μ is determined using

$$C_1 = x_1 - \mu$$

$$C_2 = (x_2 - \mu) + (x_1 - \mu) = C_1 + (x_2 - \mu)$$

and so on resulting in (8.3-1):

$$C_i = \sum_{j=1,i} (x_j - \mu). \tag{8.3-1}$$

These values are displayed on a chart such as that in Fig. 8.3-1.

Both axes are converted to the same scale in units. The scale factor w determines the scaling of the axes. It indicates which CuSum value represents a single unit on the y-axis. In general, w is given in a multiple of the standard deviation $w = q \cdot s$ determined in the pre-period with $1 \leq q \leq 2$. When the CuSum chart is constructed by mean values obtained by n analysis, then the standard deviation of the mean s_m is used:

$$s_m = \frac{s}{\sqrt{n}}. \tag{8.3-2}$$

A single unit on the x-axis corresponds to the difference between two observations, for example, one day. The scaling is determined by the scaling factor w which corresponds the unit on the CuSum (y)-axis.

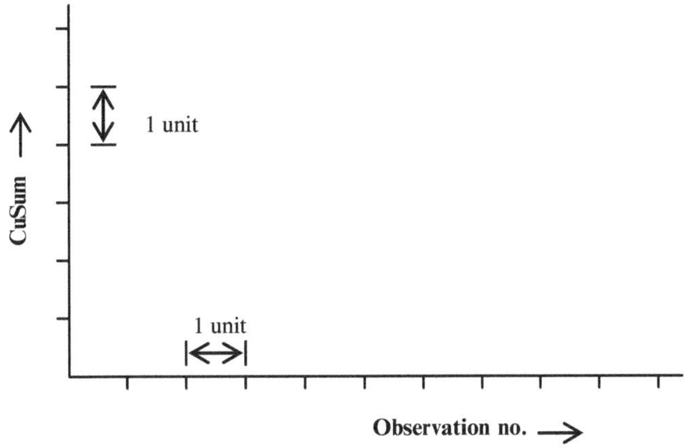

Fig. 8.3-1 Pattern of a CuSum chart

CuSum charts cannot be interpreted using warning and action limits as in the interpretation of Shewhart chart, but there are some possibilities for recognizing an out-of-control situation:

- Visual estimation of the slope of the CuSum line. An out-of-control situation can be shown by changes in the slope of the CuSum line.
- Numerical criteria.
- Use of software packages such as [6].
- Use of the V-mask as the decision criterion, which will be described below.

The dimension of the *V-mask* can be specified by two distinct parameters:

- θ, half the angle formed by the V-mask arms.
- d the distance between the origin and the vertex, as shown in Fig. 8.3-2.

The vertical distance between the origin and the upper (or lower) V-mask arm h corresponds to an interval of one unit on the horizontal axis.

The V-mask is laid over the CuSum chart in such a manner that the vertex always points in the direction of the observations and the origin of the V-mask (point •) is located at the most recently plotted point and overlays each point in turn. Note that the horizontal line must always be kept parallel to the x-axis. An out-of-control condition is signaled at an observation (or at time t) if one or more of the points plotted up to time t crosses an arm of the V-mask. An upward shift is signaled by points crossing the lower arm, and a downward shift by points crossing the upper arm. The observation at which the shift occurred corresponds to the observation at which a distinct change is observed in the slope of the plotted points.

The parameters θ and d can be calculated, for example, by the (8.3-3) and (8.3-4) [2]:

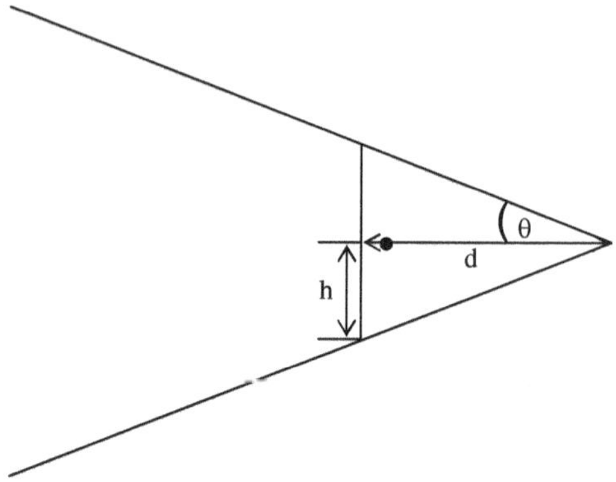

Fig. 8.3-2 Parameters of the V-mask

$$\theta = \arctan\left(\frac{D}{2w}\right) \tag{8.3-3}$$

$$d = \frac{2 \cdot s^2 \cdot \ln \alpha}{D^2}, \tag{8.3-4}$$

where D is the smallest deviation which can be recognized at a certain significance given in multiples of the standard deviation; s is the standard deviation of the target value or is determined by reference values of the pre-period; w is the scaling factor $w = q \cdot s$, for example, $w = 2s$, and α is the risk of error of the first kind; $\alpha = 0.0027$ corresponds to the three-sigma control limit of the mean value chart.

The CuSum chart is used to detect small changes between 0 and $0.5\,s$. For larger shifts $(0.5 - 2.5\,s)$, the simple Shewhart charts are just as good and easier to use.

Challenge 8.3-1

In order to control whether the validated ion chromatographic (IC) method for the determination of nitrite-N is fit-for-purpose in routine analysis, a control sample with a content of $c = 12.25\,\mathrm{mg\,L^{-1}}$ is analyzed under the same conditions. The results are shown in Table 8.3-1.

(a) Construct a Shewhart chart with warning and action limits for $P = 95.5\%$ and $P = 99.7\%$, respectively, and check whether an out-of-control situation can be detected.

(b) Construct a CuSum chart and check by a V-mask using the scaling factor $w = 1s$ and the smallest deviation $D = 1.3 \cdot s$ whether the method can be considered to be under statistical control at $P = 99.7\%$. Compare the result obtained by the Shewhart chart.

(continued)

Table 8.3-1 Results of controlling the IC method for the determination of nitrite-N in routine analysis using a control sample with $c = 12.25\,\text{mg}\,\text{L}^{-1}$

Observation no.	c in mg L^{-1}	Observation no.	c in mg L^{-1}
1	12.28	11	12.34
2	12.37	12	12.17
3	12.00	13	12.34
4	12.23	14	12.35
5	12.38	15	11.89
6	12.18	16	12.12
7	12.01	17	12.18
8	12.23	18	12.20
9	12.33	19	12.09
10	12.38	20	12.15

Solution to Challenge 8.3-1

(a) The standard deviation of the results given in Table 8.3-1 is $s = 0.140\,\text{mg}\,\text{L}^{-1}$ and the mean value is $\bar{x} = 12.21\,\text{mg}\,\text{L}^{-1}$.

Warning limits (WL) calculated as $\mu \pm 2 \cdot s = 12.21 \pm 0.28\,\text{mg}\,\text{L}^{-1}$ are 12.49 mg L^{-1} and 11.93 mg L^{-1}.

Action limits (AL) calculated as $\mu \pm 3 \cdot s = 12.21 \pm 0.42\,\text{mg}\,\text{L}^{-1}$ are 12.63 mg L^{-1} and 11.79 mg L^{-1}.

The Shewhart chart is shown in Fig. 8.3-3.

According to the Shewhart chart, no out-of-control condition can be detected. The measured value of observation no. 15 lies indeed outside the lower warning line, but inside the action line. Because the next measured value is again inside the warning line no out-of-control situation is present at observation no. 15.

(b) The CuSum-values calculated according to (8.3-1) are summarized in Table 8.3-2 and the CuSum chart is shown in Fig. 8.3-4.

In order to check for an out-of-control situation, the V-mask must be constructed using the parameters:

- Standard error of the mean $s_{\text{m}} = 0.03133\,\text{mg}\,\text{L}^{-1}$, which is calculated by (8.3-2) from the results given in Table 8.3-1.
- The scaling factor given as $w = 1 \cdot s$.
- The smallest deviation $D = 1.5 \cdot s$.

Calculation of the angle θ :

$$\theta = \arctan\left(\frac{D}{2w}\right) = \arctan\left(\frac{1.5}{2}\right) = 0.6435 \qquad (8.3\text{-}5)$$

$$\theta = 36.87°.$$

(continued)

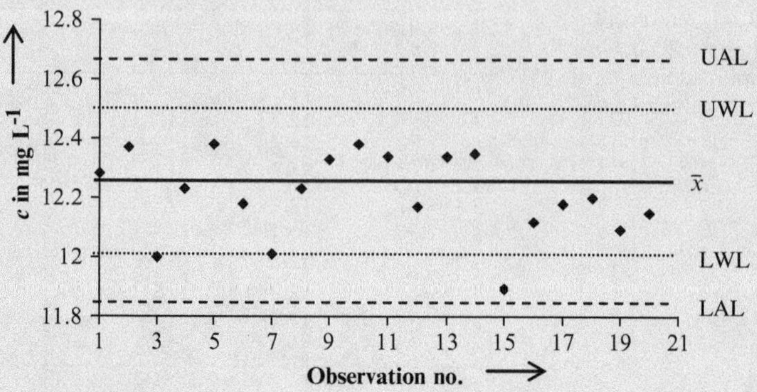

Fig. 8.3-3 Shewhart chart constructed from data in Table 8.3-1

Table 8.3-2 Calculation of the cumulative sum (CuSum) for the data given in Table 8.3-1

Observation no.	$\mu - x_i$	C_i	Observation no.	$\mu - x_i$	C_i
1	−0.03	−0.03	11	−0.09	0.02
2	−0.12	−0.15	12	0.08	0.10
3	0.25	0.10	13	−0.09	0.01
4	0.02	0.12	14	−0.10	−0.09
5	−0.13	−0.01	15	0.36	0.27
6	0.07	0.06	16	0.13	0.40
7	0.24	0.30	17	0.07	0.47
8	0.02	0.32	18	0.05	0.52
9	−0.08	0.24	19	0.16	0.68
10	−0.13	0.11	20	0.10	0.78

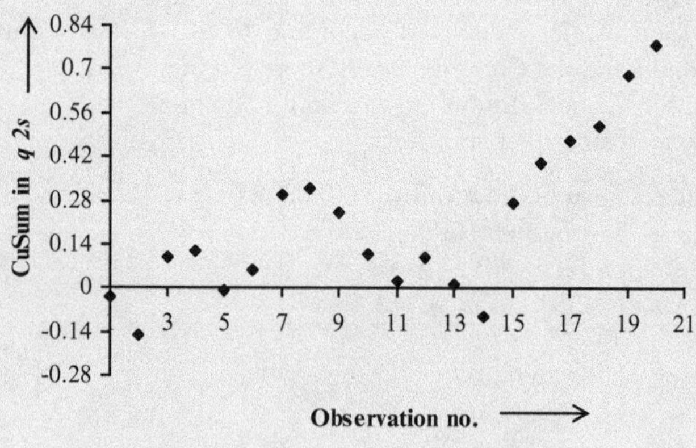

Fig. 8.3-4 CuSum chart for the data given in Table 8.3-2

Calculation of the distance d in the V-mask:

$$d = \frac{-2 \cdot s^2}{D^2} \cdot \ln \alpha = \frac{-2}{1.3^2} \cdot \ln 0.027 = 4.25 \text{ units.} \tag{8.3-6}$$

The V-mask constructed with $\theta = 37°$ and $d = 4.3$ units overlies observation no. 15. As Fig. 8.3-5 shows, observation no. 14 falls outside the lower arm of the V-mask, indicating an upward shift which is manifest at observation point 15. Note that the Shewhart mean chart does not show any out-of-control situation. This demonstrates the higher sensitivity of the CuSum chart in comparison with the Shewhart mean value chart.

The relative merits of different chart types when applied to detect gross errors, shifts in mean, and shifts in variability are summarized in Table 8.3-3.

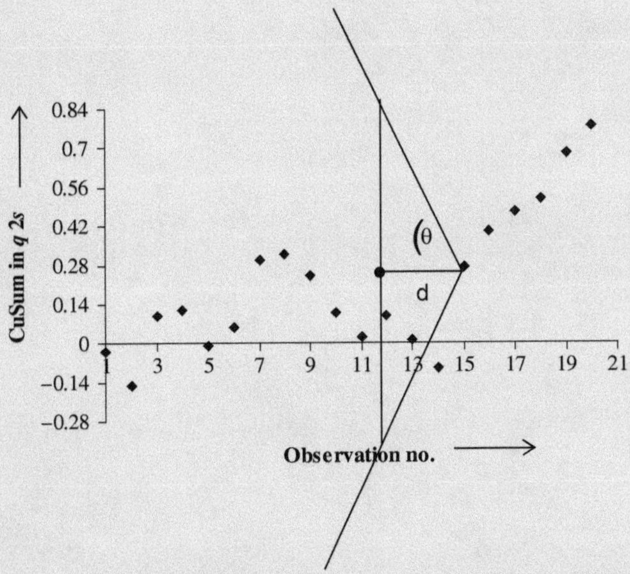

Fig. 8.3-5 CuSum chart for the data given in Table 8.3-2 with the V-mask overlying the measured value obtained by observation no. 15

Table 8.3-3 Relative merits of different chart types when applied to detect changes in the first column

Cause of change	Chart type		
	Mean	Range	CuSum
Gross error	+++	++	+
Shifts in mean	++		+++
Shifts in variability		+++	

+ suitable, ++ very suitable, +++ especially suitable for recognizing out-of-control situations

References

1. Xie M, Kuralmani V, Goh TN (2002) Statistical models and control charts for high quality processes. Springer, Berlin
2. Funk W, Dammann V, Donnevert G (2005) Qualitätssicherung in der Analytischen Chemie, 2 Aufl. Wiley-VCH, Weinheim
3. http://statsoft.com/textbook/*quality-control-charts*
4. Mullins E (2003) Statistics for the quality control chemistry laboratory. RSC, Cambridge
5. ISO 8258 (1991) Shewart control charts. International Organization for Standardization, Geneva
6. Copyright Statsoft, Inc. (1984–2003) Statistical quality control charts

Chapter 9
Interlaboratory Studies

9.1 Purpose and Types of Interlaboratory Studies

In interlaboratory studies, which are usually organized by a reference laboratory ("the organizer"), the participating laboratories receive part of a homogeneous bulk material which must be analyzed according to a given protocol. The results obtained are returned to the reference laboratory which evaluates the results and gives feedback to the participating laboratories.

Interlaboratory studies are primarily performed for one of three reasons:

1. Validation of a measurement method (method performance studies).
   Such studies are essential in order to compile DIN/CEN/ISO standards. Furthermore, the value of the precision obtained by the interlaboratory studies can be used to estimate the measurement uncertainty (see Chap. 10). The organization, accomplishment, and estimation of the results may be carried out according to DIN ISO 5725-2 [1], DIN 38402-41 [2], and DIN 38402-42 [3]. Participation in interlaboratory studies is by the laboratory's own choice.
2. Validation of a reference materials (material certification studies).
   A group of selected laboratories analyses, usually by different methods, a homogeneous material in order to determine the most probable mean value of the reference material with the smallest uncertainty. In general, interlaboratory studies for the certification of reference materials are carried out and established according to ISO Guide 35: 2006 [4]. There are statistical principles to assist in the understanding of the associated uncertainty and to establish its metrological traceability. Reference materials that undergo all the steps described in ISO Guides are usually accompanied by a certificate and called certified reference materials (CRM).
3. Assessing laboratory performance (proficiency studies).
   A proficiency test scheme comprises the regular distribution to participating laboratories for independent testing of test materials of which the true concentrations are known or have been assigned in some way, often from the interlaboratory study itself. The choice of the analytical methods is left to the laboratory itself. Proficiency testing shares two key objectives:

M. Reichenbächer and J.W. Einax, *Challenges in Analytical Quality Assurance*,
DOI 10.1007/978-3-642-16595-5_9, © Springer-Verlag Berlin Heidelberg 2011

(a) The provision of regular, objective, and independent assessments of the accuracy (which involves trueness and precision) of analytical laboratory results on routine test samples.
(b) The promotion of improvements in the quality (accuracy) of routine analytical data. Proficiency testing is described in international guidelines and standards such as ISO/IEC 17043:2010 [5] or DIN 38402-45:2003 [6].

Participation in proficiency testing is necessary for laboratories in the course of their accreditation.

Note that the operational protocols for these three types of study are quite different.

9.2 Method Performance Studies

In the following, we describe the interlaboratory method performance studies according to DIN ISO 5725-2 [1] and DIN 38402-42 [3] for estimating objectively the laboratory quality and the analytical method, respectively.

There should be at least eight participating laboratories, but 15 or more may be better. In general, each laboratory analyzes the same samples with four replicates using the same declared analytical method. The results are relayed to the organizer of the interlaboratory study to evaluate the results by the stages given below.

The evaluation of the outlier-free data set is based on the random effect model of ANOVA.

Note that the equations for the calculation of parameters are only given where necessary. Most of them should be familiar from previous chapters:

- Stage 1: Calculation of the preliminary within-laboratory parameters, mean value $\bar{x}_i^*$, and standard deviation s_i^*.
  Note that the asterisk denotes that these parameters are preliminary data because they may still contain outliers. This is also true for the following parameters.
- Stage 2: Preliminary rejection of type 1 outliers (outliers within the laboratory data) checked by the Grubbs test.
  Each laboratory is checked and the test value $\hat{r}_m$ is compared with the critical values at the two-sided significance level $P = 90\%$, which is equal to the one-sided significance level $P = 95\%$. If $\hat{r}_m$ is greater than $r_m(P = 95\%, n_k)$, the test value is provisionally rejected from the data set.
- Stage 3: Recalculation of the mean values $\bar{x}_i$ and the standard deviation s_i of the outlier-free data set of each laboratory i.
- Stage 4: Calculation of the parameters of the outlier-free laboratory mean values $\bar{x}_i$.
  To evaluate type 2 outliers (outliers of the mean values $\bar{x}_i$), the laboratories' mean value $\bar{x}_L^*$ of the means $\bar{\bar{x}}$ as well as their standard deviation $s_{\bar{x}_L}^*$ are calculated.
- Stage 5: Rejection of type 2 outliers as in stage 2, using the Grubbs test.

- Stage 6: Determination of the smallest ($\bar{x}_{\min}$) and the largest ($\bar{x}_{\max}$) of the laboratory means.
- Stage 7: Re-integration of outliers of type 1 $x^*_{\mathrm{OL}_{ik}}$ which fulfill the condition $\bar{x}_{\min} \leq x^*_{\mathrm{OL}_{ik}} \leq \bar{x}_{\max}$.
- Stage 8: Rejecting of type 3 outliers (outliers in laboratory precision) checked by the Cochran test.
  If the test value exceeds the critical value at the significance level $P = 99\%$, the laboratory with maximal variance is rejected as an outlier. The test is repeated until no more outliers can be found.
- Stage 9: Calculation of the following final parameters on the basis of the data set free of outliers of types 1–3:
  - Laboratory mean values $\bar{x}_i$
  - Standard deviation s_i
  - Total number of the analytical values n
  - Number of the remaining laboratories l
  - Grand mean $\bar{\bar{x}}$
  - Repeatability standard deviation s_r according to ANOVA

$$s_r = \sqrt{\frac{\sum_{i=1}^{l} \mathrm{SS}_{x_i x_i}}{n - l}}. \tag{9.2-1}$$

  - Coefficient of variation of repeatability CV_r

$$\mathrm{CV}_r\ \% = \frac{s_r}{\bar{\bar{x}}} \cdot 100. \tag{9.2-2}$$

  - Reproducibility standard deviation s_R

$$s_R = \sqrt{s_L^2 + s_r^2}. \tag{9.2-3}$$

with the between-laboratory variance s_L^2

$$s_L^2 = \frac{s_b^2 - s_r^2}{\bar{\bar{n}}}. \tag{9.2-4}$$

The between-group variance s_b^2 according to ANOVA

$$s_b^2 = \frac{\sum_{i=1}^{l} n_i (\bar{x}_i - \bar{\bar{x}})^2}{l - 1} \tag{9.2-5}$$

and

$$\bar{\bar{n}} = \frac{1}{l-1}\left[\sum_{i=1}^{l} n_i - \frac{\sum_{i=1}^{l} n_i^2}{\sum_{i=1}^{l} n_i}\right]. \qquad (9.2\text{-}6)$$

Note that if the values of all n_i are equal then (9.2-6) can be simplified to (9.2-7):

$$\bar{\bar{n}} = n_i. \qquad (9.2\text{-}7)$$

– Coefficient of variation of reproducibility CV_R

$$CV_R\ \% = \frac{s_R}{\bar{\bar{x}}} \cdot 100. \qquad (9.2\text{-}8)$$

- Stage 10: Calculation of the recovery rate η

$$\eta\ \% = \frac{\bar{\bar{x}}}{\mu} \cdot 100. \qquad (9.2\text{-}9)$$

- Stage 11: Estimation of bias by a t-test.
  The test value $\hat{t}$ calculated by

$$\hat{t} = \frac{|\bar{\bar{x}} - \mu|}{s_R} \cdot \sqrt{n} \qquad (9.2\text{-}10)$$

is compared with the critical value $t(P = 99\%, \mathrm{df} = n - 1)$. If the test value is greater than the critical value, a significant difference between the true value μ (or x_{ref}) and the grand mean $\bar{\bar{x}}$ is detected. In this case, the analytical method should be re-validated.

- Stage 12: Presentation of an overview of the whole parameter set obtained by the interlaboratory study as a table with comments.

Challenge 9.2-1

An interlaboratory study was carried out for the determination of inorganic anions in synthetic industrial water by the ion chromatographic method according to DIN EN ISO 10304-1:2009-07 [7], in order to check the quality of the method for a specific matrix and to obtain parameters for the calculation of the measurement uncertainty (see next Chapter). Nine laboratories took part in the interlaboratory trial and four replicates should be carried out. The results obtained for the analyte *bromide* is given in Table 9.2-1. The true value is $\mu = 15.2\ \mathrm{mg\,L^{-1}}$.

(continued)

Table 9.2-1 Data of the analyte *bromide* in mg L^{-1} obtained by an interlaboratory study

Laboratory	Replicate 1	2	3	4
1	13.8	13.9	14.0	13.7
2	15.1	15.0	14.9	15.1
3	15.2	15.0	15.1	15.1
4	15.5	15.2	15.4	15.4
5	14.8	14.9	14.8	14.9
6	15.2	15.3	15.0	15.2
7	15.4	15.4	15.2	15.3
8	15.1	15.0	15.1	14.5
9	15.3	15.2	15.3	15.1

Table 9.2-2 Overview of the symbols of all parameters for the analyte *bromide* in industrial water obtained by an interlaboratory study

Number of laboratories	l
Number of outlier-free individual analytical values	n
Number of outliers	n_{OL}
Percentage of outlier values	$n_{\mathrm{OL}}\%$
Grand mean value	$\bar{\bar{x}}$
True value	μ
Recovery rate	$\eta\%$
Reproducibility standard deviation	s_{R}
Coefficient of variation of the reproducibility	$\mathrm{CV_R}\%$
Repeatability standard deviation	s_{r}
Coefficient of variation of the repeatability	$\mathrm{CV_r}\%$
Degrees of freedom of the repeatability standard deviation	df_{s_r}

Evaluate the interlaboratory study according to the procedure given in DIN 38402-42 [3] and fill in Table 9.2-2 with the results obtained by the interlaboratory study.

Solution to Challenge 9.2-1

The results of the interlaboratory study are evaluated according to the DIN-procedure given above.

Stages 1 and 2: The intermediate quantities and results of the Grubbs test for type 1 outliers are summarized in Table 9.2-3. The test values $\hat{r}_{m,i}$ for each laboratory are calculated by (3.2.3-2). The critical value is $r_m(P = 95\%, n_k = 4) = 1.463$. As Table 9.2-3 shows, the lowest value of laboratory 8 ($x_{8,4} = 14.5\ \mathrm{mg\,L^{-1}}$) must be provisionally rejected as an outlier because $\hat{r}_{m,x_{\min}} = 1.480$ exceeds the critical value.

Stages 3, 4, and 5: The intermediate quantities and results of the Grubbs test for type 2 outliers calculated by

(continued)

Table 9.2-3 Intermediate quantities and results of the Grubbs test for type 1 outliers

Laboratory	Intermediate quantities				Test values	
	$\bar{x}$	s	$x^*_{\max}$	$x^*_{\min}$	$\hat{r}_{m,\,x_{\max}}$	$\hat{r}_{m,\,x_{\min}}$
1	13.85	0.1291	14.0	13.7	1.162	1.162
2	15.025	0.0957	15.1	14.9	0.783	1.306
3	15.100	0.0816	15.2	15.0	1.225	1.225
4	15.375	0.1258	15.5	15.2	0.993	1.391
5	14.850	0.0577	14.9	14.8	0.866	0.866
6	15.175	0.1258	15.3	15.0	0.993	1.391
7	15.325	0.0957	15.4	15.2	0.783	1.306
8	14.925	0.2872	15.1	14.5	0.609	1.480
9	15.225	0.0957	15.3	15.1	0.783	1.306

Table 9.2-4 Intermediate quantities and results of the Grubbs test for type 2 outliers using the data set free of type 1 outliers

Lab	$\bar{x}_i$	$\hat{r}_{m,\bar{x}_i}$
1	13.850	2.502
2	15.025	0.056
3	15.100	0.220
4	15.375	0.819
5	14.850	0.325
6	15.175	0.383
7	15.325	0.710
8	15.067	0.147
9	15.225	0.492
$\bar{x}^*_{\mathrm{L}}$	14.999	
$s^*_{\bar{x}_{\mathrm{L}}}$	0.45929	

$$\hat{r}_m = \frac{|\bar{x}_i - \bar{x}_L|}{s_{\bar{x}l}} \tag{9.2-11}$$

are given in Table 9.2-4.

As Table 9.2-4 shows, the data set of laboratory 1 must be rejected because the mean value $\bar{x}_1$ is confirmed as an outlier. The test value $\hat{r}_{m,\bar{x}_1} = 2.502$ is greater than the critical value $r_m(P = 95\%, l = 9) = 2.110$.

Stage 6: The extreme mean values of the remaining means, after rejection of type 2 outliers, are: $\bar{x}_{\min} = 14.850$ and $\bar{x}_{\max} = 15.375$.

Stage 7: Because $x_{8,4} = 14.5 < \bar{x}_{\min} = 14.85$, the condition $\bar{x}_{\min} \leq x^*_{\mathrm{OL}_{ik}} \leq \bar{x}_{\max}$ is not fulfilled, and therefore the outlier $x_{8,4} = 14.5$ cannot be re-integrated into the data set.

Stage 8: The test value for the check on outliers of the laboratory precision (type 3 outliers) is calculated by (9.2-12) using the data set which is free of outliers of types 1 and 2:

$$\hat{C} = \frac{\left(s^*_{i,\max}\right)^2}{\sum\limits_{i=1}^{l^*} s_i^2}. \tag{9.2-12}$$

(continued)

According to the results of the Cochran test given in Table 9.2-5, the variances of the eight laboratories checked are homogeneous, because the test value $\hat{C}$ does not exceed the critical value $C(P = 99\%, n_k = 4, l = 8) = 0.521$. Thus, there are no outliers of type 3.

Stages 9 and 10: The intermediate quantities and results of the calculation of the parameters of the interlaboratory study are listed in Table 9.2-6. Further parameters are listed in the tables given above.

(continued)

Table 9.2-5 Intermediate quantities and results of the Cochran test for type 3 outliers

Laboratory	$\bar{x}_i$	n_k	s_i^2
2	15.03	4	0.00917
3	15.10	4	0.00667
4	15.38	4	0.01583
5	14.85	4	0.00333
6	15.18	4	0.01583
7	15.33	4	0.00917
8	15.07	3	0.00333
9	15.23	4	0.00917
$\sum_{i=1}^{l^*} s_i^2$	0.07250	s^2_{max}	0.01583
$\hat{C}$	0.218		

Table 9.2-6 Intermediate quantities for the calculation of the parameters of the interlaboratory study

Lab	s_i	$SS_{x_ix_i}$	$n_i(\bar{x}_i - \bar{\bar{x}})^2$
2	0.0957	0.0275	0.0554
3	0.0816	0.0200	0.0073
4	0.1258	0.0475	0.2158
5	0.0577	0.0100	0.3427
6	0.1258	0.0475	0.0042
7	0.0957	0.0275	0.1329
8	0.0577	0.0067	0.0173
9	0.0957	0.0275	0.0271
n	31	l	8
$\bar{\bar{x}}$	15.143		
Repeatability standard deviation s_r and $CV_r\%$			
$\sum_{1=1}^{l} SS_{x_ix_i}$	0.2142	$df_r = n - l$	23
s_r	0.0965	$CV_r\%$	0.64
Reproducibility standard deviation s_R and $CV_R\%$			
$\sum_{i=1}^{l} n_i$	31	$\sum_{i=1}^{l} n_i^2$	121
$\bar{\bar{n}}$	3.871	$\sum_{i=1}^{l} n_i(\bar{x}_i - \bar{\bar{x}})^2$	0.8028
s_b^2	0.1147	s_L^2	0.0272
s_R	0.1911	$CV_R\%$	1.26
Recovery rate $\eta\%$			99.62

Table 9.2-7 Overview of the whole parameter set for the analyte *bromide* in industrial water obtained by an interlaboratory study

Number of laboratories	l	8
Number of outlier-free individual analytical values	n	31
Number of outliers	n_{OL}	5
Percentage of outlier values	$n_{OL}\%$	86.1
Grand mean value	$\bar{\bar{x}}$ in mg L^{-1}	15.15
True value	μ in mg L^{-1}	15.20
Recovery rate	$\eta\%$	99.62
Reproducibility standard deviation	s_R in mg L^{-1}	0.911
Coefficient of variation of the reproducibility	$CV_R\%$	1.26
Repeatability standard deviation	s_r in mg L^{-1}	0.0965
Coefficient of variation of the repeatability	CV_r in %	0.64
Degrees of freedom of the repeatability standard deviation	df_{s_r}	23

Stage 11: The test value $\hat{t} = 1.669$ calculated by (9.2-10) with the data given in Table 9.2-6 does not exceed the critical value $t(P = 99\%, df = n - 1) = 2.750$, which means that no bias can be detected. Re-validation of the analytical method is not necessary.

Stage 12: The overview of the results of the interlaboratory study for the determination of the analyte *bromide* in industrial water according to DIN EN ISO 10304-1:2009-07 is presented in Table 9.2-7.

9.3 Proficiency Testing

Scores are commonly used for proficiency testing. They have the advantages that they are a simple way to compare laboratories with each other and that they can be used to eliminate laboratories from accreditation if they do not perform sufficiently well. The most common scoring is the z-score which is calculated by

$$z = \frac{x_i - x_a}{\sigma_p}, \tag{9.3-1}$$

where x_i are the results reported by the participating laboratories i, x_a is the assigned value for the test material, and σ_p is the standard deviation for proficiency assessment.

Three types of laboratories can be distinguished by the absolute values of the z-scores:

- $|z| \leq 2$ satisfactory performance
- $2 < |z| \leq 3$ questionable performance
- $|z| > 3$ unsatisfactory performance

A laboratory should take corrective action [8] if

- The z-score shows a unsatisfactory performance ($|z| > 3$)

- Two consecutive questionable results are obtained for the same measurement ($2 < |z| \leq 3$).

There are two critical steps in the organization of a proficiency testing scheme: specifying

- The assigned value (x_a)
- The standard deviation σ_p.

because both values determine the z-scores according to (9.3-1). There is a number of possibilities for an appropriate choice of these parameters [9]: some of them are given below:

Assigned value x_a:

- x_a is taken from a certified reference material (CRM).
- x_a is a reference value determined by a single expert laboratory.
- x_a is obtained from a consensus of expert laboratories which analyze the material using suitable methods.
- x_a is obtained from the results from all participants in the proficiency testing round. The assigned value is then normally based on robust estimation, i.e. using the median $\tilde{x}$.

Standard deviation σ_p

- σ_p is prescribed based on the fitness for purpose criterion; for example, σ_p is set at 10% of the median value by the organizer.
- σ_p is based on the results from a reproducibility study (see the previous chapter):

$$\sigma_p = \sqrt{\sigma_L^2 + \left(\frac{\sigma_r^2}{n}\right)} \qquad (9.3\text{-}2)$$

with

$$\sigma_L = \sqrt{\sigma_R^2 - \sigma_r^2}, \qquad (9.3\text{-}3)$$

where σ_R is the reproducibility standard deviation, σ_r is the repeatability standard deviation, and n is the number of replicates obtained from the collaborative study.

- σ_p is obtained from participants' results themselves using methods of robust statistics such as the median absolute deviation (MAD) and MAD_E, respectively. MAD is the median of absolute deviations from the data set median, calculated by:

$$\text{MAD} = \text{median}\left(|x_i - \tilde{x}|_{i=1,2,\ldots n}\right). \qquad (9.3\text{-}4)$$

For a normal deviation (9.3-5) is valid.

$$\text{MAD}_\text{E} \approx \frac{\sigma}{1.483} \tag{9.3-5}$$

Thus, (9.3-6) is an estimate for the standard deviation,

$$\text{MAD}_\text{E} = \sigma_p = 1.483 \cdot \text{MAD} \tag{9.3-6}$$

compare the values for MAD_E and the standard deviation s in the following example.

Example:
The given data set is:

3.5	3.2	3.6	2.9	3.7	3.1	3.4

Ranked data:

2.9	3.1	3.2	3.4	3.5	3.6	3.7

Median: $\tilde{x} = 3.4$ Mean value; $\bar{x} = 3.34$
Calculation von MAD according to (9.3-4) using the median $\tilde{x} = 3.4$:

0.5	0.3	0.2	0	0.1	0.2	0.5

Ranked deviations:

0	0.1	0.2	0.2	0.3	0.3	0.5

MAD	0.2	$\text{MAD}_\text{E} = \sigma_p$	0.297	s	0.288

For further details see also in [6].

Challenge 9.3-1
In order to check laboratory performance, a proficiency study was organized for the determination of *lead* in flood sediment. The material was dried at 105°C, pre-sieved to grain size <2 mm, and then sieved to grain size <100 µm. After homogenization and partition the material was send to the 21 participants. The results are listed in Table 9.3-1.

(a) Estimate the performance of the laboratories by means of the z-score method using the median and adjusted median absolute deviation (MAD_E).
(b) How many of the laboratories would have their performance judged as satisfactory, questionable performance and unsatisfactory, respectively?
(c) Show the plot of z-scores from the proficiency testing scheme for the determination of lead.

Table 9.3-1 Results from a proficiency testing round for the determination of lead in flood sediments

Laboratory no.	c_{Pb} in mg kg^{-1}	Laboratory no.	c_{Pb} in mg kg^{-1}
1	155	12	154
2	162	13	154
3	165	14	207
4	166	15	168
5	143	16	152
6	165	17	168
7	164	18	153
8	141	19	184
9	156	20	166
10	163	21	165
11	155		

Solution to Challenge 9.3-1

(a) Because no other parameters are given for the calculation of the z-scores, the assigned value x_a and the standard deviation are obtained from a robust estimation of the participants' results. Using the parameters from the robust statistics should be justified because of the obvious distribution of the values in the range 141–207 mg kg^{-1}.

Ranked data, MAD values and ranked MAD values from the results of the proficiency test for the determination of lead are presented in Table 9.3-2.

(continued)

Table 9.3-2 Ranked data, MAD values and ranked MAD values from the results of the proficiency test for the determination of lead

Laboratory no.	Ranked data	MAD		
		Values	Laboratory no.	Ranked values
8	141	22	10	0
5	143	20	2	1
16	152	11	7	1
18	153	10	3	2
12	154	9	6	2
13	154	9	21	2
1	155	8	4	3
11	155	8	20	3
9	156	7	15	5
2	162	1	17	5
10	163	0	9	7
7	164	1	1	8
3	165	2	11	8
6	165	2	12	9
21	165	2	13	9
4	166	3	18	10
20	166	3	16	11
15	168	5	5	20
17	168	5	19	21
19	184	21	8	22
14	207	44	14	44

Because n is odd, the median is $\tilde{x}_{(n+1)/2} = 163\ \text{mg kg}^{-1}$, which is similar to the mean value $\bar{x} = 162.2\ \text{mg kg}^{-1}$. The median absolute deviation calculated by (9.3-4) is MAD = 7 mg kg^{-1}. Thus, the adjusted median absolute deviation is $\text{MAD}_\text{E} = \sigma_\text{p} = 10.38\ \text{mg kg}^{-1}$ calculated according to (9.3-6) which is used for σ_p.
The z-scores calculated by

$$z = \frac{x_i - \tilde{x}}{\sigma_\text{p}} = \frac{x_i - 163\ \text{mg kg}^{-1}}{10.38\ \text{mg kg}^{-1}} \qquad (9.3\text{-}7)$$

are listed in Table 9.3-3.

(b) Estimation of the performance of the laboratories:

- Laboratory no. 14 is considered to have unsatisfactory performance, with $|z| \geqslant 3$.
- Laboratories nos. 8 and 19 are considered to have questionable performance, with $2 < |z| \leq 3$.
- The performance of all other laboratories is considered satisfactory, because $|z| \leq 2$.

(c) The plot of z-scores from the proficiency testing scheme for the determination of lead in flood sediments is shown in Fig. 9.3-1.

Table 9.3-3 z-scores for the participants from the proficiency testing scheme for the determination of lead in flooding sediment

Laboratory no.	z-score	Laboratory no.	z-score
8	−2.119	7	0.096
5	−1.927	3	0.193
16	−1.060	6	0.193
18	−0.963	21	0.193
12	−0.867	4	0.289
13	−0.867	20	0.289
1	−0.771	15	0.482
11	−0.771	17	0.482
9	−0.674	19	2.023
2	−0.096	14	4.239
10	0.000		

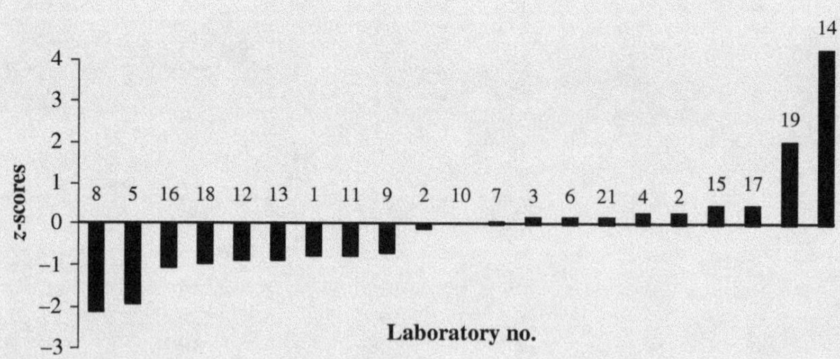

Fig. 9.3-1 Plot of z-scores from the proficiency testing scheme for the determination of lead in flood sediments

References

1. DIN ISO 5725-2 (2002) Accuracy (trueness and precision) of measurement methods and results – part 2: basic method for the determination of repeatability and reproducibility of a standard measurement method. Beuth, Berlin
2. DIN 38402-41 (1984) German standard methods for the examination of water, waste water and sludge. General information (group A) – Inter-laboratory tests; planning and organization. Beuth, Berlin
3. DIN 38402-42 (2005) German standard methods for the examination of water, waste water and sludge – General informations (group A) – part 42: Interlaboratory trials for method validation, evaluation. Beuth, Berlin
4. ISO Guide 35 (2006) Certification of reference material – general and statistical principles. International Organization for Standardization, Geneva
5. ISO/IEC 17043 (2010) Conformity assessment – general requirements for proficiency testing. International Organization for Standardization, Geneva
6. DIN 38402-45 (2009) German standard methods for the examination of water, waste water and sludge – General information – Interlaboratory comparisons for proficiency testing of laboratories. Beuth, Berlin
7. EN ISO 10304-1 (2009) Water quality – determination of dissolved anions by liquid chromatography of ions – part 1: determination of bromide, chloride, fluoride, nitrate, nitrite, phosphate and sulfate. Beuth, Berlin
8. Massart DL, Vandeginste BGM, Buydens LMC, De Jong S, Lewi PJ, Smeyers-Verbeke J (1997) Handbook of chemometrics and qualimetrics, Part A. Elsevier, Amsterdam
9. Ellison SLR, Berwick VJ, Duguit Farrant TJ (2009) Practical statistics for the analytical scientist, 2nd edn. RSC, Cambridge
10. Lawn RE, Thompson M, Walker RF (1997) Proficiency testing in analytical chemistry. RSC, Cambridge

Chapter 10
Measurement Uncertainty

10.1 Purpose, Definitions, and Terminology

Analytical results obtained for customers are usually the basis for correct decisions: for example, when looking at allowable concentration limits. But all measurements are affected by a certain error. Which error should be used as the basis of the decision?

In Fig. 10.1-1, two results with the same mean value are compared with the allowable threshold value. According to result 1 calculated solely from results in control samples, the threshold value is not exceeded, whereas result 2, obtained by the *expanded measurement uncertainty* including all random and systematical errors of the complete analytical method, clearly indicates the crossing of the legal threshold value.

The measurement uncertainty gives information as to what size the measurement error *might* have. The measurement uncertainty is therefore an important part of the reported results in order to make correct decisions. Furthermore, knowledge of the measurement uncertainty is important for the laboratory for its own quality control and to improve the required quality.

Estimation of the measurement uncertainty is required by regulatory authorities. Thus, the "General requirements for the competence of testing and calibration laboratories" [1] states "A calibration laboratory ... shall have and shall apply a procedure to estimate the uncertainty of measurement for all calibrations and types of calibrations," and furthermore "When estimating the uncertainty, all uncertainty components which are of importance in the given situation shall be taken into account using appropriate methods of analysis."

Now, what is *measurement uncertainty*?

Measurement uncertainty is defined in "ISO Guide to the Expression of Uncertainty in Measurement" (the GUM) [2] as "A parameter associated with the result of a measurement, that characterizes the dispersion of the values that could reasonably be attributed to the measurand."

There are some practical approaches for verifying the mathematical analytical approach to GUM, in particular the EURACHEM/CITAC-Guide [3] and, especially

M. Reichenbächer and J.W. Einax, *Challenges in Analytical Quality Assurance*,
DOI 10.1007/978-3-642-16595-5_10, © Springer-Verlag Berlin Heidelberg 2011

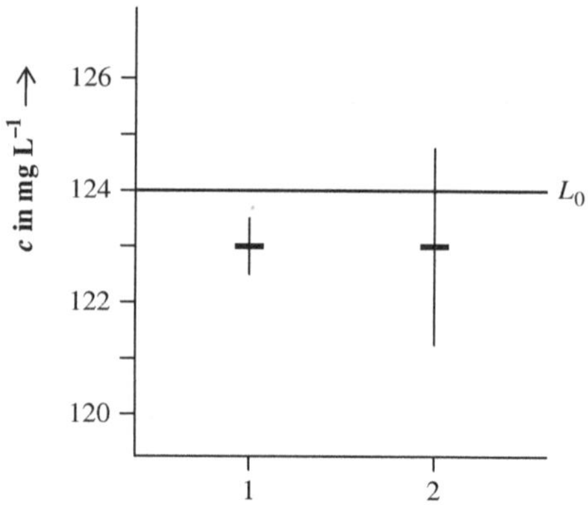

Fig. 10.1-1 Comparison of two analytical results with the threshold value $L_0 = 124\,\text{mg}\,\text{L}^{-1}$. 1 – Result obtained by control samples; 2 – result calculated with the expanded uncertainty

recommended for environmental laboratories, the Nordtest Report procedure TR 537 [4].

After explaining the required definitions, the EURACHEM/CITAC and Nordtest procedures will be described.

When expressed as a standard deviation, an uncertainty is known as a *standard uncertainty*, denoted as u. The uncertainty of the result can arise from many possible sources which must all be separately treated in estimating the overall uncertainty, which is termed *combined standard uncertainty* and denoted by $u_c(y)$ for a measurement result y calculated according to the law of propagation of uncertainty. In general, in analytical chemistry an *expanded uncertainty* U is used which gives an interval within which the value of the measurand is believed to lie with a defined level of significance level. U is obtained by multiplying $u_c(y)$ by a *coverage factor* k which is based on the level of confidence desired. For an approximate significance level $P = 95\%$, k is 2.

The GUM defines two different methods of estimating uncertainty.

- *Type A:*
  Method of evaluation of uncertainty by the statistical analysis of a series of repeated observations.
- *Type B:*
  Method of evaluation of uncertainty by other means than the statistical analysis of series.
  Sources of *Type B* uncertainties can be, for example:
  - Tolerances of the measuring devices given by the manufacturers
  - Data obtained by certificates
  - Data obtained by earlier measurements
  - Data obtained on basis of judgments

  The calculation of these kinds of uncertainty is given in the following chapters.

10.2 Steps in Measurement Uncertainty Estimation

Step 1: Specifying the measurand Write down clearly what is being measured, including the relationship between the measurand and the input quantities upon which it depends.

If sampling steps are to be included, estimation of uncertainties associated with the sampling procedure must be considered. Furthermore, it must be decided whether the measurand is a result of a so-called *empirical* or a *non-empirical* method. In contrast to a non-empirical method, where the results obtained by various methods should be independent of the method, in the case of an empirical method the result depends on the method used. The latter is the case, for example, if the method includes an extraction step, when the extracted analyte can depend on the choice of extraction conditions.

Step 2: Identifying the relevant uncertainty sources A comprehensive list of relevant sources of uncertainty can be assembled but the *cause and effect diagram* (also called a fish-bone diagram) is a very convenient way of listing the uncertainty sources, showing how they relate to each other and indicating their influence on the result. Such cause and effect diagrams are used in the presentation of the solutions to the Challenges. Figure 10.2-1 shows, for example, the cause and effect diagram for the determination of the density of ethanol after combination of similar effects (precision and temperature effects are grouped together) and cancellation (the bias of weighing is cancelled because of weighing by difference using the same balance).

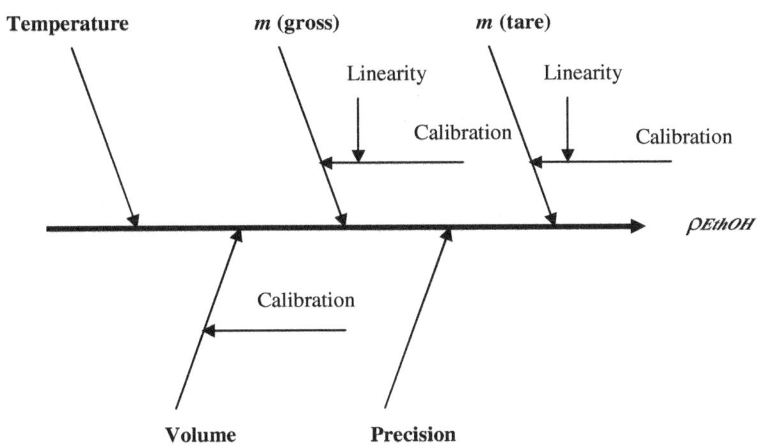

Fig. 10.2-1 Cause and effect diagram for the determination of the density of ethanol (EtOH)

Typical sources of uncertainty are, for example:

- Sampling
- Storage condition
- Instrument effects
- Reagent purity
- Measurement conditions
- Matrix effects and others.

Step 3: Quantifying uncertainty After uncertainty sources have been identified, the next step is to quantify the uncertainty arising from theses sources. This can be done by

- Evaluating of the uncertainty arising from each individual source as the basis for the calculation of the combined uncertainty

or

- Determining directly the combined contribution to uncertainty from these sources using method performance data.

The procedures which may be adopted depending on the data available and the additional information required are described in detail in [3]. Note that not all listed components of the uncertainty make a significant contribution to the combined uncertainty; such contributions should be eliminated from further estimations. After elimination of non-significant contributions, simplification by grouping sources covered by existing data, quantification of the grouped components, and quantification of the remaining components, the components must be converted into standard deviations.

Step 4: Calculating combined uncertainty Before the combined uncertainty can be calculated, all uncertainty contributions must be expressed as standard uncertainties, that is, standard deviations. This can involve conversion from some other measures of dispersion.

The *Type A* uncertainty component is evaluated experimentally from the dispersion of n repeated measurements. It is expressed as a standard deviation. For the contribution to uncertainty from single measurements, the standard uncertainty is simply the observed standard deviation s; for results subjected to averaging, the standard deviation of the mean $s_{\bar{x}}$ is used:

$$u = \frac{s_{\bar{x}}}{\sqrt{n}}. \qquad (10.2\text{-}1)$$

Example 1: The following five results given in % (w/w) are averaged to give the mean of a related substance in a drug:

1.54	1.49	1.58	1.55	1.46	1.53

The mean is $\bar{x} = 1.53$ % (w/w), and the standard deviation is $s = 0.043$ % (w/w). Because the results are averaged, the standard uncertainty in the mean value is $u = \dfrac{0.43\%}{\sqrt{5}} = 0.019$ % (w/w).

If only a single observation is made, the standard deviation is calculated from the relative standard deviation s_r obtained during validation:

$$u = x \cdot s_r. \tag{10.2-2}$$

However, *Type B* estimates of uncertainty are based on different information which is converted to an estimated uncertainty u:

- Tolerance as $x \pm a$ without specifying a level of confidence.

  The formula used depends on the kind of distribution.

  Assuming a rectangular distribution of width $2a$, symmetrical about x, and all values within the interval equally probable, the uncertainty is calculated by (10.2-3):

$$u = \frac{a}{\sqrt{3}}. \tag{10.2-3}$$

Assuming that values close to x are more likely than near the bounds, the triangular distribution should be used with the uncertainty

$$u = \frac{a}{\sqrt{6}}. \tag{10.2-4}$$

Example 2: A 10 mL grade A volumetric flask is certified within ± 0.1 mL.

  The standard uncertainty is
  $u = \frac{0.1}{\sqrt{3}} = 0.06$ mL, assuming rectangular distribution and
  $u = \frac{0.1}{\sqrt{6}} = 0.04$ mL, assuming triangular distribution.

- Confidence interval as $x \pm \Delta x$ with a significance level $P\%$.

  The uncertainty is calculated by

$$u = \frac{\Delta x}{t}, \tag{10.2-5}$$

where t is the two-sided value of the t-factor for the level of significance P and number of degrees of freedom.

Where the number of degrees of freedom for the confidence interval is not given, the t-factor with infinite degrees of freedom is used, i.e. $t = 1.96$ for $P = 95\%$.

Example 3: A specification states that a balance reading is within ± 0.2 mg with $P = 95\%$ significance.

The standard uncertainty is $u = \frac{0.2}{1.96} \approx 0.1$ mg.

- Expanded uncertainty as $x \pm U$

  The standard uncertainty is calculated by

$$u = \frac{U}{k}, \tag{10.2-6}$$

where k is the coverage factor given. But if k is not given, $k = 2$ should be used. Thus, (10.2-6) is:

$$u = \frac{U}{2}. \tag{10.2-7}$$

Calculation of the standard uncertainty from *linear least squares calibration*:

The standard uncertainty $u(\hat{x}, y)$ in a predicted value $\hat{x}$ due to variability in y can be estimated by

$$u(\hat{x}) = \frac{s_{y.x}}{a_1}\sqrt{\frac{1}{n_a} + \frac{1}{n_c} + \frac{\left(\hat{\bar{y}} - \bar{y}\right)^2}{a_1^2 \cdot \mathrm{SS}_{xx}}}, \tag{10.2-8}$$

in which the symbols are as explained in Sect. 4.2.1. For more information see [3].

The next stage is to calculate the combined standard uncertainty.

The general relationship between the combined standard uncertainty $u(y)$ of a value y and the uncertainty of the independent parameters $x_1, x_2, \ldots, x_n$ on which it depends is

$$u(y) = \sqrt{\sum_{i=1}^{n} c_i^2 u(x_i)^2} = \sqrt{\sum \left(\frac{\partial y}{\partial x}\right)^2 \cdot u(x_i)^2}, \tag{10.2-9}$$

where the sensitivity c_i is the partial differential of y with respect to x_i and $u(x_i)$ denotes the uncertainty in y arising from the uncertainty in x_i. The contribution of each variable is just the square of the associated uncertainty expressed as a standard deviation multiplied by the square of the relevant sensitivity coefficient which describes how the value of y varies with changes in the parameters $x_1, x_2, \ldots, x_n$.

For two independent input quantities x_1 and x_2, according to (10.2-9) the combined uncertainty is calculated by (10.2-10):

$$u[f(x_1, x_2)] = \sqrt{\left[\frac{\partial f(x_1)}{\partial x_1} \cdot u(x_1)\right]^2 + \left[\frac{\partial f(x_2)}{\partial x_2} \cdot u(x_2)\right]^2}. \tag{10.2-10}$$

Note that the variables are not independent, and therefore the covariances $u(x_i, x_k)$ between x_i and x_k must be considered.

The covariance is related to the correlation coefficient r_{ik} and is calculated by (10.2-12):

$$u(x_i, x_k) = u(x_i) \cdot u(x_k) \cdot r_{ik}. \tag{10.2-11}$$

According to the "law of propagation of uncertainty," (10.2-9) leads to the two simple rules for combining standard uncertainties:

Rule 1: If variables are added or subtracted, $y = p + q$ or $y = p - q$, the combined uncertainty is calculated by

$$u(y) = \sqrt{u(p)^2 + u(q)^2}. \tag{10.2-12}$$

Rule 2: For a mathematical model involving only products or quotients, $y = p \cdot q$ and $y = p/q$, respectively, their uncertainties combine *as relative* uncertainties:

$$\frac{u(y)}{y} = \sqrt{\left(\frac{u(p)}{p}\right)^2 + \left(\frac{u(q)}{q}\right)^2}. \tag{10.2-13}$$

The final stage is to multiply the combined standard uncertainty by the chosen coverage factor k in order to obtain the *expanded uncertainty U*:

$$U(y) = k \cdot u(y). \tag{10.2-14}$$

The choice of the factor k is based on the level of significance. For most purposes, k is taken as 2 for an approximate of significance level $P = 95\%$, but the t-factor can also be used for exact requirements.

10.3 Spreadsheet Method for Uncertainty Calculation

The uncertainty can be calculated by the equations given above, but the calculation can be simplified using spreadsheet software according to the procedure first described by Kragten [5]. The procedure takes advantage of an approximate numerical method, differentiation, and requires knowledge only of numerical values of the parameters and their uncertainty, which is explained below.

Given the mathematical model

$$y = f(x, w, z) \tag{10.3-1}$$

with the input quantities x, w, z and the output quantity y, according to (10.2-9) the change of the output quantity in relation to one input quantity x may be found by the tangent to the function $y = f(x)$, as shown in Fig. 10.3-1 in which $f(x)$ is the result found for the nominal or observed value of the quantity x_i and $f(x + \Delta x)$ the result for the value plus its standard uncertainty.

The difference quotient (see Fig. 10.3-1) approximates the slope of the tangent. In the limit $\Delta x \rightarrow 0$, the difference quotient converts to the differential quotient, the slope of the tangent which corresponds to the standard uncertainty according to (10.2-9).

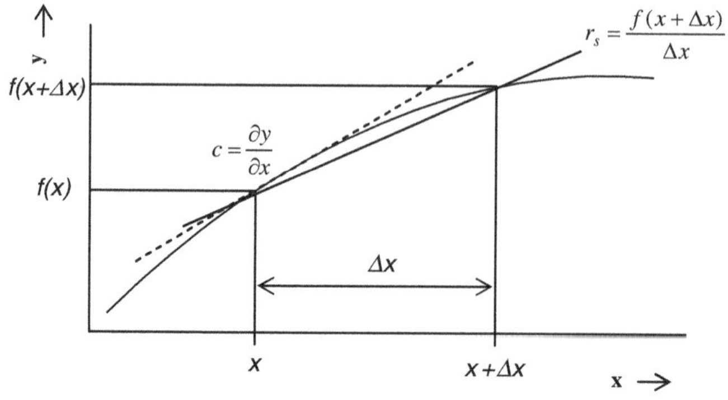

Fig. 10.3-1 Graphical illustration of the standard uncertainty for the value x

Fig. 10.3-2 Step 1 of spreadsheet calculation of combined uncertainty

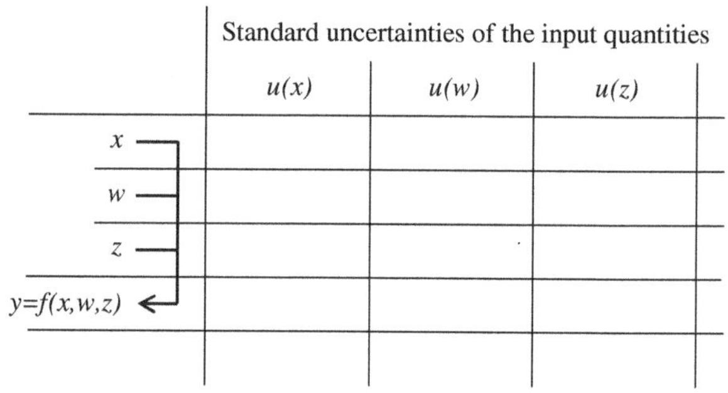

Using Δx as the value for the standard uncertainty $u(x)$ of the input quantity x, one obtains an approximate expression for the value of the standard uncertainty of the model given in (10.3-1):

$$r_s = \frac{\partial f}{\partial x} = \frac{f(x + u(x), w, z) - f(x, w, z)}{u(x)} \tag{10.3-2}$$

$$\frac{\partial f}{\partial x} \cdot u(x) = f(x + u(x), w, z) - f(x, w, z). \tag{10.3-3}$$

Equation (10.3-3) is the basis for the construction of a spreadsheet to calculate combined uncertainty, which will verified by Excel for the model given in (10.3-1).

Step 1: Enter the values of x, w, and z, in the formula for calculating y according to (10.3-1), and the standard uncertainties of the input quantities in a spreadsheet as shown in Fig. 10.3-2.

Step 2: Enter the input quantities plus their standard uncertainties in the diagonal of the $(n \times n)$ matrix of the spreadsheet and complete the matrix by entering the input quantities in the outer-diagonal positions (see Fig. 10.3-3).

Step 3: Copy the formula y for the input quantities into the positions on the right so that the input quantities plus their uncertainties y'_x, y'_w, y'_z are given in each column of the row as shown by Fig. 10.3-4.

Step 4: In order to obtain the partial differentials according to (10.3-3) the input quantity y must again be subtracted. Therefore, the input quantity y is entered in the next row and is copied into the columns to the right of it. After subtraction, the partial differentials are given in the row as shown in Fig. 10.3-5.

Step 5: To obtain the combined standard uncertainty in y, the individual contributions are squared, added together and the square root is taken (see Fig. 10.3-6).

The *result* is given by:

$$u(y) = \sqrt{\left\{\left(\frac{\partial y}{\partial x}\cdot u(x)\right)^2 + \left(\frac{\partial y}{\partial w}\cdot u(w)\right)^2 + \left(\frac{\partial y}{\partial z}\cdot u(z)\right)^2\right\}}. \qquad (10.3\text{-}4)$$

Note that the spreadsheet construction procedure can also be extended to cope with the situation that any of the variables are correlated and correlated uncertainty contributions must therefore be considered, as given in [3, 6].

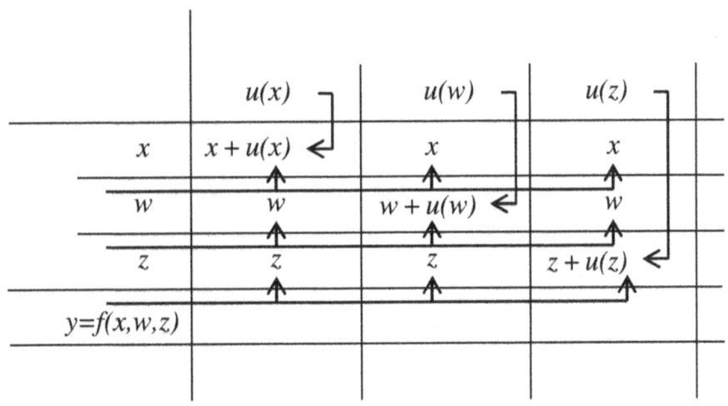

Fig. 10.3-3 Step 2 of spreadsheet calculation of combined uncertainty

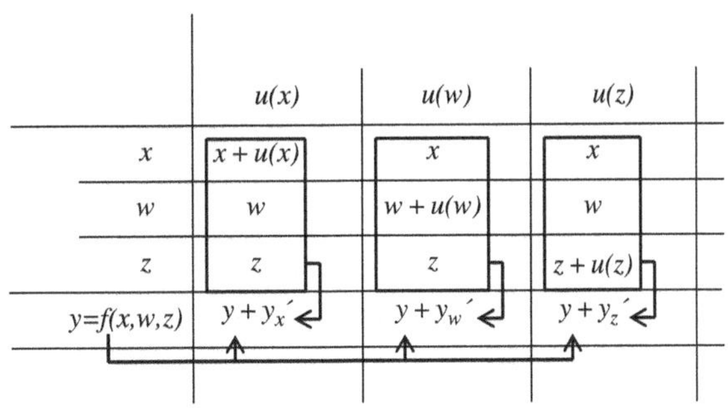

Fig. 10.3-4 Step 3 of spreadsheet calculation of combined uncertainty

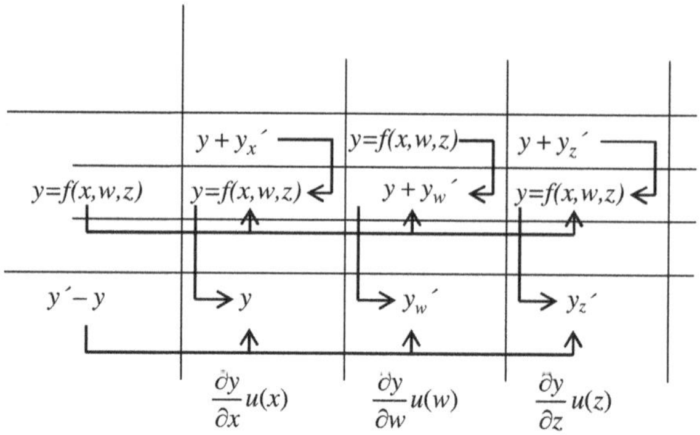

Fig. 10.3-5 Step 4 of spreadsheet calculation of combined uncertainty

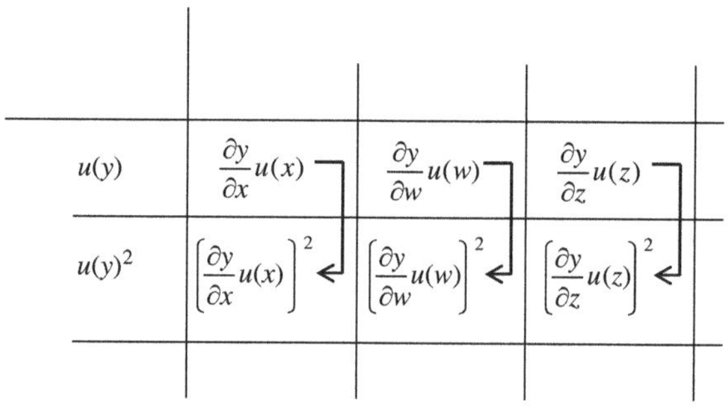

Fig. 10.3-6 Step 5 of spreadsheet calculation of combined uncertainty

Additionally, the individual contributions to the combined uncertainty can be represented by the graph formats available in Excel. Thus, one can easily and quickly recognize the main sources of uncertainty and which contributions can be rejected [3].

Challenge 10.3-1

Solve the following problems:

(a) To re-calibrate a 10 mL pipette, the volume of water ($\rho_{20°C} = 0.998207\,\mathrm{g\,cm^{-3}}$) was measured by ten replicates giving the following results in g:

99.85	99.82	99.81	99.82	99.72
99.84	99.80	99.85	99.83	99.81

What are the mean value and the standard uncertainty for a single pipetting step?

(b) The data set for the calibration of the photometric determination of nitrite-N is listed in Table 10.3-1.

The measured values of the absorbance A for a sample are:

0.4892	0.4886	0.4895

(continued)

Table 10.3-1 Calibration data set for the photometric determination of nitrite-N

Standard	1	2	3	4	5	6
c (mg L^{-1})	0.05	0.10	0.15	0.20	0.25	0.30
A_i	0.1845	0.3197	0.4603	0.5895	0.7202	0.8501

Table 10.3-2 Data obtained in a proficiency test of benzo[*a*]pyrene in drinking water

x_{true} in ng L^{-1}	$\bar{\bar{x}}$ in ng L^{-1}	s_r in %	s_{bw} in ng L^{-1}	s_{in} in ng L^{-1}
24	19.05	79.4	4.921	2.340

Calculate the predicted value $\hat{x}$ and the standard uncertainty of the sample.

(c) Table 10.3-2 shows the results obtained by a proficiency test of the determination of benzo[*a*]pyrene in drinking water according to DIN 38 407 F18 [7].

What standard uncertainty and relative standard uncertainty can be used in the own laboratory for the determination of benzo[*a*]pyrene?

(d) The specification for a 10 mL burette is quoted by the manufacturer as ± 0.02 mL.

Calculate the standard uncertainty under the conditions of a rectangular and a triangular distribution.

(e) According to the calibration certificate for a balance, the measurement uncertainty is ± 0.0005 g with a significance level of $P = 95\%$.

Calculate the standard uncertainty.

(f) The standard deviation of repeated weighing of 0.1 g is calculated to be 0.00015 g.

Calculate the standard uncertainty.

Solution to Challenge 10.3-1

(a) *Type A* uncertainties, the standard uncertainty is expressed as the standard deviation, which is calculated in the known manner: $u(x) = 0.0375$ mg, which gives, after conversion by the density, $u(x) = 0.04$ mL and $u(x)_r\% = 0.04$, respectively.

(b) The standard uncertainty of a predicted value is $u(\hat{x}) = 0.0011\,\text{mg}\,\text{L}^{-1}$ calculated by (10.2-8) with $a_1 = 2.6621\,\text{L}\,\text{mg}^{-1}$, $s_{y.x} = 0.00412$, $n_a = 3$, $n_c = 6$, $SS_{xx} = 0.043750\,\text{mg}^2\,\text{L}^{-2}$, $\bar{y} = 0.5207$, and $\hat{\bar{y}} = 0.4891$. The predicted value is $\hat{x} = 0.163\,\text{mg}\,\text{L}^{-1}$ calculated according to (5.2-15) with the intercept $a_0 = 0.0548$ and the slope given above.

(continued)

(c) The standard uncertainty used by this laboratory is equal the standard deviation in the laboratory s_{in} obtained by the interlaboratory test, which is 2.340 ng L^{-1}:

$u = 2.340\,\text{ng}\,L^{-1}$, $u_r\% = 12.3$.

(d) The *Type B* standard uncertainty calculated by (10.2-3) and (10.2-4) is:

$u = 0.012$ mL (for a rectangular distribution) and

$u = 0.008$ mL (for a triangular distribution).

(e) The uncertainty is calculated according to (10.2-5). Since the number of degrees of freedom is unknown, the *t*-factor for large degrees of freedom is used, which is 1.96.

The uncertainty is

$$u = \frac{0.0005\,\text{g}}{1.96} = 0.00026\,\text{g}.$$

(f) Because the standard uncertainty is expressed as a standard deviation, no conversion is necessary.

$u = 0.00015$ g

Challenge 10.3-2

In a laboratory, a standard solution must be prepared based on an aqueous solution of acetic acid with the concentration $c = 4\%$ (w/w) prepared using the following procedure:

40 mL of a specified stock solution of $c = 100 \pm 0.5\%$ (v/v) is pipetted into a 1 L volumetric flask using a class *A* 20 mL pipette and the flask is filled with water. The difference between the laboratory temperature and the calibration temperature of the pipette and volumetric flask is not more than $\pm$ 2°C.

Calculate the expanded uncertainty *U* at the significance level $P = 95\%$. The manufacturer's calibration data of the volumetric flask and the pipette and the standard deviation of the manual operations, obtained by earlier tests in the laboratory, are given in Table 10.3-3. The coefficient of volume expansion of water is $2.1 \cdot 10^{-4}$ °C^{-1}.

Table 10.3-3 Calibration data of flask and pipette as well as standard deviation of the manual operations

Calibration data at 20°C	
1 L volumetric flask	± 4 mL
20 mL class *A* pipette	± 0.03 mL
Standard deviation of the manual operations	
1 L volumetric flask	1.5 mL
20 mL class *A* pipette	0.016 mL

Solution to Challenge 10.3-2

Step 1: Specifying the measurand. The measurand is the concentration which is calculated by

$$c_{\mathrm{HAc}} = \frac{2 \cdot V_{\mathrm{pip}} \text{ in mL} \cdot c_{\mathrm{stock}}}{V_{\mathrm{flask}} \text{ in mL}}. \qquad (10.3\text{-}5)$$

Step 2: Identifying the relevant uncertainty sources. The relevant uncertainty sources are shown in a cause and effect diagram (Fig. 10.3-7).

Step 3: Quantifying uncertainty

- Calculation of standard uncertainties from the manufacturer's data according to (10.2-3), assuming rectangular distribution:

$$u = \frac{\text{tolerance}}{\sqrt{3}}. \qquad (10.3\text{-}6)$$

Stock solution	0.29%
1 L flask	2.31 mL
20 mL pipette	0.017 mL

- Calculation of the relative standard uncertainty of the volume of the flask and the pipette according to rule 1 using (10.2-12):

$$u(V) = \sqrt{u^2_{\mathrm{tolerance}} + u^2_{\mathrm{rep}}}, \qquad (10.3\text{-}7)$$

(continued)

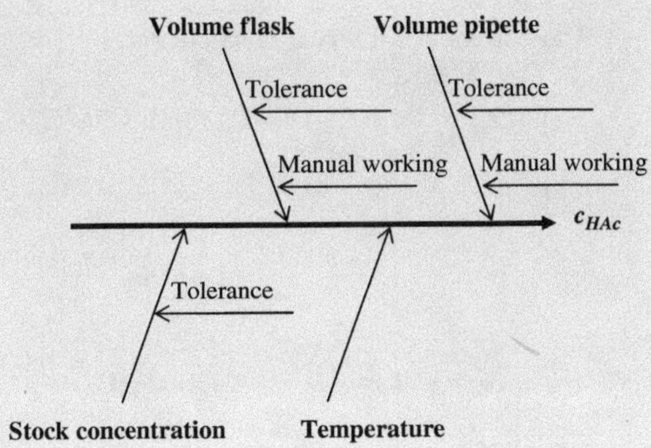

Fig. 10.3-7 Cause and effect diagram for the preparation of standard concentration

$$u_r(V) = \frac{u(V)}{V}, \tag{10.3-8}$$

$$u(V_{flask}) = \sqrt{(2.31\,\text{mL})^2 + (1.5\,\text{mL})^2} = 2.75\,\text{mL}, \tag{10.3-9}$$

$$u_r(V_{flask}) = \frac{2.75\,\text{mL}}{1,000\,\text{mL}} = 0.00275, \tag{10.3-10}$$

$$u(V_{pip}) = \sqrt{2 \cdot ((0.017\,\text{mL})^2 + (0.016\,\text{mL})^2)} = 0.033\,\text{mL}. \tag{10.3-11}$$

Note that the 20 mL pipette must be used twice.

$$u_r(V_{pip}) = \frac{0.033\,\text{mL}}{20\,\text{mL}} = 0.00165. \tag{10.3-12}$$

- Calculation of the relative standard uncertainty of the stock solution, assuming rectangular distribution (10.2-3):

$$u(c) = \frac{0.5\%(\text{v/v})}{\sqrt{3}} = 0.29\%(\text{v/v}) \tag{10.3-13}$$

$$u_r(c) = \frac{0.29\%(\text{v/v})}{100\%(\text{v/v})} = 0.0029. \tag{10.3-14}$$

- Calculation of the relative uncertainty of the temperature
  The influence of temperature on the volume is given by:

$$u(T) = \frac{CI(V)}{t(P = 95\%, \text{df} = \infty)}. \tag{10.3-15}$$

The confidence interval is calculated by

$$CI(V) = \theta \cdot \Delta T \cdot V = 0.00021°\text{C}^{-1} \cdot 2°\text{C} \cdot 1,000\,\text{mL}$$
$$= 0.42\,\text{mL} \tag{10.3-16}$$

$$u(T) = \frac{0.42\,\text{mL}}{1.96} = 0.21\,\text{mL} \tag{10.3-17}$$

$$u_r(T) = \frac{0.21\,\text{mL}}{1,000\,\text{mL}} = 0.00021. \tag{10.3-18}$$

(continued)

Step 4: Calculating the combined uncertainty Because of the multiplicative combination of the input quantities according to (10.3-5), rule 2 with (10.3-5) has to be applied using the relative standard uncertainties of the components given in (10.3-10), (10.3-12), (10.3-14), and (10.3-18):

$$u_r(c) = \sqrt{u_r^2(V_{\text{flask}}) + u_r^2(V_{\text{pip}}) + u_r^2(c_{\text{stock}}) + u_r^2(T)} \quad (10.3\text{-}19)$$

$$u_r(c) = \sqrt{0.0027^2 + 0.00165^2 + 0.0029^2 + 0.00021^2} = 0.0043. \quad (10.3\text{-}20)$$

The combined standard uncertainty is

$$u(c) = u_r(c) \cdot c = 0.0043 \cdot 4\% \text{ (v/v)} = 0.017\% \text{ (v/v)}.$$

Result. The concentration of the acetic acid is $c \pm u(c) = 4 \pm 0.02\%$(v/v).

Examples of the complex application including the complete analytical method are extensive, but they are presented in the appendix of EURACHEM [3].

10.4 Procedure of the Nordtest Report

The Nordtest handbook is written primarily for environmental testing laboratories in the Nordic countries and supports implementation of the concept of measurement uncertainty for their routine measurements, but it is also the basis of regulations in other countries; see, for example, in [5]. The practical, understandable, and common method of measurement uncertainty calculations is mainly based on already existing quality control and validation data, which means that additional determinations are, in general, not necessary. By using existing and experimentally determined quality control and method validation data, the probability of including *all* uncertainties will be maximized.

The model is a simplification of the model presented in ISO guide [4]:

$$y = m + (\delta + B) + e \quad (10.4\text{-}1)$$

where

- y is the measurement of the result
- m is the expected value for y
- δ is the method bias
- B is the laboratory bias – the uncertainties for these are combined into u_{bias}
- e is the random error under within-laboratory reproducibility conditions R_w which is the intermediate measure between the repeatability limit r and the

reproducibility limit R, where operator and/or equipment and/or time and/or calibration can vary, but only in the same laboratory. An alternative name is *intermediate precision* (see Chap. 5.2).

The flow chart for the calculation of the uncertainty, involving six defined steps, should be followed in all cases:

Step 1. Specify measurand. For example, ammonia is measured in water by photometric determination according to DIN EN ISO 11732 [8].

Step 2. Quantify the reproducibility within the laboratory u_{R_w}. This can be achieved by:

1. Stable control samples covering the whole analytical process: usually with one sample each at low and high concentration levels
   When a stable control sample is treated using the complete analytical process and it has a matrix similar to the samples, the within-laboratory reproducibility at a specific concentration level can be obtained by the *mean value chart.*

Example 4: The results of the control sample obtained by the two mean value charts for the two working ranges 0.5–5 µg L^{-1} and 10–100 µg L^{-1} are given in Table 10.4-1.

2. Stable synthetic control samples (standard samples).
   When a synthetic control sample is used for quality control, and matrix of the control sample is not the same as the natural samples, a *mean value* and a *range chart* has to be kept.
   The uncertainty is calculated by:

$$u_{R_w} = \sqrt{u^2_{R_w,\text{Standard}} + u^2_{R_w,\text{Range}}}, \qquad (10.4\text{-}2)$$

with

$$u_{R_w,\text{Standard}} = s_{\text{mean value chart}} \qquad (10.4\text{-}3)$$

$$u_{R_w,\text{Range}} = \frac{\bar{R}}{1.128} \quad \text{(for two replicates)}. \qquad (10.4\text{-}4)$$

Table 10.4-1 Data from mean value control charts

Parameter	Range	
	0.5–5 µg L^{-1}	10–100 µg L^{-1}
True value μ	2.5 µg L^{-1}	45.0 µg L^{-1}
Mean value $\bar{x}$	2.52 µg L^{-1}	45.30 µg L^{-1}
Standard deviation s_{R_w}	0.12 µg L^{-1}	1.30 µg L^{-1}
s_r %	4.8	2.9

Example 5: For the determination of ammonia, a mean value chart and a range chart are kept for various matrices. The results obtained for the determination of ammonia-N are:

- Mean value chart: $\mu = 300\ \mu\text{g L}^{-1}$, $s = 3.0\ \mu\text{g L}^{-1}$
- Range chart (for two replicates): $\bar{R}\% = 5.5$.

The uncertainty components calculated according to (10.4-3) and (10.4-4) are:

$$u_{R_w,\text{Standard}}\% = \frac{3.0}{300} \cdot 100 = 1.0 \qquad (10.4\text{-}5)$$

$$u_{R_w,\text{Range}}\% = \frac{5.5\%}{1.128} = 4.88. \qquad (10.4\text{-}6)$$

Thus, the standard uncertainty of the reproducibility of the laboratory calculated by (10.4-2) is $u_{R_w}\% = \sqrt{1 + 4.88^2} = 4.98$.

3. Unstable control samples

If the laboratory does not have access to stable control samples, e.g. for the determination of sum parameters, the reproducibility can be estimated by the data of the *range chart* (R-chart) obtained by the analysis of natural duplicate samples. However, this only gives the within-day variation (repeatability) for sampling and measurement, and there will also be a "long-term" uncertainty (the variation between the series) which is hard to measure. Therefore, to estimate the total within-laboratory reproducibility, the following approximation is used:

$$u_{R_w} = \sqrt{2} \cdot u_{R_w,\text{Range}}. \qquad (10.4\text{-}7)$$

Example 6: For the determination of oxygen in seawater, a R-chart is maintained over a long period. The differences obtained by duplicate measurements of natural samples used in the R-chart yield the value $\bar{R}\% = 5.5$. The uncertainty of the range calculated by (10.4-4) is $u_{R_w,\text{Range}}\% = 4.88$ and, thus, the total standard uncertainty is

$$u_{R_w}\% = \sqrt{2} \cdot 4.88\% = 6.90. \qquad (10.4\text{-}8)$$

Step 3. Estimate the method and laboratory bias u_{bias}. Note that sources of bias should be eliminated if possible.

For estimation of the uncertainty of the method and the laboratory bias, two components have to be estimated to obtain u_{bias} :

- The laboratory variation RMS_{bias} (root mean square) which is the bias (as % difference from the nominal or certified value) and its deviation
- The uncertainty of the nominal/certified value $u_{C_{\text{ref}}}$ or u_{recovery}(method variation).

The general formula for the calculation of the total systematical deviation u_{bias} is:

$$u_{\text{bias}} = \sqrt{\text{RMS}^2_{\text{bias}} + u^2_{c_{\text{ref}}}}. \tag{10.4-9}$$

In order to estimate u_{bias} there are three possibilities:

1. Using certified reference material CRM.

Regular analysis of a CRM can be used to estimate the bias. The material should be analyzed in at least five different analytical series, e.g. on five different days, and the results used for a mean value chart.

If only *one* CRM is used, the laboratory deviation RMS_{bias} required for (10.4-9) can be estimated by

$$\text{RMS}_{\text{bias}} = \sqrt{\text{bias}^2 + \left(\frac{s_{\text{bias}}}{\sqrt{n}}\right)^2}, \tag{10.4-10}$$

where bias is the relative difference between the nominal or certified value and the laboratory mean value obtained by the mean value chart, and s_{bias} is the standard deviation of the bias also obtained by the mean value chart with n replicates.

Example 7: A certificate value is $c_{\text{ref}} = 11.5 \pm 0.5\,\text{mg}\,\text{L}^{-1}$ with a 95% confidence interval. The mean value and its standard deviation obtained by a mean value chart are $\bar{x} = 11.9\,\text{mg}\,\text{L}^{-1}$ and $s_{\text{bias}} = 0.27\,\text{mg}\,\text{L}^{-1}$, obtained by $n = 12$.

What is the value of the total standard uncertainty of the bias u_{bias}?

Steps of the solution:

- Converting the confidence interval to uncertainty $u_{c_{\text{ref}}}$ according to (10.2-5):

$$u_{c_{\text{ref}}} = \frac{0.5\,\text{mg}\,\text{L}^{-1}}{1.96} = 0.26\,\text{mg}\,\text{L}^{-1}. \tag{10.4-11}$$

- Converting the uncertainty to relative uncertainty:

$$u_{c_{\text{ref}}}\% = \frac{0.26\,\text{mg}\,\text{L}^{-1}}{11.5\,\text{mg}\,\text{L}^{-1}} \cdot 100 = 2.26. \tag{10.4-12}$$

- Relative standard deviation of the bias from the mean value chart:

$$s_{\text{bias}}\% = \frac{0.27\,\text{mg}\,\text{L}^{-1}}{11.9\,\text{mg}\,\text{L}^{-1}} \cdot 100 = 2.27. \tag{10.4-13}$$

- Calculation of bias:

$$\text{bias}\% = \frac{(11.9 - 11.5)\,\text{mg}\,\text{L}^{-1}}{11.5\,\text{mg}\,\text{L}^{-1}} \cdot 100 = 3.48. \tag{10.4-14}$$

– Calculation of the total uncertainty u_{bias} according to (10.4-9) and (10.4-10):

$$u_{\text{bias}}\% = \sqrt{\text{bias}^2 + \left(\frac{s_{\text{bias}}}{\sqrt{n}}\right)^2 + u_{C_{\text{ref}}}^2} = \sqrt{3.48^2 + \left(\frac{2.27}{\sqrt{12}}\right)^2 + 2.26^2} = 4.2. \tag{10.4-15}$$

2. Interlaboratory comparison.

In this case the results from interlaboratory comparisons are used in the same way as a reference material, i.e. estimating the bias. A laboratory should participate at least six times within a reasonable time interval, for example 3 years, in order to correctly evaluate the bias.

The procedure is similar to that for reference materials. But, because the certified value of a CRM is normally better defined than a nominal or assigned value in an interlaboratory comparison, the calculated uncertainty $u_{C_{\text{ref}}}$ can be too high and is not valid for estimation of u_{bias}.

3. Recovery tests.

Recovery tests, for example the recovery rate of a standard addition to a sample in the validation process, can also be used to estimate the systematic error. In this way, validation data can provide a valuable input to uncertainty estimation.

The recovery rate of spiked samples is determined with at least five samples. The uncertainty is then given by two components:

- The bias $\text{RSM}_{\text{bias}}\%$ as the difference from the value 100.
- The uncertainty of the spiking.

The total uncertainty u_{bias} is calculated according to the following equations:

$$u_{\text{bias}} = \sqrt{\text{RMS}_{\text{bias}}^2 + u_{\text{spike}}^2}, \tag{10.4-16}$$

$$\text{RMS}_{\text{bias}} = \sqrt{\frac{\sum (\text{bias}_i)^2}{n}}, \tag{10.4-17}$$

$$u_{\text{spike}} = \sqrt{u_V^2 + u_c^2}, \tag{10.4-18}$$

$$u_V = \sqrt{u_{V_{\text{bias}}}^2 + u_{V_{\text{rep}}}^2}, \tag{10.4-19}$$

$$u_{V_{\text{bias}}} = \frac{\text{max.deviation}}{\sqrt{3}}, \tag{10.4-20}$$

where

– RMS_{bias} is the root mean square of the deviations obtained by n replicates
– u_{spike} is the uncertainty of spiking

- u_V is the uncertainty of the spiked volume
- u_c is the uncertainty of the spiked concentrations
- $u_{V_{\text{bias}}}$ is the systematic uncertainty of the stock volume
- $u_{V\text{rep}}$ is the random uncertainty of the stock volume.

Example 8: For the determination of ammonia-N, the component of u_{bias} was evaluated by recovery experiments.

1. Uncertainties in the manufacture's data:

   - Uncertainty of the concentration u_c of the stock solution with the certified confidence interval $\Delta x = \pm 1.5\%$ at the significance level $P = 95\%$:

$$u_c = \frac{1.5\%}{1.96} = 0.77\%. \tag{10.4-21}$$

   - Systematic uncertainty of the stock volume by the maximal deviation of 1%:

$$u_{V_{\text{bias}}} = \frac{1\%}{\sqrt{3}} = 0.58\%. \tag{10.4-22}$$

   - Random uncertainty of the stock volume $u_{V_{\text{rep}}} = 0.5\%$.

2. Measurement of the stock solutions by six replicates gave deviations of 100% recovery (bias_i) in %:

5	3	2	4	1	4

3. Calculations:
   - Uncertainty of the spiking according to (10.4-9)

$$u_{spike} = \sqrt{(0.76\%)^2 + (0.77\%)^2} = 1.1\% \tag{10.4-23}$$

with (10.4-10)

$$u_V = \sqrt{(0.58\%)^2 + (0.5\%)^2} = 0.76\%. \tag{10.4-24}$$

   - Root mean square of the bias_i values

$$\text{RMS}_{\text{bias}}\% = \sqrt{\frac{(5\%)^2 + (3\%)^2 + (2\%)^2 + (4\%)^2 + (1\%)^2 + (4\%)^2}{6}} = 3.44. \tag{10.4-25}$$

– Total systematic uncertainty according to (10.4-16) using the intermediate results of (10.4-23) and (10.4-26):

$$u_{\text{bias}}\% = \sqrt{(3.44\%)^2 + (1.1\%)^2} = 3.61. \tag{10.4-26}$$

Step 4. Calculate the combined uncertainty u_{comb}. The uncertainties obtained by the reproducibility within the laboratory u_{R_w} and the systematic bias u_{bias} estimated in steps 2 and 3, respectively, are summed to give the combined uncertainty, which is calculated by:

$$u_{\text{comb}} = \sqrt{u_{R_w}^2 + u_{\text{bias}}^2}. \tag{10.4-27}$$

Example 9: The combined uncertainty for the analytical determination of ammonia-N calculated by (10.4-27) using the uncertainties calculated in examples 7 and 8 is

$$u_{\text{comb}}\% = \sqrt{(4.2\%)^2 + (3.61\%)^2} = 5.54. \tag{10.4-28}$$

Step 5. Calculate the expanded uncertainty U(P). As described above the expanded uncertainty is obtained by multiplying the combined uncertainty by the coverage factor k. For the significance level $P = 95\%$, $k = 2$:

$$U\% = 2 \cdot u_{\text{comb}}. \tag{10.4-29}$$

Example 10: The expanded uncertainty for the photometric determination of ammonia-N calculated by the combined uncertainty of example 9 is:

$$U\% = 2 \cdot 5.54 = 11.1. \tag{10.4-30}$$

Let us assume the mean value is $\bar{x} = 50\,\text{mg}\,\text{L}^{-1}$; the true value μ then lies within the boundaries 44.45 mg L^{-1} and 55.55 mg L^{-1} at a significance level $P = 95\%$, with the corresponding risk of $\alpha = 5\%$ that the true value lies outside these values.

Table 10.4-2 Data of interlaboratory studies for the IC determination of sulfate

Exercise	Nominal value x_{ref} in mg L^{-1}	Laboratory result x_i in mg L^{-1}	s_R in mg L^{-1}	Participants
2006	75	77	7.1	26
	258	253	20.1	42
2007	135	139	12.4	33
	214	211	15.9	31
2008	186	190	20.4	35
	98	100	8.8	38

Challenge 10.4-1

In a laboratory, the concentration of sulfate in industrial water is determined by the ion chromatographic method according to DIN EN ISO 10304-1 [9].

(a) Calculate the expanded uncertainty at the significance level $P = 95\%$ using the following data obtained by the method validation and by control charts.

The customer's requirement for expanded uncertainty is $\pm 10\%$. Can this limit be achieved?

In the course of method validation, the laboratory has taken part in interlaboratory tests over the last 3 years, and the results are given in Table 10.4-2.

The standard deviation obtained by the *mean value* control chart with the nominal value 200 mg L^{-1} is $s = 2.2\,\text{mg}\,L^{-1}$.

The mean value $\bar{R}$ obtained by a *range* control chart constructed with data of stable synthetic control samples in various matrices is $\bar{R}\% = 4.5$.

(b) Let us assume that the allowable sulfate concentration of a specific industrial water is 190 mg L^{-1}. A control sample gives the mean value $\bar{x} = 175\,\text{mg}\,L^{-1}$. Is the limit value exceeded?

(c) The pure analytical error obtained by the method validation is $s_r\% = 2.2$. Calculate the uncertainty solely on the basis of the analytical error and decide whether the allowable limit value is exceeded.

Solution to Challenge 10.4-1

(a) The solution is presented according to the steps given in the Nordtest documents

Step 1: Specify measurand. The measurand is *sulfate* which should be determined in industrial water by DIN EN ISO 10304-1 [9].

Step 2: Quantify the reproducibility within the laboratory u_{R_w}. The reproducibility within the laboratory u_{R_w} is estimated by the second method given above: data obtained by control charts constructed using

(continued)

data of stable synthetic control samples in various matrices according to (10.4-2) – (10.4-4).

$$u_{R_w,\text{Standard}} = s_{\text{mean value chart}} = \frac{2.2\,\text{mg}\,\text{L}^{-1}}{200\,\text{mg}\,\text{L}^{-1}} \cdot 100\% = 1.1\% \qquad (10.4\text{-}31)$$

$$u_{R_w,\text{Range}}\% = \frac{\bar{R}\%}{1.128} = \frac{4.5}{1.128} = 3.99 \qquad (10.4\text{-}32)$$

$$u_{R_w}\% = \sqrt{u^2_{R_w,\text{Standard}} + u^2_{R_w,\text{Range}}} = \sqrt{(1.1)^2 + (3.99)^2} = 4.14. \qquad (10.4\text{-}33)$$

Step 3: Estimate the method and laboratory bias u_{bias}. The estimation of u_{bias} is verified by data obtained by the interlaboratory comparison given in Table 10.4-2. Intermediate quantities for calculation of the uncertainty are shown in Table 10.4-3.

The components required for the estimation of the uncertainty u_{bias} according to (10.4-16) are calculated) using the data given in Table 10.4-3:

$$\text{RMS}_{\text{bias}}\% = \sqrt{\frac{\sum (\text{Bias})^2}{n}} = \sqrt{\frac{30.40}{6}} = 2.25. \qquad (10.4\text{-}34)$$

The uncertainty component from the nominal value $u_{c_{\text{ref}}}$ is calculated according to the standard error of the mean using the mean value of the reproducibility standard deviation $\bar{s}_{\text{R}}$ and the mean value of the number of participants in the interlaboratory exercises $\bar{n}_{\text{interlab}}$:

$$u_{c_{ref}}\% = \frac{\bar{s}_R}{\sqrt{\bar{n}_{\text{interlab}}}} = \frac{8.97}{\sqrt{34}} = 1.53 \qquad (10.4\text{-}35)$$

(continued)

Table 10.4-3 Intermediate quantities for calculation of the uncertainty u_{bias}

Exercise	Participants	Bias in %	$(\text{Bias})^2$ in $\%^2$	s_{R} in %
2006	26	2.67	7.11	9.47
	42	1.94	3.76	7.79
2007	33	2.96	8.78	9.19
	31	1.40	1.97	7.43
2008	35	2.15	4.62	10.97
	38	2.04	4.16	8.98
Mean	34			8.97
Sum			30.40	

Now, the uncertainty of the bias u_{bias} is calculated by (10.4-9) using the results of (10.4-34) and (10.4-35):

$$u_{\text{bias}}\% = \sqrt{\text{RMS}_{\text{bias}}^2 + u_{C_{\text{ref}}}^2} = \sqrt{(2.25)^2 + (1.53)^2} = 2.72. \qquad (10.4\text{-}36)$$

Step 4: Calculate the combined uncertainty u_{comb}. The combined uncertainty is calculated by (10.4-14):

$$u_{\text{comb}}\% = \sqrt{u_{R_w}^2 + u_{\text{bias}}^2} = \sqrt{(4.14)^2 + (2.72)^2} = 4.95. \qquad (10.4\text{-}37)$$

Step 5: Calculate the expanded uncertainty $U(P = 95\%)$. The expanded uncertainty at the significance level $P = 95\%$ calculated by (10.4-30) is:

$$U\% = 2 \cdot u_{\text{comb}} = 2 \cdot 4.95 = 9.9. \qquad (10.4\text{-}38)$$

The customer's requirement for expanded uncertainty ($\pm 10\%$) can thus be achieved.

(b) According to the results given in (10.4-38), the true value lies within the boundaries $\bar{x} \pm U(x)\%$ which for the measured mean value is $\bar{x} = 175 \pm 17.3\,\text{mg}\,\text{L}^{-1}$. The upper value (192.3 mg L^{-1}) lies above the allowable limit value (190 mg L^{-1}), and thus the limit is exceeded.

(c) The uncertainty calculated according to (10.2-2) is

$$u(\hat{x}) = RSD \cdot \hat{x} = \frac{2.2 \cdot 175\,\text{mg}\,\text{L}^{-1}}{100} = 3.9\,\text{mg}\,\text{L}^{-1}. \qquad (10.4\text{-}39)$$

The upper limit of the analytical result is $\hat{x} + u(\hat{x}) = 175\ \text{mg}\,\text{L}^{-1} + 3.9\,\text{mg}\,\text{L}^{-1} = 178.9\,\text{mg}\,\text{L}^{-1}$, which is smaller than the allowable limit value. According to the result obtained by (10.4-39) the limit value is *not* exceeded, a practical example of the situation demonstrated in Fig. 10.1-1.

References

1. DIN EN ISO/IEC 17025 (1999) General requirements for the competence of testing and calibration laboratories, Beuth, Berlin
2. ISO guide 98 (GUM) (1995) International Organization for Standardization, Geneva, 1995
3. Ellison SLR, Rösslein M, Williams A (eds) (2004) Quantifying uncertainty in analytical measurements, EURACHEM/CITAC guide, 2nd edn. hppt://www.measurementuncertainty.org
4. Nordtest Report TR 537 (2003) Handbook for calculation of measurement uncertainty in environmental laboratories, 2nd edn. http://www.nordicinnovation.net/NTtec537
5. DEV A0–4 (2006) Leitfaden zur Abschätzung der Messunsicherheit aus Validierungsdaten. Beuth, Berlin

6. Kragten J (1994) Calculating standard deviations and confidence intervals with a universally applicable spreadsheet techniques. Analyst 119:216–66
7. DIN EN 38 407 F18 (1996) Determination of polycyclic aromatic hydrocarbons (PAH). Beuth, Berlin
8. DIN EN ISO 11 732 (2005) Water quality – Detection of ammonium nitrogen – Method by flow analysis (CFA and FIA) and spectrophotometric determination. Beuth, Berlin
9. DIN EN 10304-1 (2009) Determination of anions by liquid chromatography of ions, Part 1: Determination of bromide, chloride, fluoride, nitrate, nitrite, phosphate, and sulfate. Beuth, Berlin

Appendix

Table A-1 Area of the standard normal variable *z* according to Fig. 2.2.1-3

z	0.00	0.01	0.02	0.03	0.04	0.05	0.06	0.07	0.08	0.09
0.0	0.0000	0.0040	0.0080	0.0120	0.0160	0.0199	0.0239	0.0279	0.0319	0.0359
0.1	0.0398	0.0438	0.0478	0.0517	0.0557	0.0596	0.0636	0.0675	0.0714	0.0753
0.2	0.0793	0.0832	0.0871	0.0910	0.0948	0.0987	0.1026	0.1064	0.1103	0.1141
0.3	0.1179	0.1217	0.1255	0.1293	0.1331	0.1368	0.1406	0.1443	0.1480	0.1517
0.4	0.1554	0.1591	0.1628	0.1664	0.1700	0.1736	0.1772	0.1808	0.1844	0.1879
0.5	0.1915	0.1950	0.1985	0.2019	0.2054	0.2088	0.2123	0.2157	0.2190	0.2224
0.6	0.2257	0.2291	0.2324	0.2357	0.2389	0.2422	0.2454	0.2486	0.2518	0.2549
0.7	0.2580	0.2612	0.2642	0.2673	0.2704	0.2734	0.2764	0.2794	0.2823	0.2582
0.8	0.2881	0.2910	0.2939	0.2967	0.2995	0.3023	0.3051	0.3078	0.3106	0.3133
0.9	0.3159	0.3186	0.3212	0.3238	0.3264	0.3289	0.3315	0.3340	0.3365	0.3389
1.0	0.3413	0.3438	0.3461	0.3485	0.3508	0.3531	0.3554	0.3577	0.3599	0.3621
1.1	0.3643	0.3665	0.3608	0.3708	0.3729	0.3749	0.3770	0.3790	0.3810	0.3830
1.2	0.3849	0.3869	0.3888	0.3907	0.3925	0.3944	0.3962	0.3980	0.3997	0.4015
1.3	0.4032	0.4049	0.4066	0.4082	0.4099	0.4115	0.4131	0.4147	0.4162	0.4177
1.4	0.4192	0.4207	0.4222	0.4236	0.4251	0.4265	0.4279	0.4292	0.4306	0.4319
1.5	0.4332	0.4345	0.4357	0.4370	0.4382	0.4394	0.4406	0.4418	0.4429	0.4441
1.6	0.4452	0.4463	0.4474	0.4484	0.4495	0.4505	0.4515	0.4525	0.4535	0.4545
1.7	0.4554	0.4564	0.4573	0.4582	0.4591	0.4599	0.4608	0.4616	0.4625	0.4633
1.8	0.4641	0.4649	0.4656	0.4664	0.4671	0.4678	0.4686	0.4693	0.4699	0.4706
1.9	0.4713	0.4719	0.4726	0.4732	0.4738	0.4744	0.4750	0.4756	0.4761	0.4767
2.0	0.4772	0.4778	0.4783	0.4788	0.4793	0.4798	0.4803	0.4808	0.4812	0.4817
2.1	0.4821	0.4826	0.4830	0.4834	0.4838	0.4842	0.4846	0.4850	0.4854	0.4857
2.2	0.4861	0.4864	0.4868	0.4871	0.4875	0.4878	0.4881	0.4884	0.4887	0.4890
2.3	0.4893	0.4896	0.4898	0.4901	0.4904	0.4906	0.4909	0.4911	0.4913	0.4916
2.4	0.4918	0.4920	0.4922	0.4925	0.4927	0.4929	0.4931	0.4932	0.4934	0.4936
2.5	0.4938	0.4940	0.4941	0.4943	0.4945	0.4946	0.4948	0.4949	0.4951	0.4952

From Otto M (2007) Chemometrics. Wiley-VCH, Weinheim, p 304

M. Reichenbächer and J.W. Einax, *Challenges in Analytical Quality Assurance*,
DOI 10.1007/978-3-642-16595-5, © Springer-Verlag Berlin Heidelberg 2011

Table A-2 Limits of the one-sided and two-sided t-distribution $t(\bar{P}, \mathrm{df})$ and $t(P, \mathrm{df})$, respectively

df	$\bar{P}_{\text{one-sided}} = 95\%$	$P = 95\%$	$\bar{P}_{\text{one-sided}} = 99\%$	$P = 99\%$
1	6.314	12.706	31.821	63.657
2	2.920	4.303	6.965	9.925
3	2.353	3.182	4.541	5.841
4	2.132	2.776	3.747	4.604
5	2.015	2.571	3.365	4.032
6	1.943	2.447	3.143	3.707
7	1.895	2.365	2.998	3.499
8	1.860	2.306	2.896	3.355
9	1.833	2.262	2.821	3.250
10	1.812	2.228	2.764	3.169
11	1.796	2.201	2.718	3.106
12	1.782	2.179	2.681	3.055
13	1.771	2.160	2.650	3.012
14	1.761	2.145	2.624	2.977
15	1.753	2.131	2.602	2.947
16	1.746	2.120	2.583	2.921
17	1.740	2.110	2.567	2.898
18	1.734	2.101	2.552	2.878
19	1.729	2.093	2.539	2.861
20	1.725	2.086	2.528	2.845
25	1.708	2.060	2.485	2.787
30	1.697	2.042	2.457	2.750
40	1.684	2.021	2.423	2.704
50	1.676	2.009	2.403	2.678
	...	...	...	...
∞	1.645	1.960	2.327	2.576

From Excel function = TINV(α, df)

Table A-3 Limits of the one-sided F-distribution for the significance level $P = 95\%$

df	1	2	3	4	5	6	7	8	9	10	12	20	∞
1	161	199	216	225	230	234	237	239	241	242	244	6,209	254
2	18.513	19.000	19.164	19.247	19.296	19.330	19.353	19.371	19.385	19.396	19.413	99.449	19.496
3	10.128	9.552	9.277	9.117	9.013	8.941	8.887	8.845	8.812	8.786	8.745	26.690	8.526
4	7.709	6.944	6.591	6.388	6.256	6.163	6.094	6.041	5.999	5.964	5.912	14.020	5.628
5	6.608	5.786	5.409	5.192	5.050	4.950	4.876	4.818	4.772	4.735	4.678	9.553	4.365
6	5.987	5.143	4.757	4.534	4.387	4.284	4.207	4.147	4.099	4.060	4.000	7.396	3.669
7	5.591	4.737	4.347	4.120	3.972	3.866	3.787	3.726	3.677	3.637	3.575	6.155	3.230
8	5.318	4.459	4.066	3.838	3.687	3.581	3.500	3.438	3.388	3.347	3.284	5.359	2.928
9	5.117	4.256	3.863	3.633	3.482	3.374	3.293	3.230	3.179	3.137	3.073	4.808	2.707
10	4.965	4.103	3.708	3.478	3.326	3.217	3.135	3.072	3.020	2.978	2.913	4.405	2.538
12	4.747	3.885	3.490	3.259	3.106	2.996	2.913	2.849	2.796	2.753	2.687	3.858	2.296
20	4.351	3.493	3.098	2.866	2.711	2.599	2.514	2.447	2.393	2.348	2.278	2.938	1.843
∞	3.841	2.996	2.605	2.372	2.214	2.099	2.010	1.938	1.880	1.831	1.752	1.878	1.008

From Excel function = FINV(5%, df_1, df_2)

Table A-4 Limits of the one-sided F-distribution for the significance level $P = 99\%$

df	1	2	3	4	5	6	7	8	9	10	12	20	∞
1	4,052	4,999	5,403	5,625	5,764	5,859	5,928	5,981	6,022	6,056	6,106	6,209	6,366
2	98.503	99.000	99.166	99.249	99.299	99.333	99.356	99.374	99.388	99.399	99.416	99.449	99.499
3	34.116	30.817	29.457	28.710	28.237	27.911	27.672	27.489	27.345	27.229	27.052	26.690	26.125
4	21.198	18.000	16.694	15.977	15.522	15.207	14.976	14.799	14.659	14.546	14.374	14.020	13.463
5	16.258	13.274	12.060	11.392	10.967	10.672	10.456	10.289	10.158	10.051	9.888	9.553	9.021
6	13.745	10.925	9.780	9.148	8.746	8.466	8.260	8.102	7.976	7.874	7.718	7.396	6.880
7	12.246	9.547	8.451	7.847	7.460	7.191	6.993	6.840	6.719	6.620	6.469	6.155	5.650
8	11.259	8.649	7.591	7.006	6.632	6.371	6.178	6.029	5.911	5.814	5.667	5.359	4.859
9	10.561	8.022	6.992	6.422	6.057	5.802	5.613	5.467	5.351	5.257	5.111	4.808	4.311
10	10.044	7.559	6.552	5.994	5.636	5.386	5.200	5.057	4.942	4.849	4.706	4.405	3.909
12	9.330	6.927	5.953	5.412	5.064	4.821	4.640	4.499	4.388	4.296	4.155	3.858	3.361
20	8.096	5.849	4.938	4.431	4.103	3.871	3.699	3.564	3.457	3.368	3.231	2.938	2.421
∞	6.635	4.605	3.782	3.319	3.017	2.802	2.639	2.511	2.407	2.321	2.185	1.878	1.011

From Excel function = FINV(1%, df_1, df_2)

Table A-5 Chi-squared distribution for different degrees of freedom and significance levels $\chi^2(P, \mathrm{df})$

df	$P = 90\%$	$P = 95\%$	$P = 99\%$
1	2.706	3.841	6.635
2	4.605	5.991	9.210
3	6.251	7.815	11.345
4	7.779	9.488	13.277
5	9.236	11.070	15.086
6	10.645	12.592	16.812
7	12.017	14.067	18.475
8	13.362	15.507	20.090
9	14.684	16.919	21.666
10	15.987	18.307	23.209
11	17.275	19.675	24.725
12	18.549	21.026	26.217
13	19.812	22.362	27.688
14	21.064	23.685	29.141
15	22.307	24.996	30.578
16	23.542	26.296	32.000
17	24.769	27.587	33.409
18	25.989	28.869	34.805
19	27.204	30.144	36.191
20	28.412	31.410	37.566
21	29.615	32.671	38.932
22	30.813	33.924	40.289
23	32.007	35.172	41.638
24	33.196	36.415	42.980
25	34.382	37.652	44.314

From Excel function = CHIINV(α, df)

Table A-6 Significance table for testing outliers according to Grubbs

df	$\bar{P} = 95\%$	$\bar{P} = 99\%$
3	1.153	1.155
4	1.463	1.492
5	1.672	1.749
6	1.822	1.944
7	1.938	2.097
8	2.032	2.221
9	2.110	2.323
10	2.176	2.410
11	2.234	2.485
12	2.285	2.550
13	2.331	2.607
14	2.371	2.659
15	2.409	2.705
16	2.443	2.747
17	2.475	2.785
18	2.504	2.821
19	2.532	2.854
20	2.557	2.884
21	2.580	2.912
22	2.603	2.939
23	2.624	2.963

(continued)

Table A-6 (continued)

df	$\bar{P} = 95\%$	$\bar{P} = 99\%$
24	2.644	2.987
25	2.663	3.009
26	2.681	3.029
27	2.698	3.049
28	2.714	3.068
29	2.730	3.085
30	2.745	3.103
31	2.759	3.119
32	2.773	3.135
33	2.786	3.150
34	2.799	3.164
35	2.811	3.178
36	2.823	3.191
37	2.835	3.204
38	2.846	3.216
39	2.857	3.228
40	2.866	3.240
41	2.877	3.251
42	2.887	3.261
43	2.896	3.271
44	2.905	3.282
45	2.914	3.292
46	2.923	3.302
47	2.931	3.310
48	2.940	3.319
49	2.948	3.329
50	2.956	3.336
60	3.025	3.411
70	3.082	3.471
80	3.130	3.521
90	3.171	3.563

From DIN EN 53 804-1 (2002) Statistical evaluation – part 1: Continuous characteristics. Beuth, Berlin
Funk W, Dammann V, Donnevert G. (2005) Qualitätssicherung in der Analytischen Chemie 2, Aufl., Wiley-VCH, Weinheim

Table A-7 Critical one-sided Q-values for testing outliers according to Dixon

n	$P = 95\%$	$P = 99\%$
3	0.941	0.988
4	0.765	0.889
5	0.642	0.780
6	0.560	0.698
7	0.507	0.637
8	0.554	0.683
9	0.512	0.635
10	0.477	0.597
11	0.576	0.679
12	0.546	0.642
13	0.521	0.615

(*continued*)

Table A-7 (continued)

n	$P = 95\%$	$P = 99\%$
14	0.546	0.641
15	0.525	0.616
16	0.507	0.595
17	0.490	0.577
18	0.475	0.561
19	0.462	0.547
20	0.450	0.535
21	0.440	0.524
22	0.430	0.514
23	0.421	0.505
24	0.413	0.497
25	0.406	0.489
26	0.399	0.482
27	0.393	0.475
28	0.387	0.469
29	0.381	0.463

From DIN EN 53 804-1 (2002) Statistical evaluation –part 1: Continuous characteristics. Beuth, Berlin
n number of observations

Table A-8 Significance table for testing normal distribution according to David

n	Lower limit		Upper limit	
	$P = 95\%$	$P = 99\%$	$P = 95\%$	$P = 99\%$
5	2.15	2.02	2.753	2.803
6	2.28	2.15	3.012	3.095
7	2.40	2.26	3.222	3.338
8	2.50	2.35	3.399	3.543
9	2.59	2.44	3.552	3.720
10	2.67	2.51	3.685	3.875
11	2.74	2.58	3.80	4.012
12	2.80	2.64	3.91	4.134
13	2.86	2.70	4.00	4.244
14	2.92	2.75	4.09	4.34
15	2.97	2.80	4.17	4.44
16	3.01	2.84	4.24	4.52
17	3.06	2.88	4.31	4.60
18	3.10	2.92	4.37	4.67
19	3.14	2.96	4.43	4.74
20	3.18	2.99	4.49	4.80
25	3.34	3.15	4.71	5.06
30	3.47	3.27	4.89	5.56
35	3.58	3.38	5.04	5.42
40	3.67	3.47	5.16	5.56
45	3.75	3.55	5.26	5.67
50	3.83	3.62	5.35	5.77
55	3.90	3.69	5.43	5.86
60	3.96	3.75	5.51	5.94

From Sachs L (1991) Angewandte Statistik: Anwendung statistischer Methoden, 7. Aufl. Springer, Berlin
n – number of observations

Table A-9 Critical *C*-values for testing homogeneity of variances at the significance level *P* = 95% and *P* = 99% (*given in italics*) according to Cochran

	Degrees of freedom *df*									
k	1	2	3	4	5	6	7	8	9	10
2	0.9985	0.9750	0.9392	0.9057	0.8772	0.8534	0.8332	0.8159	0.8010	0.7880
	0.9999	*0.9950*	*0.9794*	*0.9582*	*0.9373*	*0.9172*	*0.8988*	*0.8823*	*0.8674*	*0.8539*
3	0.9669	0.8709	0.7977	0.7457	0.7071	0.6771	0.6530	0.6333	0.6167	0.6025
	0.9933	*0.9423*	*0.8831*	*0.8335*	*0.7933*	*0.7606*	*0.7335*	*0.7107*	*0.6912*	*0.6743*
4	0.9065	0.7679	0.6841	0.6287	0.5895	0.5598	0.5365	0.5175	0.5017	0.4884
	0.9676	*0.8643*	*0.7814*	*0.7212*	*0.6761*	*0.6410*	*0.6129*	*0.5897*	*0.5702*	*0.5536*
5	0.8412	0.6838	0.5981	0.5441	0.5065	0.4783	0.4564	0.4387	0.4241	0.4118
	0.9279	*0.7885*	*0.6957*	*0.6329*	*0.5875*	*0.5531*	*0.5259*	*0.5037*	*0.4854*	*0.4697*
6	0.7808	0.6161	0.5321	0.4803	0.4447	0.4184	0.3980	0.3817	0.3682	0.3568
	0.8828	*0.7218*	*0.6258*	*0.5635*	*0.5195*	*0.4866*	*0.4608*	*0.4401*	*0.4229*	*0.4084*
7	0.7271	0.5612	0.4800	0.4307	0.3947	0.3726	0.3535	0.3384	0.3259	0.3154
	0.8376	*0.6644*	*0.5685*	*0.5080*	*0.4659*	*0.4347*	*0.4105*	*0.3911*	*0.3751*	*0.3616*
8	0.6798	0.5157	0.4377	0.3910	0.3595	0.3362	0.3185	0.3043	0.2926	0.2829
	0.7945	*0.6152*	*0.5209*	*0.4627*	*0.4226*	*0.3932*	*0.3704*	*0.3522*	*0.3373*	*0.3248*
9	0.6385	0.4775	0.4027	0.3548	0.3286	0.3067	0.2901	0.2768	0.2659	0.2568
	0.7544	*0.5727*	*0.4810*	*0.4251*	*0.3870*	*0.3592*	*0.3378*	*0.3207*	*0.3067*	*0.2950*
10	0.6020	0.4450	0.3733	0.3311	0.3029	0.2823	0.2666	0.2541	0.2439	0.2353
	0.7175	*0.5358*	*0.4469*	*0.3934*	*0.3572*	*0.3308*	*0.3106*	*0.2945*	*0.2813*	*0.2704*
12	0.5410	0.3924	0.3264	0.2880	0.2624	0.2439	0.2299	0.2187	0.2098	0.2020
	0.6528	*0.4751*	*0.3919*	*0.3428*	*0.3099*	*0.2861*	*0.2680*	*0.2535*	*0.2419*	*0.2320*
15	0.4709	0.3346	0.2758	0.2419	0.2195	0.2034	0.1911	0.1815	0.1736	0.1671
	0.5747	*0.4069*	*0.3317*	*0.2882*	*0.2593*	*0.2386*	*0.2228*	*0.2104*	*0.2002*	*0.1918*
20	0.3894	0.2705	0.2205	0.1921	0.1137	0.1602	0.1501	0.1422	0.1357	0.1303
	0.4799	*0.3297*	*0.2654*	*0.2288*	*0.2048*	*0.1877*	*0.1748*	*0.1646*	*0.1567*	*0.1501*
24	0.3434	0.2354	0.1907	0.1656	0.1493	0.1374	0.1286	0.1216	0.1160	0.1113
	0.4247	*0.2871*	*0.2295*	*0.1970*	*0.1759*	*0.1608*	*0.1495*	*0.1406*	*0.1338*	*0.1283*
30	0.2929	0.1980	0.1593	0.1377	0.1237	0.1137	0.1061	0.1002	0.0958	0.0921
	0.3632	*0.2412*	*0.1913*	*0.1653*	*0.1454*	*0.1327*	*0.1232*	*0.1157*	*0.1100*	*0.1054*

From internet: http://www.watpon.com/table/cochran

k – number of the levels/samples, *df* – degrees of freedom of the replilcates in each level/sample

Table A-10 Critical limits for testing trends according to Neumann

n	$P = 95\%$	$P = 99\%$
4	0.7805	0.6256
5	0.8204	0.5379
6	0.8902	0.5615
7	0.9359	0.6140
8	0.9825	0.6628
9	1.0244	0.7088
10	1.0623	0.7518
11	1.0965	0.7915
12	1.1276	0.8280
13	1.1558	0.8618
14	1.1816	0.8931
15	1.2053	0.9221
16	1.2272	0.9491
17	1.2473	0.9743
18	1.2660	0.9979
19	1.2834	1.0199
20	1.2996	1.0406
21	1.3148	1.0601
22	1.3290	1.0785
23	1.3425	1.0958
24	1.3552	1.1122
25	1.3671	1.1287
26	1.3785	1.1426
27	1.3892	1.1567
28	1.3994	1.1702
29	1.4091	1.1830
30	1.4183	1.1951

From Sachs L (1999) Angewandte Statistik: Anwendung statistischer Methoden, 9. Aufl. Springer, Berlin
n number of observations

Index

FSC
www.fsc.org
MIX
Papier | Fördert
gute Waldnutzung
FSC® C083411

Kolibri 360 GmbH
Ferdinand-Jühlke-Straße 7
99095 Erfurt, Deutschland
produktsicherheit@kolibri360.de